Emergency Response Management of Offshore Oil Spills

Scrivener Publishing
3 Winter Street, Suite 3
Salem, MA 01970

Scrivener Publishing Collections Editors

Publishers at Scrivener

Martin Scrivener (martin@scrivenerpublishing.com)
Phillip Carmical (pcarmical@scrivenerpublishing.com)

Emergency Response Management of Offshore Oil Spills

Guidelines for Emergency Responders

Nicholas P. Cheremisinoff
and Anton Davletshin

Scrivener

WILEY

Co-published by John Wiley & Sons, Inc. Hoboken, New Jersey, and Scrivener Publishing LLC, Salem, Massachusetts.
Published simultaneously in Canada.

Cover design by Kris Hackerott.

Library of Congress Cataloging-in-Publication Data:

ISBN 978-0-470-92712-0

Printed in the United States of America

10 9 8 7 6 5 4 3 2 1

Contents

Preface

The Macondo well is located in Mississippi Canyon Block 252 of the Gulf of Mexico. British Petroleum (BP) is the operator and principal developer of the oil field with 65% of interest, while 25% is owned by Anadarko Petroleum Corporation, and 10% by MOEX Offshore 2007, a unit of Mitsui. The prospect may have held as much as 50 million barrels (2.1×10^9 gallons) of producible reserves of oil. On October 7, 2009 the Transocean Marianas semi-submersible rig commenced drilling, but operations were halted at 4,023 feet below the sea floor on November 29, 2009, when the rig was damaged by Hurricane Ida. The Transocean's Deepwater Horizon rig was leased by the British Petroleum Corporation and resumed drilling operations in February 2010.

An explosion on the drilling Deepwater Horizon rig occurred on April 20, 2010. The Deepwater Horizon sank on April 22, 2010, in water approximately 5,000 feet deep, and has been located resting on the seafloor approximately 1,300 feet (about a quarter of a mile) northwest of the well.

The spill is an unfolding event, and while the runaway well has been capped nearly 205 million gallons of oil have been released into the Gulf with the scientific community claiming that nearly 75% has been unaccounted for. The impact of the spill touches almost every fabric of life. The damage to marine life, wildlife, coral reefs and disruption to the ecosystem food chain cannot even begin to be taken into consideration because we simply do not have a sense of how extensive this ongoing event is. Literally hundreds and possibly, in the end, thousands of miles of shoreline and wetlands will be lost not simply for decades, but some sensitive wetland zones will disappear forever. Businesses and livelihoods for all practical purposes are gone. For families that have made their living in the fishing industry, they will simply have to find other means by which to make livings to support their families. While the younger population may retool to another profession or perhaps wind up working on oil rigs in the region, a question for our government is what do people who are approaching or are in senior citizen age brackets do especially when they do not have sufficient savings to support themselves and are unable to retool to another profession so late in their lives? Indeed, what additional social-economic burden does this place on our country?

The impacts of the spill do not simply affect businesses, property values, wildlife, and marine life, as well as raise a big question on the health and safety of communities – the impacts include denying citizens of state and local services and access to public lands, a large part of which were no less than national treasures. BP's oil has in essence annexed thousands of miles of shoreline along the coast of the United States. It is in fact one of the largest acts of trespassing in U.S. history.

Since the Exxon Valdez oil spill of 1989 the oil industry has sat idle. It has not developed technologies that make deep offshore drilling safe. While technologies for exploration are 21st century, emergency and post emergency response technologies to oil spills at sea have remained frozen in time with an infrastructure that is now proving to be woefully inadequate to the response in the Gulf.

This volume was written for two reasons. First, we think the public needs to be better educated about industry practices. The volume explores the events leading up to the mega spill in the Gulf. While there are seemingly endless advertisements on the part of British Petroleum on how the company is taking full responsibility for the catastrophe, congressional hearings have uncovered extensive wrongdoing on the part of the company and major flaws both in the technologies being used and in the oversight, or rather lack thereof, on the part of the former Minerals Management Services, which remains responsible for regulating offshore oil operations.

A second reason we have written this book is that we believe many of the practices being employed in the post-emergency response activities potentially place workers at high risk of exposures to a mix of chemicals. Crude oil contains carcinogens and numerous toxins. Chemical dispersants and shoreline cleaning agents contain a host of different chemicals including carcinogens and concentrated toxins. Apart from these chemicals being deadly poisons to marine and wild life, they pose serious health risks to workers and potentially to the public if the spill is not eventually contained. The volume provides extensive information on safe work practices and takes considerable effort to provide comparative product information on the arsenal of chemicals that are being applied in nothing short of chemical warfare to combat the spill.

There are seven chapters to the volume. The first chapter provides an overview of crude oil properties and explores marine spills from a historical standpoint. Offshore oil spills are a more common event than the industry has often admitted to. Despite it being a rather frequent event in exploration and recovery, no technology innovations have been made within the last 30 years to prevent or respond to such events. Chapter 2 goes further into the origins of spills and also provides a case study. The case study focuses on a former Gulf of Mexico spill, which has become BP's game plan in responding to its own spill. We question why the failed procedures used in a 1979 mega spill in the Gulf of Mexico would be expected to have any better results in 2010 when no new innovations or technologies have improved over the old approaches.

Chapter 3 addresses chemical dispersants. The intent of chemical dispersants is to break the oil up and accelerate the natural biodegradation process. But in doing so, the oil enters into a dispersed phase, which causes it to sink into the water column. BP has applied unprecedented amounts of chemical dispersants in such a way that scientists are now reporting that the oxygen and methane levels in the Gulf are reaching the tipping point, whereby parts of the Gulf may be incapable of supporting marine life. While dispersants have proven effective in addressing small to mid-size oil spills, they have never been applied to the extent currently being applied in the Gulf. The practices suggest that the use of these chemicals is simply masking the problems created by this catastrophic event and concealing from the public the true damages. The chapter provides extensive product information and comparisons of dispersant products in terms of their effectiveness, toxicity, and chemical ingredients.

Chapter 4 addresses the arsenal of technologies and tools available for combating spills that impact shorelines. These tools and technologies are both manpower intensive and primitive. At best, this can be described as chemical and trench warfare. Shoreline cleaning chemicals, sorbents, and the use of booms and skimmers are covered in detail.

Chapter 5 discusses several emerging technologies. This is not a large chapter that is testament to the level of effort placed by the industry sector in developing and investing in post-emergency response clean-up equipment. Still, there are several technologies that can be immediately applied to the Gulf disaster, and we wonder why they were not fully commercialized and endorsed by the industry.

Chapter 6 focuses on spill response and worker protection. Much of what is addressed in this chapter deals with post-emergency response, which is concerned with clean-up. The level of training and skills required by professionals and workers differs considerably between emergency and post-emergency forces. These skills and job functions are explained in detail. Further, the type of training and levels of protection required for responders varies depending on the nature of the spill countermeasures being employed. The chapter begins by reviewing many of the countermeasures covered in earlier chapters, but it devotes considerable discussions to in-situ burning. This is a practice that was partially effective in the Valdez incident, but its origins date back to World War II. Despite a very long history as a countermeasure, we find surprisingly that the impacts to air pollution and the extent of possible worker exposure have been poorly defined by the industry. This chapter also provides considerable information on how the National Contingency Plan works, or is supposed to work, and then provides an extensive compendium of information and data that post-emergency responders need to be trained on and aware of. The last part of this chapter covers the essential elements of an oil spill response plan.

Chapter 7 is titled "Standard of Care and the BP Oil Spill." The chapter examines the events and decisions made by the British Petroleum Corporation weeks, days, and hours prior to the April 20th fire and explosion. The public record to date shows that BP placed cost and schedule for the Macondo above public safety. The record to date further shows a breakdown in government services, whereby the Mineral Management Services was remiss in performing inspections of the Deepwater Horizon rig, allowed BP to self-certify its own safety and oils spill response plan, and took no actions to regulate design flaws in blowout preventers.

The opinions expressed in this volume are those of the authors. We have applied best practices in performing the research relied upon in preparing the book and believe that opinions stated at various points are reasonable and based on factual data and sources. The events in the Gulf are changing and as new information becomes available, some of our opinions may change; however, this remains to be seen.

We wish to thank both Scrivener Publishing and John Wiley & Sons, Inc. for their joint efforts in the production of this volume.

Nicholas P. Cheremisinoff, Ph.D.
Anton R. Davletshin

August 2010

1

Toxic Nature of Crude Oil

1.1 High Risk Areas

In general, spilled oil is most harmful when shallow, productive waters, porous sediments, low-energy aquatic environments, or special-use habitats are impacted. Examples of high-risk locations include wetlands, sheltered tidal flats, shallow bays, coarse sand and gravel beaches, and sites with concentrated reproductive and migratory activities.

1.2 Potential Impacts

The impacts on local shore regions from a crude oil spill can be economically devastating. There are impacts on the fishing and tourism industries, shipping, recreation, and property values, to mention a few. In addition to the economic impacts on the coastal communities, the environmental damages can be devastating to the local ecosystems. The BP Gulf oil spill threatens to destroy an entire ecosystem, a way of life that has existed for generations, and national treasures and can possibly have global implications.

Recent computer models suggest that the ongoing spill is likely to contaminate the entire eastern shoreline and move on to the shores of foreign countries.

Crude oil and petroleum products vary in their toxicity, and the sensitivity of fish to petroleum varies according to species. The water-soluble fractions of crude oil can stunt fish growth. Negative impacts on fish are primarily seen in the eggs, larvae, and early juveniles, with varied effects on the adults.

The following are some negative impacts from oil spills:

- Pink salmon fry are affected by exposure to water-soluble fractions of crude oil, while pink salmon eggs are tolerant of benzene and water-soluble petroleum.
- Fish rapidly metabolize aromatic hydrocarbons due to their enzyme system.
- Depressed feeding, decreased swimming activity, and increased mortality occur in fish.
- The mortality of eggs and larvae increases (such as after the Argo Merchant No.6 fuel oil discharge where 20% of the cod eggs and 46% of the Pollock eggs in the discharge zone were dead). During the Torrey Canyon (Bunker C) discharge, 90% of the pilchard eggs in the discharge area were killed. Following the Amoco Cadiz (crude oil) discharge, a one-year old class of flat-fish was thought to have been reduced.
- Exclusion of fishermen from the fishing grounds and other disruptions of fishing can change the population balance to date (e.g., salmon over-escapement in Prince William Sound after the Exxon Valdez (crude oil) spill).
- The fouling of costly fishing gear during the spill can set the fisherman back even after the fishing restrictions are lifted.
- The tainting of fish (such as change in flavor or smell) and the public's fear of tainting, mortality, or other effects of non-motile inshore species, such as rockfish, occurs.
- Mortality and the tainting of fish maintained in mariculture enclosures, where the escape of fish is prevented, is common (e.g., the Braer oil discharge off the Shetlands affected salmon in mariculture enclosures).

- Sublethal effects occur, such as fin erosion, ulceration of the integument, liver damage, lesions in the olfactory tissue, reduced hatching success, reduced growth, change in egg buoyancy, malformations that interfere with feeding, arrest of cell division, and genetic damage.
- The oiling of feathers is considered to be the primary cause of most bird deaths following oil spills. Oil disrupts and destroys the fine strand structure of the feathers, resulting in the loss of water repellency and body insulation. As the oiled plumage becomes matted, water penetrates the feathers and chills the animal's body. The combined results are a loss of buoyancy and hypothermia. The natural response to oil matted plumage is preening, subsequent to which oiled birds ingest the oil while attempting to remove the petroleum from their feathers. The ingestion of petroleum results in anemia, pneumonia, kidney and liver damage, stunted growth, altered blood chemistry, and decreased egg production.
- Chicks are exposed to petroleum by ingesting food regurgitated by impacted adults, and they may also be poisoned.
- Recovery from the effects of oil spills on local populations of invertebrates can require up to 10 years depending on the type of oil, the circumstances of the spill, and the organisms affected. Invertebrates (zooplankton) in the water column of large bodies of water return to pre-spill conditions much faster than invertebrates in small bodies of water (fresh-water lakes, streams); however, this is a broad generalization.
- Gulls, storm petrels, and guillemots experience elevated corticosterone, thyroxin, and increased size of adrenal glands after ingesting a single dose of 0.1% (of diet) crude oil.
- Exposure to oil by birds has been shown to lead to changes in behavior that ultimately cause reduced reproductive success. Effects include cessation or delay of egg laying, increased nesting phenology, nest abandonment, reduced feeding of young, mate switching, interruption of courtship behavior, egg rejection,

parental rejection of chicks, impairment of incubation behavior, and reduced nest attentiveness.
- Epifauna, such as mussels and bivalves, generally survive oiling as adults due to their protective shells, but they have no enzymatic system for purging. Therefore, bioaccumulation occurs, resulting in reduced feeding absorption efficiency followed by growth reductions.
- Oiling of chicks by the externally treated adult has been reported to result in the rejection of the chicks.
- Wedge-tailed shearwaters orally exposed to Santa Barbara crude oil had a laying and incubation frequency significantly lower than controls.
- Oil concentrations as low as 1 uL/egg (1.3% of the surface of a mallard egg) are toxic. This is attributed to a function of the aromatic component of crude oil rather than impaired gas exchange.
- Mallards that ingest crude oil experience delayed laying, decreased oviposition, and decreased shell thickness.
- The hatching success of herring and black-backed gull eggs decreased in response to 10 uL of crude or weathered crude oil applied externally to eggshells.
- Heron, tern, and brown pelican eggs experience reduced hatchability when oiled either directly or via the adult's feathers.
- Following the Santa Barbara oil spill, a large number of premature births were observed in sea lions.

Crude oil is toxic to humans. The industry downplays this and argues crude oil is of low toxicity to humans. Table 1.1 tabulates compositions of crude oil as reported on Material Safety Data Sheets from different suppliers. Crude oil contains both chemical toxins and carcinogens. Among the chemicals are benzene, toluene, ethylbenzene, xylene, polyaromatic hydrocarbons (PAHs), and toxic heavy metals not listed on MSDSs. While concentrations are low, when there is a spill of the magnitude of the Deepwater Horizon spill (the BP disaster of the Gulf) covering thousands of square miles of surface area, the volatile organic compounds (VOCs) rapidly evaporate accelerated by constant surface area renewal caused by waves. Personnel working on vessels that skim up the surface oil or combat the spill along shorelines and marshes are being exposed to carcinogens and other toxic chemical ingredients that are known teratogens, or systemic poisons, which are known to impair liver and kidney functions. Crude oils contain similar chemicals to coal tar pitch volatiles, or PAHs, which are a large family of toxic chemicals that are suspected and confirmed human carcinogens. When oil is burned at the surface in order to combat the spill, more carcinogenic PAHs are created as these are products of incomplete combustion.

The toxicity of crude oil involves an incredibly complex mixture of inorganic and organic chemicals. There is uncertainty in the application of dose-response relationships based on crude oil as a whole mixture to both humans and aquatic life. A simplified approach to determining toxicity is an "indicator chemical approach" which involves selecting a subset of chemicals from the whole mixture that represents the "worst case" in terms of mobility and toxicity. This approach has been used with crude oil with the subsets of chemicals being volatile organics such as benzene, toluene, ethylbenzene, and xylenes (collectively referred to as BTEX) and polycyclic aromatic hydrocarbons (PAHs).

The BTEX group is of significance because the chemicals in the family are soluble in water, highly mobile in the environment, and represent the more volatile and soluble components of crude oil. In addition, benzene is an EPA defined Class A carcinogen.

In contrast, PAHs are not highly mobile. These chemicals tend to have low solubility and vapor pressures (hence, they are referred to as semi-volatile organic chemicals or SVOCs) but they are of interest because they range from toxic to carcinogenic. These chemicals are prevalent in crude oil, representing the heavier or less volatile crude oil components. PAHs and their transformation products are among the most hazardous constituents of crude oil.

The effects of petroleum and individual PAHs on living organisms such as mammals include impaired immune systems, impaired reproduction, and reduced growth and development. Overall, the effects are immunological, reproductive, fetotoxic, and genotoxic.

The toxicity of crude oil may be affected by various factors such as "weathering" time or the addition of oil dispersants. Weathered and "fresh" crude oil may have different toxicities depending on oil type and weathering time. Oil and dispersant mixtures can be equally as toxic as crude oil alone.

Table 1.1 Compositions of crude oil reported on various MSDSs.

Concentrations as Reported on MSDS	**Toluene**					100 ppm	0.1–5 wt.%
	Xylene					100 ppm	0.1–5 wt.%
	Coal Tar Pitch Volatiles				0.2 wt.%		
	Pentane				1.5–2.5 vol.%		
	Isopentane				0.3–1.5 vol.%		
	Butane				0.8–1 vol.%	10 ppm	
	Hydrogen Sulfide			Varies (<1) wt.%		10 ppm	<0.5 wt.%
	n Hexane			0–1.4 wt.%	0.8–1 vol.%		
	Polycyclic aromatic hydrocarbons (PAHs)		Trace				
	Ethylbenzene						0.1–5 wt.%
	Benzene	0–0.1 wt.%	Trace	0.0–0.1 wt.%			0.1–5 wt.%
	Asphaltene Content	0–2 wt.%					
	Polars	1–5 wt.%					
	Aromatics	8–15 wt.%					
	Saturates	80–90 wt.%					
	Sulfur Compounds	0–2 wt.%	Varied				
Effective Date		7-Aug	11-Mar-08	No Date	5/13/2002	No Date	30-Jun-08
Supplier		Martin Marietta Materials	Irving	Mark West	BP West Coast Products LLC	Gibson Energy ULC	Plains Marketing, L.P.

1.3 Definitions

The following definitions are given in the NOAA HMRAD Shoreline Countermeasures Manual for tropical coastal environments:

- Light oils (diesel, No. 2 fuel oils, light crudes):
  - Moderately volatile; will leave residue (up to 1/3 of spilled amount)
  - Moderate concentrations of toxic (soluble) compounds
  - Well "oil" intertidal resources with long-term contamination potential
  - Has potential for subtidal impacts (dissolution, mixing, sorption onto suspended sediments)
  - No dispersion necessary
  - Cleanup can be very effective

- Medium oils (most crudes):
  - About 1/3 will evaporate within 24 hours
  - Maximum water-soluble fraction is 10–100 ppm
  - Oil contamination of intertidal areas can be severe and long-term
  - Impact to waterfowl and fur-bearing mammals can be severe
  - Chemical dispersion is an option within 1–2 days
  - Clean-up is most effective if conducted quickly

- Heavy oils (heavy crude oils, No. 6 fuel, bunker C):
  - Heavy oils with little or no evaporation or dissolution
  - Water-soluble fraction is likely to be <10 ppm
  - Heavy contamination of intertidal areas is likely
  - Severe impacts to waterfowl and fur-bearing mammals (coating and ingestion)
  - Possible long-term contamination of sediments
  - Weathers very slowly
  - Dispersion is seldom effective
  - Shoreline clean-up is difficult under all conditions

The following are common synonyms. Although the term "petroleum" is sometimes considered a synonym for crude oil, it more commonly refers to petroleum products as a whole and not just the crude oil from which these products are derived. Oil, both crude and refined, is a very broad topic. It has been divided into several different entries (although there is still some overlap on certain topics):

- Base oil
- Coal liquid
- Coal oil
- Crude oil, petroleum
- Crude petroleum
- Petrol
- Petroleum crude
- Petroleum oil
- Rock oil
- Seneca oil
- Naphtha

The following are terms concerning chemical composition. These terms are important from the standpoint of identifying the chemicals of exposure, environmental sampling, and assessing toxicity.

1.3.1 Polycyclic Aromatic Hydrocarbons (PAHs)

Polycyclic aromatic hydrocarbons (PAHs) are a group of over 100 different chemicals that are formed during the incomplete burning of coal, oil, gas, garbage, or other organic substances like tobacco or charbroiled meat, but they are also found in oil. The simplest PAHs, as defined by the International Union on Pure and Applied Chemistry (IUPAC) (G.P. Moss, IUPAC nomenclature for fused-ring systems), are phenanthrene and anthracene. Both of these chemicals contain three fused aromatic rings. Smaller molecules, such as benzene, are not PAHs. PAHs may contain four-, five-, six-, or seven-member rings, but those with five or six are most common. PAHs composed only of six-membered rings are called alternant PAHs. Certain alternant PAHs are called "benzenoid" PAHs. The name is derived from benzene, which is an aromatic hydrocarbon with a single, six-membered ring. These can be benzene rings that are interconnected with each other by single carbon-carbon bonds and with no rings remaining that do not contain a complete benzene ring. The set of alternant PAHs is closely related to a set of mathematical entities called polyhexes, which are planar figures composed by conjoining regular hexagons of identical sizes.

PAHs that contain up to six fused aromatic rings are often known as "small" PAHs, and those containing more than six aromatic rings are called "large" PAHs. Due to the availability of samples of the various small PAHs, the bulk of research on PAHs has been on those of up to six rings. The biological activity and occurrence of the large PAHs do appear to be a continuation of the small PAHs. They are believed to derive from combustion products but at lower levels than the small PAHs due to the kinetic limitation of their production through the addition of successive rings. Also, with many more isomers possible for larger PAHs, the occurrence of specific structures is smaller.

Another aromatic compound is Naphthalene ($C_{10}H_8$ constituent of mothballs). This chemical consists of two coplanar six-membered rings sharing an edge and is another aromatic hydrocarbon. By formal convention it is not a true PAH, though it is referred to as a bicyclic aromatic hydrocarbon.

Note that aqueous solubility decreases approximately one order of magnitude for each additional ring.

Other toxic PAHs include the following:

- Benz(a)anthracene
- Benzo(a)pyrene (B(a)P)
- Benzo(b)fluoranthene
- Benzo(k)fluoranthene
- Chrysene
- Dibenz(a,h)anthracene
- Indeno(1,2,3-cd)pyrene
- Acenaphthylene
- Acenaphthene
- Anthracene
- Fluorene
- Phenanthrene
- Fluoranthene
- Pyrene
- Benzo(g,h,i)perylene

1.3.2 Total Petroleum Hydrocarbons (TPH)

Individual compounds of crude oil can be classified into the following categories:

- Hydrocarbons, which include alkanes (normal and branched chains), cycloalkanes, alkenes, aromatics, and naphthenoaromatics
- Non-hydrocarbons, which include nitrogen, sulfur and oxygen (NSO) compounds, asphaltenes and resins (including NSO heterocyclics), metallo-organics, and inorganic metal salts

The toxicity of individual chemicals differs widely. Generally, the most toxic compounds are benzene, ethyl benzene, toluene, and xylenes.

1.4 Examples of Historical Oil Spills and Their Impacts

Table 1.2 lists 26 spills that have occurred over the years. The list is a sampling of the spills that have been reported. For relative comparison, while many consider the 1989 Exxon Valdez spill in Alaska one of the largest spills in history, it was actually quite small in comparison to many other spills. The Valdez spill ranks as the 34th largest spill in the world.

The Gulf War oil spill, resulting from the 1990 Iraqi invasion of Kuwait, is regarded as the largest oil spill in history. Iraqi forces opened valves at the Sea Island oil terminal and dumped oil from several tankers into the Persian Gulf in order to thwart a potential landing by U.S. Marines. The immediate reports from Baghdad said that American air strikes had caused a discharge of oil from two tankers. Coalition forces determined the main source of oil to be the Sea Island terminal in Kuwait. American air strikes destroyed the pipelines in order to prevent further spillage into the Persian Gulf. The oil spill, which began on January 23, 1991, caused considerable damage to wildlife in the Persian Gulf, especially in areas surrounding Kuwait and Iraq. Estimates on the volume spilled usually range around 11 million barrels (462 million gallons). The oil slick reached a maximum size of 101 by 42 miles (4,242 square miles) and was 5 inches thick in some areas. Marshlands and mud tidal flats still contain large quantities of oil, and full recovery is likely to take decades. The long-term effects were very significant. The salt marshes, which occur along almost 50% of the coastline, have shown the heaviest impact compared to the other ecosystem types for nearly two decades. Sand beaches are on the best way to complete recovery. The main reason for the delayed recovery of the salt marshes is the absence of physical energy (wave action) and the mostly anaerobic milieu of the oiled substrates.

Of the 26 spills noted in Table 1.2, nearly 1.2 billion gallons of oil was spilled. This is equivalent to the combined capacities of 1,848 Olympic-size swimming pools.

There are no reliable estimates of the BP oil spill in the Gulf of Mexico at the moment. This disaster began April 20, 2010 when the rig exploded and sank. News media reports as of June 19, 2010 have reported official numbers of between 35,000 and 60,000 barrels per day being released into the Gulf. As of this writing, those estimates represent a release of between 89 and 153 million gallons of oil. The BP Gulf disaster (known as the Deepwater Horizon spill) may prove to match the 1979 Ixtoc I oil rig disaster (see Table 1.2).

The damage to ecosystems and the way of life caused by oil spills is both far reaching and not fully defined because neither governments nor the industry have acted to provide adequate research funding in order to define the impacts and to devise safe and reliable technologies for offshore oil recovery. The Exxon Valdez spill of nearly 11 million gallons killed as many as 700,000 sea birds and 5,000 sea otters initially. Twenty-one years after the spill, the populations of sea otters in the areas of Prince William Sound have not recovered. The Pacific herring population all but collapsed in the aftermath of the spill. Two intensely studied pods of killer whales in the sound suffered heavy losses from the spill and never fully recovered. One of the pods no longer has a reproductive female and is therefore doomed to extinction. The oil that was spilled in Prince William Sound is in fact still sitting there.

The degradation of oil takes decades. The authors have worked on oil pipeline releases that occurred in the 1960s. The daughter products of degradation follow a pathway that is strongly dependent on the ecosystem it is attacking. While oil industry ads and government officials seek to calm public outcry and anger over the chronic impacts from the Deepwater Horizon spill by reporting that ecosystem damages will be minimized by natural attenuation, the fact remains that this is nonsense. Microbial action in a rich marine environment like the Gulf will accelerate degradation but only to a point. The microbes move on leaving large complex chemical compounds in their aftermath. These are the asphaltenes that are too difficult to digest. What remains after decades is the dense tarry material that serves one useful purpose – surfacing roadways. Estimates from 2003 show that there was as much as 21,000 gallons of crude oil remaining in Prince William Sound decades after the Exxon Valdez spill. After all of these years, the oil, the legal disputes, and the negative impacts to the environment are still there.

The Ixtoc I spill was even more devastating and even more mysterious because less documentation exists. The oil drifting north from the spill destroyed hundreds of millions of crabs on Mexican beaches and killed 80% of the segmented worms and shrimp-like crustaceans in the sand along Texas beaches.

Since 1970, there have been 1,700 spills on record from tankers alone. The smallest recorded spill is 2,100 gallons of oil. There seems to be no reliable records on how much toxic chemical dispersants have been applied over the years in response to some of these spills. After the Deepwater Horizon spill, the fact still remains that there is a lack of oversight and complacency in issuing licenses for offshore drilling without ensuring appropriate emergency response infrastructure and plans.

The International Tanker Owners Pollution Federation (ITOPF) maintains a database[1] of oil spills from tankers, combined carriers, and barges. The database reports information on accidental spillages since 1970, excluding those resulting from acts of war. The data reported includes the type of oil spilt, the spill amount, the cause and location of the incident, and the vessel involved. For historical reasons, spills are generally categorized by size: <7 tonnes, 7–700 tonnes, and >700 tonnes (<50 bbls, 50–5,000 bbls, >5,000 bbls). The actual amount spilled is also recorded. The database shows nearly 10,000 incidents, the vast majority of which (82%) fall into the smallest category, i.e., < 50 bbls (bbls stands for barrels).

In general, the incidence of large spills is relatively low. However, ITOPF has noted that detailed statistical analysis is rarely possible. Regardless, this extensive database does show that the number of

[1]http://www.itopf.com/information-services/data-and-statistics/statistics/#background

Table 1.2 Examples of major oil spills ranked from largest to smallest.

Date	Location	Incident	Amount of Oil Spilled		Equivalent Number of Olympic Size Swimming pools
			In Gallons	In Barrels	
19-Jan-91	Persian Gulf, Kuwait	During the Gulf War, Iraqi forces, attempting to thwart a potential landing of American soldiers, opened the valves at an offshore oil terminal and dumped oil from several tankers. The oil they released created a 4-inch thick oil slick that covered 4,000 square miles. That's enough oil to cover the entire state of Rhode Island one-foot deep in oil.	520,000,000	12,380,952	802.5
3-Jun-79	Bay of Campeche off Ciudad del Carmen, Mexico (Incident lasted from June 3, 1979-March 23, 1980)	Ixtoc 1 Oil Spill – This oil spill did not involve a tanker, but rather an offshore oil well. Pemex, a state-owned Mexican petroleum company was drilling an oil well when a blowout occurred. The oil ignited causing the drilling rig to collapse. Oil began gushing out of the well into the Gulf of Mexico at a rate of 10,000 to 30,000 barrels a day for almost an entire year before workers were finally able to cap the well and stop the leak.	140,000,000	3,333,333	216.0

Table 1.2 (cont.) Examples of major oil spills ranked from largest to smallest.

Date	Location	Incident	Amount of Oil Spilled		Equivalent Number of Olympic Size Swimming pools
			In Gallons	In Barrels	
8-Sep-94	Kolva River, Russia	Kolva River Oil Spill – A ruptured pipeline caused this enormous oil spill. The pipeline had been leaking for eight months, but the oil was contained by a dike. When the dike collapsed, it sent millions of gallons of oil into the Russian Arctic.	84,000,000	2,000,000	129.6
18-Sep-83	Persian Gulf, Iran	Nowruz Oil Field Spill – This spill was the result of a tanker collision with an oil platform. The platform tilted and was closed, but the weakened platform collapsed sending oil spewing into the Persian Gulf. Delays in getting the leak capped were caused by the ongoing Iran-Iraq War.	80,000,000	1,904,762	123.5
6-Aug-83	Saldanha Bay, South Africa	Castillo de Bellver Oil Spill – The Castillo de Bellver caught fire approximately 70 miles northwest of Cape Town, South Africa. The ship drifted before breaking in two 25 miles off the coast. The ships stern sank in deep	79,000,000	1,880,952	121.9

		water still carrying approximately 31 million gallons of oil. The bow section was towed away and deliberately scrapped.			
16-Mar-78	Portsall, France	Amoco Cadiz Oil Spill – The Amoco Cadiz was caught in a fierce winter storm that damaged its rudder. The ship put out a distress call that it was no longer able to maneuver. Several ships responded, but none were able to stop the massive ship from running aground. On March 17th, the gigantic supertanker broke in two sending all of its 69 million gallons of oil into the English Channel. The ship was later sunk by the French.	69,000,000	1,642,857	106.5
11-Apr-91	Genoa, Italy	M/T Haven Tanker Oil Spill – This oil tanker exploded and sank off the coast of Italy killing six people. The source of the explosion was alleged to be the poor state of repair the ship was in. Supposedly, the Haven was scrapped after being hit by a missile during the Iran-Iraq War and then put back into operation. The vessel continued leaking its remaining oil into the Mediterranean for 12 years after the sinking.	45,000,000	1,071,429	69.4

Table 1.2 (cont.) Examples of major oil spills ranked from largest to smallest.

Date	Location	Incident	Amount of Oil Spilled		Equivalent Number of Olympic Size Swimming pools
			In Gallons	In Barrels	
10-Nov-88	Off the coast of Nova Scotia, Canada – This spill occurred approximately 700 nautical miles off the coast of Newfoundland.	Odyssey Oil Spill – No information on the incident was uncovered from research.	40,700,000	969,048	62.8
18-Mar-67	Scilly Isles, UK	The Torrey Canyon Oil Spill – The spill created an oil slick measuring 270 square miles. It contaminated approximately 180 miles of coastland and killed over 15,000 sea birds and enormous numbers of aquatic animals before the spill was finally contained.	36,000,000	857,143	55.6

		approximately 180 miles of coastland and killed over 15,000 sea birds and enormous numbers of aquatic animals before the spill was finally contained.			
19-Dec-72	Gulf of Oman	The Sea Star Oil Spill – The South Korean supertanker, Sea Star, collided with a Brazilian tanker, the Horta Barbosa, off the coast of Oman.	35,300,000	840,476	54.5
Feb-96	Coast of Southwest Wales	Sea Empress Grounding – The grounding of the Sea Empress caused one of the largest and most environmentally damaging oil spills in European history. More than 100 km of coastline became seriously polluted by oil. Ecosystems of conservation, fishery, and recreational importance were affected. A massive clean-up operation was launched both at sea and onshore. The pollution was at its height during late February and early March of 1996, at which time huge slicks were at sea, and many shores were experiencing large-scale bulk oil pollution. By April, as a result of natural dispersion and the clean-up operation, little bulk oil remained at sea and many shores had regained a semblance of normality. Wave-exposed rocky shores improved quickly	21,097,436	502,320	32.6

Table 1.2 (cont.) Examples of major oil spills ranked from largest to smallest.

Date	Location	Incident	Amount of Oil Spilled		Equivalent Number of Olympic Size Swimming pools
			In Gallons	In Barrels	
		through natural cleansing, followed by the intensively cleaned sands of the main bathing beaches, most of which were reopened for Easter. However, many shores were affected by residual oil pollution through the summer of 1996. Contamination persisted out of sight within shingle, sands, and muds and within surviving wildlife. Autumnal storms remobilized buried oil, causing the temporary reappearance of sheens and tar balls on many beaches. This storm action accelerated the natural cleansing of the coast. By spring of 1997, few shores showed visible evidence of oiling. Birds at sea were hit hard during the early weeks of the spill, resulting in thousands of casualties. The grey seal population appeared little affected and impacts to sub			

		tidal wildlife were limited. Considerable damage was caused to shorelines affected by bulk oil. Shore seaweeds and invertebrates were killed in large quantities. Mass strandings of cockles and other shellfish occurred on sandy beaches. Rock pool fish were also affected. On many shores the process of wildlife recovery began soon after the bulk oil dispersed naturally or was removed. Surviving wildlife was supplemented by plant and animal recolonizers. Recovery is likely to be protracted in the case of sheltered shores that continue to harbor deep-seated lingering residues of oil (e.g., some muddy shores in Milford Haven). As a result of the incident, fishing activity was banned. However, following intensive government testing of shellfish and finfish, the ban was lifted in stages.			
24-Mar-89	Coast of Alaska	Exxon Valdez oil – spilled 10.8 million gallons of oil along 1,100 miles of Alaskan coast. The spill could have been much worse – the Valdez was carrying 53 million gallons.	10,800,000	257,143	16.7

Table 1.2 (cont.) Examples of major oil spills ranked from largest to smallest.

Date	Location	Incident	Amount of Oil Spilled		Equivalent Number of Olympic Size Swimming pools
			In Gallons	In Barrels	
1-Nov-79	Galveston Entrance in the Gulf of Mexico	Burmah Agate – Burmah Agate collided with the freighter Mimosa southeast of Galveston Entrance in the Gulf of Mexico. The collision caused an explosion and a fire on the Burmah Agate that burned until January 8, 1980. An estimated 2.6 million gallons of oil were released into the environment, and another 7.8 million gallons were consumed by the fire.	10,400,000	247,619	16.0
27-Apr-86	Cativa Bay – Caribbean coast of Panama	Cavity Bay Release – On May 4th, a storm broke the containment booms, releasing ~150,000 barrels of oil into the Atlantic Ocean. Winds, tides, and rain runoff washed part of the oil onto exposed shorelines. Some of the oil was carried back into Cativa Bay, and some was washed into adjacent	10,080,000	240,000	15.6

		embayments with mangrove shorelines. By May 15th, oil had spread along the coast and washed across fringing reefs and into mangrove forests and small estuaries. A total of 82 km of coastline (=11 linear km) was oiled to varying degrees. Approximately 75 ha of mangroves, primarily the red mangrove Rhizophora mangle, were killed by the discharge. Oil slicks were observed frequently in Bahia Les Minas during the 4 years following the discharge. The slicks primarily originated from fringing mangrove areas that had been impacted by the discharge. As dead red mangroves decayed and their wooden structures disappeared, erosion of the associated oiled sediment occurred, releasing trapped oil. Oil was present in mangrove sediments and continued to appear on mangrove roots during the three years following the discharge, with the highest levels of continued oiling occurring in stream habitats and the lowest levels along the open coast. Total oil concentrations remained high, up to 20% of dry weight in surface sediments, for at least the first 4 years following the discharge. Residual pools of oil in mangrove			

Table 1.2 (cont.) Examples of major oil spills ranked from largest to smallest.

Date	Location	Incident	Amount of Oil Spilled		Equivalent Number of Olympic Size Swimming pools
			In Gallons	In Barrels	
		sediments were sufficiently fluid to flow out when sediments were cored or disturbed 5 years after the discharge. The release of oil from pools under and around the collapsed Refineria Panama storage tank and from mangrove sediments caused chronic re-oiling for at least 5 years following the discharge, and undegraded oil residues were found in some heavily oiled sediments 6 years after the discharge. Thus, the discharge site, initially injured by a single point source of oil, became a chronic source of oil contamination. Hydrocarbon chemistry confirmed the long-term persistence of crude oil in mangrove sediments, with pools of trapped oil maintaining consistent hydrocarbon composition. The frequency and amount of re-oiling differed among			

		habitats. Secondary re-oiling was heaviest in sheltered drainage systems of the mangrove environment, where oil continuously leaked from the sediment, but also occurred along the open coast and along channels. Seasonal variation in weather, water levels, and tidal flushing affected the amount of oil released. The greatest amount of re-oiling occurred between February and August 1989 and appeared to be related to the collapse and cutting of dead mangroves and to replanting efforts by the Refineria Panama.			
15-Dec-76	Southeast of Nantucket Island, Massachusetts	Argo Merchant ran aground on Fishing Rip (Nantucket Shoals) 29 nautical miles southeast of Nantucket Island, Massachusetts in high winds and ten foot seas. Six days later, the vessel broke apart and spilled its entire cargo.	7,700,000	183,333	11.9
30-Jul-84	Calcasieu River Bar Channel, LA	U.K. Tank Vessel Alvenus grounded in the Calcasieu River Bar Channel 11 miles southeast of Cameron, Louisiana. The vessel suffered a structural failure that ruptured the hull near the No. 2 tanks. Offshore recovery was hampered by rough weather and the magnitude of the spill. The spill moved slowly westward, coming ashore near High	2,751,000	65,500	4.2

Table 1.2 (cont.) Examples of major oil spills ranked from largest to smallest.

Date	Location	Incident	Amount of Oil Spilled		Equivalent Number of Olympic Size Swimming pools
			In Gallons	In Barrels	
		Island along the Bolivar Peninsula and into Galveston Bay, Texas between August 2nd and 5th. Oil impacts were severe at Rollover Inlet and Crystal Beach, and on August 4th more oil was pushed ashore further south along the Galveston Island coast. The spill affected 90% of Galveston's West Beach, including 80% of the Galveston seawall and the associated rock groins and pilings. Oil in the seawall area was a concern due to the oil's smothering effect on marine organisms. Tourism losses were estimated at $1 million per day. Commercial shrimp fishermen filed a suit for $10 million. By October 1st, the oiled beaches of Bolivar Peninsula and Galveston Island had been cleaned by the removal of oiled sand.			

Jan-83	Louisiana	A well blow-out – Comparisons of control and affected sites one year after the discharge revealed that oil effects on vegetation were species-specific. Areas with high shading by mature trees had little or no understory, and few effects of the oil were observed on the dominant woody vegetation. Perennial plants were returning to the sunlit areas. In contrast, oiled areas formerly covered with floating vascular vegetation were devoid of any vegetation.	1,260,000	30,000	1.9
Mar-73	La Parguera, Puerto Rico	Liberian tanker Zoe Colocotronis ran aground and pumped crude oil overboard. The wind drove about 60% of the oil into Bahia Sucia in southwestern Puerto Rico, where it affected a number of marine habitats including red and black mangrove swamps. The discharge site and an un-oiled reference site were evaluated qualitatively one week, 13 weeks, and 3 years after the discharge. Observations were made of the degree of prop root oiling, of the prop root invertebrate community, and of oil in swamp sediments. After 3 years, dead mangroves were evident and oil remained in sediments. It was noted in 1990 that,	1,118,803	26,638	1.7

Table 1.2 (cont.) Examples of major oil spills ranked from largest to smallest.

Date	Location	Incident	Amount of Oil Spilled		Equivalent Number of Olympic Size Swimming pools
			In Gallons	In Barrels	
		although most petroleum released at sea in tropical environments degrades rapidly, contamination reaching intertidal sediments may persist for decades. They observed discrete subsurface layers of petroleum hydrocarbons in intertidal sediment cores collected from the discharge site in 1990, 13 years after the discharge. The uppermost such layer contained petroleum hydrocarbon concentrations greater than 200 mg/g. A deeper layer with less concentrated petroleum hydrocarbons was believed to correspond to the earlier Argea Prima discharge in 1962. Sediments above, between, and below these layers had low concentrations of typical biogenic hydrocarbons.			

7-Jan-94	Escambron Beach in San Juan, Puerto Rico	Barge Morris J. Berman – The barge grounded on a hard bottom consisting of rocky substrate with scattered coral after its towing cable parted. The barge had a capacity of three million gallons but was reportedly only half full. The cargo, a heavy #6 fuel oil, began spilling and impacted nearby shoreline and shallow intertidal habitats immediately. No estimated leakage rate was available. Due to strong northerly winds, the surf at the grounding site was quite strong, creating a hazardous situation as they pounded the deck of the vessel. The responsible party initially assumed responsibility for the spill but very quickly expended the ten million dollar limit of their insurance policy. Full federal funding of the spill occurred at 0600 on January 14th, and it became a United States Coast Guard (USCG) directed response. Oil, in the form of large mats, accumulated on the surface and on the bottom of the lagoons. Submerged oil posed a major clean-up problem during the response. On January 15th, the barge was re-floated, towed to a scuttling site 20 miles northeast of San Juan, and sunk. Impacts were primarily along 12 miles	743,400	17,700	1.1

Table 1.2 (cont.) Examples of major oil spills ranked from largest to smallest.

Date	Location	Incident	Amount of Oil Spilled		Equivalent Number of Olympic Size Swimming pools
			In Gallons	In Barrels	
		of shoreline from Isabella to Borinquen. This oil was buried as oily sand layers and submerged as oil and sand mats in the protected areas or crenulate bays.			
13-Dec-68	Galeta Island, Canal Zone, Panama	The oil tanker Witwater broke up in heavy seas off the Atlantic coast of Panama. The breakup of the vessel and the continued leaking of the wreck spilled 14,000 barrels of bunker C and diesel oil into the water 5 miles from Galeta Island. The oil eventually impacted Galeta Island. Strong seasonal winds pushed the slick towards Galeta Island. Sand beaches, rocky coasts, and mangroves along the island were oiled. Oil collected in a small bay on the island. An estimated 8,000 barrels of oil remained in the bow section of the wreck and leaked at a rate of 50 barrels per day.	588,000	14,000	0.9

23-Oct-07	Kab Field, Bay of Campeche, Gulf of Mexico	The Usumacinta was brought into position alongside the Kab-101 platform to finish drilling the Kab-103 well. A cold weather front passed through the Gulf of Mexico bringing storm winds of 130 km/hr with waves of 6–8 m. The adverse weather conditions caused oscillating movements of Usumacinta jack-up. These movements caused the cantilever deck of the Usumacinta to strike the top of the production valve tree on the Kab-101 platform, resulting in a leak of oil and gas. The subsurface safety valves of wells 101 and 121 were closed by PEMEX personnel, but the valves were unable to seal completely, allowing the continued leaking of oil and gas. 81 personnel on the Usumacinta were evacuated by lifeboat. There were 21 reported deaths during the evacuation of the Usumacinta with one worker missing, presumed dead. Regarding the spill, PEMEX undertook measures to contain, recover, and disperse the spill. The initial spill was reported by PEMEX at circa 422 barrels per day of mostly light crude, of which 40% was estimated to have evaporated. Chemical dispersants were used	575,400	13,700	0.9

Table 1.2 (cont.) Examples of major oil spills ranked from largest to smallest.

Date	Location	Incident	Amount of Oil Spilled		Equivalent Number of Olympic Size Swimming pools
			In Gallons	In Barrels	
		alongside sea recovery operations, with a total of 8,701 barrels recovered from the sea. By early December 2007, PEMEX estimated that around 5,000 bbls oil had been lost from the well and not recovered. There was some criticism over the use of dispersants simply causing the oil to sink to the seabed.			
6-Mar-91	Nevis Island (British)	Vista Bella – the Vista Bella sank in the Atlantic Ocean approximately 12 miles northeast of Nevis Island (British). The barge, carrying 13,300 barrels of No. 6 fuel oil, sank in approximately 2,000 feet of water. The wind speed on the day of the incident was approx. 6–10 knots. The cause of the sinking was not determined. The barge was owned by Offshore Marine Limited and operated under the Trinidad flag. Dispersant was applied from March	558,600	13,300	0.9

		9th–15th within a two-nautical mile area of the source. The U.S. Coast Guard (USCG) Atlantic Strike Team (AST) was on-scene from March 14th to the 23rd. Contractor beach clean-up began on March 27.			
20-Mar-01	Rio de Janeiro, Brasil	Ten personnel were killed when the world's largest offshore oil rig exploded and sank.	395,000	9,405	0.6
10-Aug-93	Tampa Bay, Florida	Barge Bouchard 155 – Three ships collided in Tampa Bay, Florida – the barge Bouchard 155, the freighter Balsa 37, and the barge Ocean 255.	336,000	8,000	0.5
Jul-75	Western edge of the Florida Current	The tanker Garbis discharged crude oil-water emulsion. Prevailing easterly winds drove the oil ashore along a 30 mile stretch of the Florida Keys from Boca Chica to Little Pine Key. Red mangroves with >50% of their leaves oiled were killed, and red mangrove propagules with >50% oil coverage died within 2 months. Black mangroves with >50% of pneumatophores oiled were killed. Thin oil coating left chemical burn scars, and the germination of oiled seeds decreased by 30%. Batis and Salicornia spp. died when oil coated their leaves, stems, or substrate. Lightly oiled mangrove	126,000	3,000	0.2

Table 1.2 (cont.) Examples of major oil spills ranked from largest to smallest.

Date	Location	Incident	Amount of Oil Spilled		Equivalent Number of Olympic Size Swimming pools
			In Gallons	In Barrels	
		areas appeared to exhibit normal growth 6 months after the discharge. However, young and dwarf mangroves apparently suffered permanent injury, indicated by deformed leaves, roots, and stems. The discharge site was visited in May 1980, 5 years after the discharge, and it was reported that the oil had weathered significantly.			
1993-Oct-14	Rose Island, approximately 150 miles east of America Samoa	Jin Shiang Fa ran aground on Rose Island. The crew abandoned ship and were rescued by another fishing vessel in the area. The Jin Shiang Fa, a 137-foot, Taiwanese flagged fishing vessel, was carrying 10,000 gallons of diesel in its internal fuel tanks as well as an additional 100,000 gallons (2,390 barrels) of diesel in its forward cargo holds when	110,000	2,619	0.2

		it went aground. A Coast Guard overflight reported the vessel went aground on the west side of Rose Island, south of the narrow opening into the inner lagoon. A sheen, two miles long by 200 yards wide, trailed from the vessel in a northwesterly direction away from the island. An overflight the following day indicated that the vessel had a 30 degree list to port and an 11-mile by 200-yard wide sheen trailing to the northwest.			

large spills (> 5000 bbls) has decreased significantly during the last 40 years, such that the average number of major spills for the decade (2000–2009) is about three. Excluding the BP Gulf catastrophe, the average for the decade of the 2000s is less than half of the average for the 1990s and a mere eighth of the average for the 1970s. The same is true for medium-sized spills from tankers (50–5,000 bbls) where the average number of spills occurring in the last decade was 14, and half of that was experienced during the previous decade.

It is notable that the number of large spills in the 1970s is more than half of all the spills recorded in the 40 years between 1970 and 2009. Furthermore, the average number of large spills per year during the 1990s was less than a third of that witnessed during the 1970s. This downward trend continued during the 2000s, during which only 7% of all recorded spills occurred. The ITOPF along with the industry sector report in numerous sources that oil spills, in general, have become considerably less frequent compared to a few decades ago and, in fact, further argue that the statistics reflect a safety record that has improved over time. They imply and even claim that this is a reflection of the industry acting more responsibly because, along with tracking the oil spill statistics, the industry has made efforts to identify the root causes for spill incidents and generally has attempted to devise corrective actions that reduce the risks of such incidents from occurring.

We took the data reported by the ITOPF and created the chronological bar chart of recorded spills shown in Figure 1.1. On the chart we have included the oil release into the Gulf from BP's ill-fated Macondo well blowout of April 2010. There are no reliable estimates of how much oil has been released into the Gulf from this one single event. *The New York Times*[2] has reported that the range is somewhere between BP's reported estimate of 5,000 bbls/day to as high as 60,000 bbls/day. With potential fines from the Clean Water Act, which sets a civil fine of $1,100 for every barrel of oil released in a spill, in addition to the responsibility for paying for environmental clean-up and economic loss, to as much as $4,300 a barrel in fines if the U.S. government determines that the spill was the result of gross negligence, we have no confidence that BP's own estimate is accurate. Based on the information currently available, we have assumed that the truth lies somewhere in between. Therefore, the

[2]http://www.nytimes.com/2010/07/20/us/20flow.html

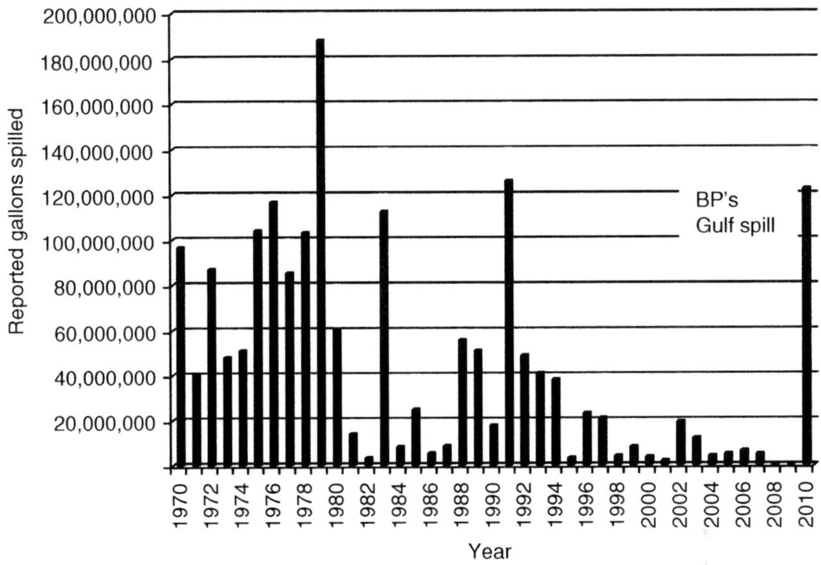

Figure 1.1 Bar chart showing reported oil spills. (Data converted from barrels to gallons from ITOPD database.)

data for the Gulf spill is shown as an average after 90 days of uncontrolled release into the Gulf.

Figure 1.1 leads one to a very different impression of the industry's safety record. While the frequency of oil spills has shown a decreasing trend since the 1970s, the chart clearly shows that for every decade since 1970 there have been catastrophic releases of oil into the environment. We don't share the same opinion as the industry sector in that it has a good track record. Nor do we believe that the industry has acted responsibly in taking steps to reduce such incidents from occurring. As explained in detail in Chapter 7, we have concluded that BP's Macondo well disaster was preventable. And while the industry sector as a whole may argue that BP's actions are not representative of the sector's checks and balances that are intended to prevent catastrophic releases of oil and the subsequent environmental damages and long-term health risks from such releases, we see more rhetoric than technical safeguards.

2

Origins of Spills

2.1 Offshore Drilling

The conventional process of drilling oil and gas wells is based on the use of a rotary drill bit that is lubricated by drilling fluids or muds. During the course of drilling, the bit grinds downward through rock layers thereby generating large amounts of ground-up rock known as drill cuttings.

The wells are constructed with multiple layers of pipe known as casing. Traditional wells are not drilled from top to bottom at the same diameter but in a series of progressively smaller-diameter intervals. The top interval is drilled starting at the surface and has the largest diameter hole. Drill bits of varying sizes are used to drill different diameter holes. The diameter can be 20″ or larger for the uppermost sections of the well, followed by different combinations of progressively smaller diameters. Common hole diameters are 17.5″, 14.75″, 12.25″, 8.5″, 7.875″, and 6.5″.

After a suitable depth has been reached, the hole is lined with casing that is slightly smaller than the diameter of the hole, and cement is then pumped into the space between the wall of the drilled hole and the outside of the casing. This surface casing is cemented. Next, a smaller diameter hole is drilled to a lower depth, and another casing string is installed to that depth and cemented. This process may be repeated several more times. The final number of casing strings depends on the regulatory requirements in place at that location and reflects the total depth of the well and the strength and sensitivity of the formations through which the well passes. Figure 2.1 shows a photograph of an offshore oil rig.

Figure 2.1 Offshore oil rig. (Source: "Environmental Benefits of Advanced Oil and Gas Exploration and Production Technology" (DOE 1999)).

Less than a generation ago, most wells were drilled to be relatively vertical and were completed at a depth to intersect a single formation. One full well was required for each completion. However, more modern technology enables more oil and gas production with less drilling and less waste generation. Drilling can be done to intersect multiple targets from the same main wellbore.

The oil and gas industry employs high-level technology in order to extract crude oil under extremely adverse conditions. Logistical barriers that the industry faces include extremely low temperatures, freezing and thawing of the tundra, and operating in remote locations in the Arctic or in fragile environments such as the Gulf of Mexico.

Horizontal drilling technology was introduced in the 1990s and currently is employed in about 90 percent of the wells drilled in some regions of the world. Oil and gas wells traditionally have been drilled vertically at depths ranging from a few thousand feet to as deep as 5 miles. Depending on subsurface geology, some wells have deviated from the strictly vertical orientation by anywhere from a few degrees to being completely horizontal or even inverted toward the surface. About 90 percent of all horizontal wells have been drilled into carbonate formations. These geological formations account for about 30 percent of all U.S. reserves.

Directional and horizontal drilling has enabled producers to reach reservoirs that are not located directly beneath the drilling rig. The advantage of this technology has been described as beneficial in avoiding sensitive surface and subsurface environmental features.

Horizontal drilling may allow a producer to contact more of the reservoir, so that more resources can be recovered from a single well. The Department of Energy has noted as an example that in the Mississippi Black Warrior Basin horizontal wells provide six times as much natural gas deliverability compared to conventional vertical wells at the Goodwin natural gas storage field. DOE further notes that in growing numbers of operations, the benefits from this increased production far outweigh the added cost for these wells. This was the case in the remote South China Sea where Phillips China, Inc. has employed advanced horizontal drilling and completion technologies to complete a 5-mile-long, extended-reach well. The use of directional drilling has also facilitated multilateral drilling and completion, enabling multiple offshoots from a single wellbore to radiate in different directions or contact resources at

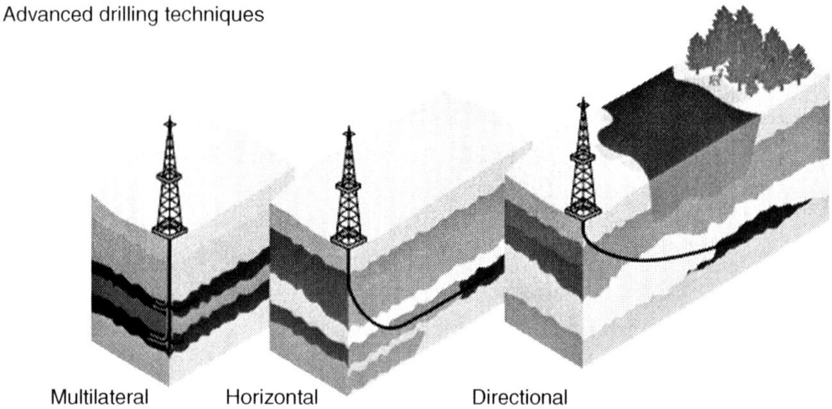

Figure 2.2 Illustration of drilling technique options. (Source: "Environmental Benefits of Advanced Oil and Gas Exploration and Production Technology" (DOE 1999)).

different depths. The development of radial drilling technology has spurred a boom in horizontal drilling. Since the mid-1980s, the drilling of horizontal wells has grown from a few to more than 2,700 wells per year worldwide. In the United States, horizontal drilling now accounts for 5 to 8 percent of the land well count at any given time. See Figure 2.2 for a conceptual illustration of drilling technologies.

Another drilling innovation is slimhole drilling. This technology is being applied to in order to tap into reserves in mature fields and is described as one that reduces waste volumes. The DOE notes as an example that a slimhole drilled to 14,760 feet and ending with a 4-⅛ inch bottomhole produces one-third less volume of cuttings than a standard well at the same depth. Operational footprints are also reduced since equipment for slimhole drilling is smaller than that used in conventional operations.

Coiled tubing technology is another advanced drilling technology. This technique is employed for drilling in re-entry, underbalanced, and highly deviated wells. It also has similar benefits of reducing drilling wastes and minimizing equipment footprints. A coiled tubing layout requires a working space that is about half of that of a conventional light workover hoist. The drilling site is easier to restore when operations are completed, and the impact of equipment mobilization on the environment is reduced. In addition, coiled tubing and slimhole drilling enable less disruptive and quieter drilling operations, thus minimizing the noise impact on wildlife or humans near the well site. Since coiled tubing is a continuous pipe, most noises associated with conventional drilling pipes are avoided. Efficient insulation and the equipment's smaller size further reduce noise levels.

All drilling operations carry high risks. Regardless of how efficient or advanced the technology is, systems fail, there is human error, and there is the potential for catastrophic failures, explosions, fires, loss of life, and releases of oil and gas into the environment. Redundant fail-safe mechanisms are employed and are supposed to prevent these events from occurring. These devices are an Acoustics Switch and Blow Out Preventers (BOPs).

A BOP is a large valve that is intended to seal off an oil or natural gas well. If underground pressure forces oil or gas into the wellbore, operators may close the valve remotely using hydraulic actuators in order to forestall a blowout from occurring, thereby regaining control of the well. Once this is accomplished, it is believed that the drilling mud density within the hole can be increased until adequate fluid pressure is placed on the influx zone, and the BOP can be opened for operations to resume pumping.

There are two general types of BOPs known as ram and annular configurations. Both types may be employed together in a stack arrangement.

A ram type BOP employs horizontally opposed hydraulic rams that can be fitted out to 1) close around the drill string, 2) shear through the drill string and then seal it, or 3) close off a wellbore when no drill pipe or tubing is in it.

An annular or spherical BOP is a more complex part, and there are several configurations. A common configuration is a flanged bottom (or clamp hub) connection and a studded top flange connection. Hydraulic operation is performed by connection to open and close (NPT) ports located on the body, which are connected to the control system. A spherical, steel-reinforced, rubber-packing element is forced inward by an upward movement of the piston. A spherical, inverted bowl in the upper housing causes steel inserts in the packing element to rotate inward toward the wellbore. The steel inserts prevent extrusion of the rubber and feed inward to close on any pipe or tubular in the wellbore. Replaceable bearings are used in the operating system to prevent metal contact of moving parts. An adapter ring located below the upper housing prevents contamination of the operating system during packing element replacement.

The BOP stack is the main wellhead pressure control equipment for oil and gas drilling, both onshore on offshore. It consists of an annular blowout preventer, a single ram blowout preventer, a double ram blowout preventer, and drilling spool.

The BOP sits directly on top of the wellhead on the ocean floor. In the event that a blowout or "kick" occurs, giant valves inside the BOP are intended to seal off the well, keeping any oil, gas, and excessive pressure contained within the ground. The sole purpose of the BOP is to safeguard against oil and gas escaping directly to the sea and causing tremendous damage to the environment. Ensuring the BOPs are properly maintained and tested is supposed to be one of the highest priorities of both the drilling contractor and the oil company involved in drilling the well. BOP testing is supposed to be closely watched by the Minerals Management Service (MMS), which apparently was not done in the Deepwater Horizon Spill incident.

The BOP is rigidly connected to the drilling rig by way of the marine riser. Drill pipe can be lowered down through the marine riser, through the BOP, into the wellhead, and then down into the well in order to drill deeper into the ocean floor. As the drilling fluid or mud is pumped through the drill pipe and out through the drill bit, it circulates all the way around, up through the marine riser, and back to the oil rig.

Drilling fluid helps clear the rock bits or "cuttings" that are constantly being chipped away as the drill bit drills deeper into the ground. The marine riser allows these "cuttings" to be brought back up to the oil rig to be collected and disposed of.

Once the mud returns to the drilling rig, the cuttings are filtered out and the mud is reused. Drilling mud can be costly, so being able to reuse it is very important. Oil companies can spend millions of dollars for drilling fluid or "mud" on every well drilled.

Oil well blowouts are uncontrolled releases of pressure from underground formations into the well hole or wellbore. Blowouts and kicks are devastating events. They can damage the drilling rig and endanger personnel on board if they are not properly dealt with. In a number of instances reported in Table 1.2 of Chapter 1, oil rig workers lost their lives to explosions and fires when uncontrolled gases from blowouts escaped from the well up to the surface of the water.

Drilling fluid is much heavier than seawater, with specific gravities that can be twice that of seawater. This is helpful when drilling a well because the weight of the drilling fluid creates pressure to keep pressure in the oil or gas formation from escaping back up through the well. The heavier the drilling fluid employed when drilling a well, the lower the probability of a formation of high pressure escaping back up into the well and up the marine riser. On the other hand, if the drilling fluid is too heavy, there is a risk of breaking or cracking the well casing. When this happens, the drilling fluid can leak out into the underground formation. This may also lead to a serious problem because it will not be possible to drill any deeper without being able to circulate the mud back up through the marine riser. Note that as the well is drilled deeper and deeper, the mud weight operating window becomes smaller.

Wells are drilled in sections. Each section of the well is drilled the same way with the cuttings and mud circulating back up to the drillship to be processed. The drill crews will drill deeper into the ocean floor so that the next section of casing can be run and cemented into place. The length of each section of casing is a decision that is left up to the drilling engineers that have closely analyzed the survey data and any data from nearby or "offset" wells.

As the drill bit continues to make its way towards the oil (known as the "pay zone"), the drilling crew closely monitors the amount of fluid in the storage tanks as well as the pressure of the formation to ensure that the well is not experiencing a blowout or kick.

If it is determined that the well is experiencing a "blowout" or kick, the BOP control valves are supposed to be closed off and the drill crew must take measures to stabilize the well. The conventional approach to stabilizing a well that has experienced a kick is to control the well by pumping heavier drilling fluids into the well in order to "force" the kicked pressure back into the formation.

Once the geologists are satisfied with the depth of the well, a series of tests called "logs" are conducted to determine how much oil and gas is present in the formation. The process is called logging because the information is logged into a database as it is collected. This is accomplished by employing high tech measurement devices that are sent into the hole to detect various features of the formation. The technology employs a combination of sound wave tools, electrical wave tools, and radiation measurement instruments whereby geologists are able to obtain readings inside the well that will assist them in determining the presence of oil.

2.2 Case Study

This case study concerns the Ixtoc I oil rig disaster which occurred in 1979. Mexico's government-owned oil company Pemex (Petróleos Mexicanos) was drilling a 1.9 mile-deep oil well when the drilling rig Sedco 135F lost drilling mud circulation.

As noted above, rotary drilling technology involves circulating the drilling mud down the drill pipe and back up the casing to the surface. The goal behind the mud circulation is to equalize the pressure through the shaft and to monitor the returning mud for any gas. When the rig lost the mud circulation, the well lost the counter-pressure. The pressure in the formation caused oil to fill the well column, subsequently blowing out the well. The oil caught fire, and Sedco 135F burned and collapsed into the sea.

At the time of the accident, Sedco 135F was drilling at a depth of about 11,800 feet below the seafloor. The day before the Ixtoc I experienced the blowout and resulting fire that caused the rig to sink, the drill bit encountered a region of soft strata. Subsequently, the circulation of drilling mud was lost resulting in a loss of hydrostatic pressure. Rather than returning to the surface, the drilling mud sought the paths of least resistance and began escaping into fractures that had formed in the rock at the bottom of the hole. Pemex officials decided to remove the bit, run the drill pipe back into the hole, and pump materials down this open-ended drill pipe in an effort to seal off the fractures that were causing the loss of circulation. This action turned out to be a fatal mistake. While removing the pipe on Sedco 135F, the drilling mud suddenly began to flow up towards the surface. By removing the drill-string, the well was "swabbed," thereby reducing frictional resistance and accelerating the upward flow from the reservoir.

An older practice that is employed to reduce pressure in a wellbore is to move pipe and wireline tools or rubber-cupped seals up the wellbore. In doing so, pressure is reduced sufficiently, and reservoir fluids may flow into the wellbore and towards the surface. The process of swabbing is dangerous in drilling operations because, as Ixtoc I discovered, it can lead to kicks and wellbore instability.

As in the case of the recent Deepwater Horizon incident in the Gulf, the rig employed a BOP, which the operator believed would stop the flow by activating the shear rams contained in the blowout preventer. These rams are intended to sever and seal off the well on the ocean floor. However, in this case drill collars had been brought in line with the BOP, and the BOP rams were not able to sever the thick steel walls of the drill collars, thus leading to a catastrophic blowout.

In the case of the Ixtoc I disaster, the drilling mud was followed by a large quantity of oil and gas at an increasing flow rate. The oil and gas fumes exploded on contact with the operating pump motors due to spontaneous combustion. This started a raging fire, which led to the collapse of the Sedco 135F drilling tower. The collapse caused damage to the underlying well structures. The damage to the well structures led to the release of vast quantities of oil into the Gulf.

During the early stages of the spill, about 30,000 barrels of oil per day (1.26 million gallons per day) were flowing unrestricted from the well. In July 1979, the pumping of mud into the well reduced the flow to 20,000 barrels per day (about 840,000 gallons per day), and early in August the pumping of nearly 100,000 steel, iron, and lead balls into the well reduced the flow to 10,000 barrels per day (420,000 gallons per day). Pemex claimed that half of the released oil burned when it reached the surface, a third of it evaporated, and the rest was contained or dispersed. Mexican authorities also drilled two relief wells into the main well in order to lower the pressure of the blowout, but the oil continued to flow for three months following the completion of the first relief well.

Pemex contracted Conair Aviation to spray the chemical dispersant COREXIT 9527 on the oil. A total of 493 aerial missions were flown, in which 1,100 square miles of oil slick were attacked. Dispersants were not used in the U.S. area of the spill because of the dispersants' inability to treat weathered oil. Eventually, the on-scene coordinator (OSC) requested that Mexico stop using dispersants north of 25°N.

COREXIT 9527 is a concentrated oil spill dispersant intended for use on a wide range of oils. A product bulletin describes the product as a self-mixing dispersant that is biodegradable and of low toxicity.[1] The following is the MSDS (Material Safety Data Sheet) for this product.

---

[1]Product Bulletin PB–EC9527A

EXXON CHEMICAL AMERICAS DIV EX – COREXIT 9527
==
MSDS Safety Information
===
FSC: 6850
MSDS Date: 06/14/1992
MSDS Num: BRYBG
LIIN: 00F030320
Product ID: COREXIT 9527
MFN: 01
Responsible Party
Cage: 72190
Name: EXXON CHEMICAL AMERICAS DIV EXXON CORP
Box: 3272
City: HOUSTON TX 77001
Info Phone Number: 713-870-6000
Emergency Phone Number: 713-870-6000
Review Ind: Y
Published: Y

===
Preparer Co. when other than Responsible Party Co.
===
Cage: 72190
Name: EXXONMOBIL CHEMICAL CO, DIV OF EXXON MOBILCORP
Address: 13501 KATY FWY
Box: 3272
City: HOUSTON TX 77253-3272

===
Contractor Summary
===
Cage: 72190
Name: EXXONMOBIL CHEMICAL CO, DIV OF EXXON MOBILCORP
Address: 13501 KATY FWY
Box: 3272
City: HOUSTON TX 77253-3272
Phone: 281-870-6000 / 800-726-2015

==
Ingredients
==
Cas: 111-76-2
RTECS #: KJ8575000
Name: 2-BUTOXYETHANOL (ETHYLENE GLYCOL MONOBUTYL ETHER), BUTYL CELLOSOLVE, BUTYL GLYCOL, GLYCOL ETHER EB
% Wt: 38
Other REC Limits: 25 PPM (SKIN)
OSHA PEL: S, 50 PPM
ACGIH TLV: S, 25 PPM; 9293

==
Health Hazards Data
==
Route of Entry Inds – Inhalation: YES
Skin: YES
Ingestion: YES
Carcinogenicity Inds – NTP: NO
IARC: NO
OSHA: NO

Effects of Exposure: EYES: IRRITATING, MAY INJURE TISSUE IF NOT REMOVED PROMPTLY. SKIN: IRRITATING. ABSORPTION MAY CAUSE HEMOLYTIC ANEMIA & KIDNEY DAMAGE EVIDENCED BY PALENESS & RED COLORATION OF URINE. INHALATION: RESPIRATORY TRACT IRRITATION. SYSTEMIC EFFECTS. INGESTION: LOW TOXICITY. INHALATION/SKIN OVEREXPOSURE: BLOOD/KIDNEY DAMAGE.

Explanation of Carcinogenicity: NONE

First Aid: EYES: FLUSH W/PLENTY OF WATER FOR AT LEAST 15 MINS. SKIN: FLUSH W/PLENTY OF WATER. USE SOAP IF AVAILABLE. REMOVE CONTAMINATED CLOTHING, INCLUDING SHOES, AFTER FLUSHING HAS BEGUN. INHALATION: REMOVE TO FRESH AIR. ADMINISTER ARTIFICIAL RESPIRATION IF BREATHING IS STOPPED. KEEP AT REST. INGESTION: INDUCE VOMITING. DON'T ATTEMPT TO GIVE ANYTHING BY MOUTH TO AN UNCONSCIOUS PERSON. OBTAIN MED ATTN.

===
Handling and Disposal
===

Spill Release Procedures: ELIMINATE IGNITION SOURCES. SMALL/LARGE SPILL:

IMPLEMENT CLEANUP PROCEDURES. FOR LARGE IF IN PUBLIC AREA, EVACUATE AREA. CONTAIN LIQUID W/SAND/EARTH. DON'T USE COMBUSTIBLES. RECOVER BY PUMPING (EXPLOSION PROOF/HAND PUMP) OR W/A SUITABLE ABSORBENT.

Waste Disposal Methods: DISPOSE OF IN ACCORDANCE W/ LOCAL, STATE, & FEDERAL REGULATIONS. PREVENT LIQUID FROM ENTERING SEWERS, WATERCOURSES OR LOW AREAS. COMPLETELY DRAIN, PROPERLY BUNG & PROMPTLY RETURN EMPTY DRUMS TO A DRUM RECONDITIONER OR PROPERLY DISPOSED OF NA 1993.

Other Precautions: EMPTY CONTAINERS RETAIN RESIDUE & CAN BE DANGEROUS. DON'T PRESSURIZE/CUT/WELD/BRAZE/SOLDER/DRILL/GRIND/EXPOSE SUCH CONTAINERS TO HEAT/FLAME/SPARKS/ELECTRICITY/OTHER IGNITION SOURCES; THEY MAY EXPLODE & CAUSE INJURY OR DEATH.

===
Fire and Explosion Hazard Information
===

Flash Point Method: TCC
Flash Point Text: 163F
Lower Limits: 1.1%
Upper Limits: 10.6%

Extinguishing Media: ALCOHOL TYPE FOAM, DRY CHEMICAL OR WATER SPRAY

Fire Fighting Procedures: USE WATER SPRAY TO COOL FIRE EXPOSED SURFACES & TO PROTECT PERSONNEL. ISOLATE FUEL SUPPLY FROM FIRE. AVOID SPRAYING WATER DIRECTLY INTO STORAGE CONTAINERS.

Unusual Fire/Explosion Hazard: COMBUSTIBLE LIQUID, CAN FORM COMBUSTIBLE MIXTURES AT TEMPS AT OR ABOVE THE FLASHPOINT.

==

Control Measures

==

Respiratory Protection: IF THE TLV HAS BEEN EXCEEDED, USE A NIOSH/MSHA APPROVED RESPIRATOR.

Ventilation: LOCAL EXHAUST. STORE & HANDLE LABORATORY SAMPLES IN A LAB HOOD. PROVIDE MECHANICAL VENTILATION IN CONFINED AREAS.

Protective Gloves: CHEMICAL RESISTANT

Eye Protection: CHEMICAL SAFETY GOGGLES W/FACE SHIELD

Other Protective Equipment: CHEMICAL SUIT, RUBBER BOOTS, LONG SLEEVE CLOTHING.

==

Physical/Chemical Properties

==

B.P. Text: 340F
M.P/F.P Text: –45F
Vapor Pres: 5
Spec Gravity: 1
Evaporation Rate & Reference: (N-BU AC = 1): 0.1
Solubility in Water: COMPLETE
Appearance and Odor: AMBER LIQUID W/GLYCOL ETHER ODOR.

==

Reactivity Data

==

Stability Indicator: YES
Stability Condition To Avoid: HEAT, FLAME, SPARKS, STATIC ELECTRICITY OR OTHER SOURCES OF IGNITION.
Materials to Avoid: STRONG OXIDIZING AGENTS
Hazardous Decomposition Products: SMOKE FUMES, CO, CO2
Hazardous Polymerization Indicator: NO

==

Toxicological Information

==

==

Ecological Information

==

==

MSDS Transport Information

==

==

Regulatory Information

==

==

Other Information

==

==

HAZCOM Label

==

Product ID: COREXIT 9527
Cage: 72190
Company Name: EXXONMOBIL CHEMICAL CO, DIV OF EXXON MOBILCORP
Street: 13501 KATY FWY
PO Box: 3272
City: HOUSTON TX
Zipcode: 77253-3272
Health Emergency Phone: 713-870-6000
Label Required IND: Y
Date Of Label Review: 09/30/1993
Status Code: C
Label Date: 09/30/1993
Origination Code: F
Chronic Hazard IND: Y
Eye Protection IND: YES
Skin Protection IND: YES
Signal Word: DANGER
Respiratory Protection IND: YES
Health Hazard: Severe
Contact Hazard: Severe
Fire Hazard: Moderate

Reactivity Hazard: Slight
Hazard and Precautions: EYES: IRRITATING, MAY INJURE TISSUE IF NOT REMOVED PROMPTLY. SKIN: IRRITATING. ABSORPTION MAY CAUSE HEMOLYTIC ANEMIA & KIDNEY DAMAGE EVIDENCED BY PALENESS & RED COLORATION OF URINE.INHALATION: RESPIRATORY TRACT IRRITATION. SYSTEMICEFFECTS.INGESTION:LOWTOXICITY.INHALATION/SKIN OVEREXPOSURE: BLOOD/KIDNEY DAMAGE. TARGET ORGANS: EYES, SKIN, KIDNEYS, RESPIRATORY TRACT, BLOOD. TARGET ORGANS: EYES, SKIN, LIVER, KIDNEYS, RESPIRATORY & DIGESTIVE TRACTS, HEART.
=======================================

Disclaimer (provided with this information by the compiling agencies): This information is formulated for use by elements of the Department of Defense. The United States of America in no manner whatsoever expressly or implied warrants, states, or intends said information to have any application, use or viability by or to any person or persons outside the Department of Defense nor any person or persons contracting with any instrumentality of the United States of America and disclaims all liability for such use. Any person utilizing this instruction who is not a military or civilian employee of the United States of America should seek competent professional advice to verify and assume responsibility for the suitability of this information to their particular situation regardless of similarity to a corresponding Department of Defense or other government situation.

Not everything in product bulletins is truthful. According to the manufacturer's MSDS, the product contains 38 wt. % 2-butoxyethanol (ethyleneglycol monobutyl ether, EGBE), butyl cellosolve, butyl glycol, and glycol ether EB. These are not exactly non-toxic chemicals. EGBE causes disorders of the central nervous system, blood, the kidneys, and the liver[2], not to mention that it is highly toxic to marine life. But there are even greater discrepancies that place workers and possibly the public at risk with this product. Another more recent NALCO Co. MSDS[3] reports that the product contains

[2]Toxicological Review of Ethylene Glycol Monobutyl Ether (EGBE), EPA/635/R-08/006F, U.S. EPA, Washington, D.C., March 2010.
[3]Nalco Company MSDS for COREXIT® EC9527A, Version Number: 2.0, Date issued : 05/11/2010

30–60 wt. % 2-butoxyethanol and states, "None of the substances in this product are listed as carcinogens by the International Agency for Research on Cancer (IARC), the National Toxicology Program (NTP) or the American Conference of Governmental Industrial Hygienists (ACGIH)." In contrast, a New Jersey Right to Know Fact Sheet[4] states that "2-Butoxy Ethanol should be handled as a CARCINOGEN – WITH EXTREME CAUTION." The chemical manufacturer and supplier may not be responsible when they do not rely on the most conservative health risk information. The reader may refer to additional discussions in Chapter 6 (see Section 6.3.1).

In Texas, an emphasis was placed on coastal countermeasures to protect the bays and lagoons formed by the barrier islands. The impacts of oil on the barrier island beaches were ranked as second in importance to protecting inlets to the bays and lagoons. This was done with the placement of skimmers and booms. Efforts were concentrated on the Brazos-Santiago Pass, Port Mansfield Channel, Aransas Pass, and Cedar Bayou, which during the course of the spill was sealed with sand. Economically and environmentally sensitive barrier island beaches were cleaned daily. Laborers used rakes and shovels to clean beaches rather than heavier equipment that would remove too much sand. Ultimately, 71,500 barrels of oil impacted 162 miles of U.S. beaches, and over 10,000 cubic yards of oiled material were removed.

This case study is most relevant to the ongoing Gulf environmental disaster. Thirty-one years after the Ixtoc disaster, the oil industry is employing the identical procedures applied in the 1970s to bring a deep drilling well blowout under control. The Mexican national oil company attempted to kill the well with drilling mud and then with steel and lead balls forced into the wellhead. They tried to contain the oil with a containment dome that was nicknamed the Sombrero. After 290 days, a relief well plugged the well with cement, and the spill finally came to an end. Barring the absence of acts of war, the Ixtoc catastrophe was the worst accidental oil spill in history, releasing 138 million gallons of oil into the Gulf. It would seem logical that the British Petroleum Co. and the rest of the industry might have learned something from the Ixtoc disaster and have developed more reliable technologies. Instead it has buried its head in the sand

---

[4]New Jersey Right to Know Fact Sheet for 2-Butoxy Ethanol, Date: February 2001 Revision: August 2008

all these years while reaping billions of dollars in profits. Clearly the federal government has failed its citizens because if there were proper oversight on the part of the Minerals Management Service, there would not be a repeat of Ixtoc.

BP's game plan to bring its Deepwater Horizon kick under control is a 31-year-old response plan borrowed from the Mexican national oil company in which attempt after attempt failed. BP's approach has included a top kill approach, in which an attempt was made to force mud and heavy objects down the well and then two containment domes, the first of which was a dismal failure and the second only partially working.

A June 5, 2010 investigative report by *USA Today*[5] reported that personnel of the Minerals Management Service accepted tickets to sports events, lunches, and other gifts from oil and gas companies. The violations in a report occurred between 2000 and 2008, long before the April 20th explosion on the Deepwater Horizon rig that caused the massive oil spill in the Gulf of Mexico. *The Washington Post* also reported flaws with federal oversight of offshore drilling. In a story on the same day, it stated that regulators "repeatedly ignored warnings from government scientists about environmental risks in its push to approve energy exploration activities quickly." It added, "Minerals Management Service officials, who can receive cash bonuses in the thousands of dollars based in large part on meeting federal deadlines for leasing offshore oil and gas exploration, frequently changed documents and bypassed legal requirements aimed at protecting the marine environment..."

Should these claims turn out to be true, then it is reasonable to expect and hope for criminal investigations to be launched into the behavior of personnel that are/were government officials paid to serve citizens and protect the environment as well as British Petroleum representatives.

The emergency remedial action response to the Deepwater Horizon spill itself is equally disturbing. The publication "Environmental Benefits of Advanced Oil and Gas Exploration and Production Technology" prepared by the U.S. Department of Energy writes, "in the past three decades, the petroleum business has transformed itself into a high-technology industry. Dramatic advances

---

[5]http://content.usatoday.com/communities/greenhouse/post/2010/05/new-us-report-shows-lax-oversight-of-offshore-drilling/1

in technology for exploration, drilling and completion, production, and site restoration have enabled the industry to keep up with the ever-increasing demand for reliable supplies of oil and natural gas at reasonable prices. The productivity gains and cost reductions attributable to extending the frontiers of technology... But public awareness of the significant and impressive environmental benefits from new exploration and production (E&P) technology advances remains limited.... Greater awareness of the industry's achievements in environmental protection will provide the context for effective policy, and for informed decision making by both the private and public sectors...Today's advanced exploration and production operations are models of high technology and resource efficiency. They incorporate high-speed computing, remote sensing and imaging, geologic interpretation, and visualization technology. They use global positioning systems, the latest geographical information systems, and 3-D seismic and 4-D imaging capabilities."

While the oil and gas industry has transformed itself into a high-tech industry, enabling it to find oil in remote locations and drill to offshore depths that are unprecedented, it has ignored remedial response technologies and has created a gap leaving both response infrastructure and technologies back in the 1980s. The emergency response techniques and infrastructure being used to respond to the Deepwater Horizon spill are identical to those relied upon to address the Exxon Valdez spill in 1989. Shockingly, there have been no advances at all. The techniques described in the chapters to follow represent the best available practices and technologies we have to address what is the nation's worst environmental disaster of all times. While these technologies are not ineffective, they are limited, and public officials who are responding to news media pressure have repeatedly admitted that they are marginal and that available resources and infrastructure affordable to address the Gulf disaster are being overwhelmed by the magnitude of the release. While the public anger mounts against BP, frustration is best focused not simply on a company that has failed to act responsibly towards protecting the environment but towards the industry as a whole and government. Elected officials have failed to establish mandatory standards by which the industry is required to act in a responsible manner to protect coastlines, preserve the environment, respect and preserve sensitive ecosystems, protect neighboring properties, and sustain Earth.

3

Use of Chemical Dispersants

3.1 Dispersants

Dispersants are chemicals that are sprayed onto oil slicks to accelerate the process of natural dispersion. Significant environmental and economic benefits can be achieved by applying these chemicals to oil spills, particularly when other at-sea response techniques are hampered by weather conditions or the limited availability of resources. In certain situations, dispersants may provide the only means of removing significant quantities of surface oil quickly. The application of these chemicals is intended to minimize the damage caused by floating oil to birds, for example, or before the oil impacts sensitive shorelines. However, the use of dispersants has limitations, and applications require careful planning since they too bring negative consequences to ecosystems and humans.

Natural attenuation or dispersion of an oil slick occurs when waves and other forms of turbulence (mixing energy) at the sea surface cause segments of the slick to break up into droplets and enter into the water column. By adding chemicals, this natural process is accelerated.

There has been considerable laboratory testing, along with a large number of field trials and successful applications, to confirm the effectiveness of chemical dispersants. Various trials, field exercises, and actual applications have been conducted in the following regions over the time periods indicated below:

- Canada, 1975–1981
- United Kingdom (U.K.), 1978
- U.S. East Coast, 1978–1979
- U.S. West Coast. 1979
- France, 1979–1983
- Norway, 1984, U.K. and Norway, 1994–1999

Two factors that are critical to successful applications are the degree of mixing energy applied and the type of crude that the dispersants are applied to.

There are two main features common to all dispersants: a surfactant and a solvent. Surfactants are comprised of molecules that are made up of two parts: an oleophilic part (with an attraction to oil) and a hydrophilic part (a component with an attraction to water). When dispersants are sprayed onto an oil slick, the solvent will transport and distribute the surfactants throughout the oil slick to the oil/water interface where the molecules arrange themselves so that the oleophilic part of the molecule is in the oil and the hydrophilic part is in the water. This alignment creates a reduction in the surface tension at the oil/water interface, causing small oil droplets to break away from the oil slick aided by the ocean's wave energy. The droplets formed will be of varying sizes, and although the larger ones may rise back to the surface some will remain in suspension and will drift apart undergoing degradation by naturally occurring microbial action. If the dispersion is successful, a characteristic brown plume can be observed to slowly spread down from the water surface shortly after the initial treatment.

Dispersants have not been used successfully on very viscous, floating oils because they tend to run off the oil into the water before the solvent can penetrate the oil layer. A rule of thumb is that dispersants are capable of dispersing liquid oils and emulsions with viscosities of less than 2,000 centistokes, which is equivalent to a medium fuel oil at 10–20°C. They are unsuitable for dealing with viscous emulsions (mousse) or oils that have a pour point near to or above that of the ambient temperature. Even oils that appear to be dispersed readily at first tend to become resistant after a period of time because the viscosity increases as a result of evaporation and emulsification. For a particular oil, the time available before the dispersant stops being effective depends on a number of variables including the surface conditions of the sea and temperature. Generally, dispersants are not effective for periods of 1 to 2 days after their application. Dispersants have shown good success with viscous oils on shorelines due to prolonged contact time, which allows for better penetration of the dispersant into the oil layer.

Dispersants are generally much more effective on seawater than freshwater. Part of this has to do with the differences in the water structures of fresh and saline waters. With freshwater, surface tension forces are a result of attractive forces between water molecules. Because of the presence of the sodium and chloride ions in seawater, there are stronger internal attractive forces between water molecules (refer to Figure 3.1).

The old expression that "oil and water don't mix" stems from the fact that when oil comes into contact with seawater it has a tendency to relax surface tension forces. See Figure 3.2. As noted earlier, surface active agents can be applied to make the oil compatible with water. This is due to the presence of soluble ends of the molecular structure (i.e., "lipophillic," meaning oil loving, and "hydrophyilic," meaning water loving). Our chemistry is sufficiently advanced that it is possible to prepare customized surfactant blends for different applications, thereby achieving a hydophillic/lipophillic balance

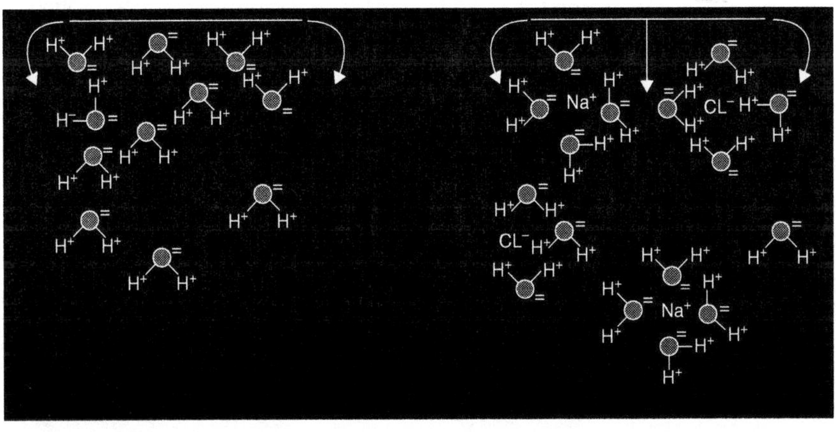

Figure 3.1 Comparison of surface tension effects from weak attractive forces in freshwater (left) and strong attractive forces in salt water (right).

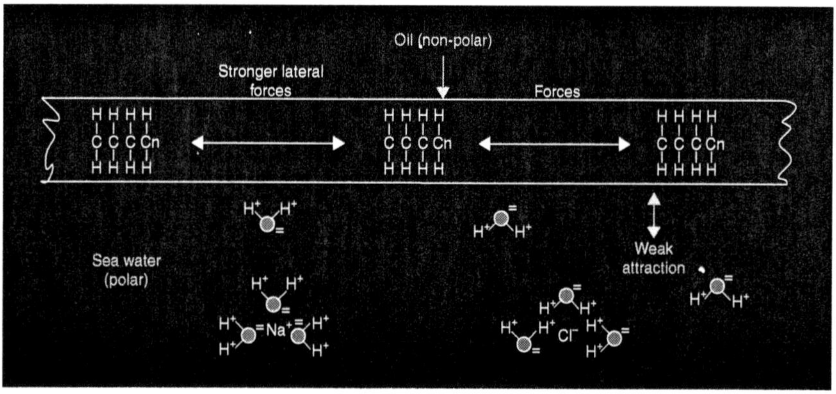

Figure 3.2 A relation of surface water surface tension forces from the introduction of oil into seawater.

(HLB balance). Generally, this balance lies between 5 and 15% for an optimum formulation for oil dispersants.

The goal of any dispersant is to reduce oil concentrations to below levels of impact on target organisms and ecosystems as rapidly as possible. When dispersants are applied to an oil slick, they disperse the oil into droplets with minimal energy. The surfaces of the oil droplets repel each other, thereby eliminating coalescence. This has the advantage of minimizing shoreline impacts. For example, dispersed oil will tend to settle out and, therefore, will not impact vegetation such as mangrove trees along the shoreline. However, there is a huge trade-off because the dispersed oil will sink and harm the neighboring coral reefs and sea grass along with many other organisms. In contrast, oil layers floating on the surface that have not been attacked by dispersants will not impact the coral or sea grasses, but they can kill the mangroves. This is why dispersants are arguably a technology that simply mask the problems created by an oil spill. It is in some ways the oil industry's approach to keeping spills out of sight and out of mind for the public.

There are several problems with the application of chemical dispersants. Among these are the following:

- There are not many products commercially available. The ones listed in the next section are essentially the only products we have been able to identify that have been tested in the field to some degree and have been applied to spills of varying sizes.

- All of the chemicals are toxic to varying degrees. Some are less toxic than others, but their toxicity to marine life varies significantly as reported under laboratory test conditions and even orders of magnitude among species. A number of the products are toxic to humans, animals, and plant and aquatic life and must be handled in dilute forms.
- Dispersants mask the problem. This means that oil is essentially herded into tiny droplets that settle out from the sea surface into large underwater plumes. While not quite a dissolved phase, the underwater plume, which extends for miles and perhaps a thousand feet or more in depth, is a dispersed phase of oil in water that cannot be separated out. Microbial action will feed on these plumes, but in their wake they remove oxygen from the seawater, thereby stressing larger animals and even creating kill zones.

Note that some products (e.g., COREXIT) require some concentration of sea salt in order to be effective dispersants. Seawater on the average contains 34 ppt (parts per trillion) of salt. The effectiveness of salt-sensitive dispersants begins to decline around 10 ppt and some products like COREXIT 9500 are ineffective at concentrations below 3 ppt.

Both Canada and France have conducted field tests with dispersants in freshwater and have applied commercially available dispersants to actual spills with reported low effectiveness. France is the only country that has devised an approval process for dispersants for freshwater applications. The following is a list of products approved under the French program:

- Dasic Freshwater
- Disperep 8
- Enersperse 1037
- Inipol IPF
- Gamlen OD 4500
- Petrotech 25

While dispersants have been developed in the United States for freshwater spill applications, none are listed on the U.S. EPA National Contingency Plan Product Schedule.

3.2 Methods of Application

3.2.1 Application at Sea

Dispersants can be applied to open water by several methods. In general, workboats are more suitable for treating minor spills in harbors or confined waters. Large multi-engine planes are best equipped for handling large offshore spills. Small, single-engine aircrafts and helicopters have been employed for treating smaller spills and near-shore areas.

The droplet size of the dispersant is important as it needs to be sufficiently large enough to overcome the effects of wind and evaporative loss but not so large that it will result in the droplets' ability to pierce through the oil slick. A uniform spray pattern of larger droplets, or "rain drops," is required rather than a fog or a mist. The key to a successful response using chemical dispersants is the ability to target the thickest part of the oil slick within a short time and before weathering or the state of the sea renders the oil unable to be dispersed.

3.2.1.1 Vessel Spraying

Dispersants are usually applied from boats equipped with spray arms. In a typical spray arm system, pumps are used to pump dispersants from a storage tank through a set of nozzles calibrated to produce a uniform spray pattern of droplets. Spray units can be portable or permanently installed on a vessel, and systems are available that deliver the dispersant either undiluted or diluted with seawater. Spray arms are mounted as far forward on the vessel as possible to avoid the effect of the bow wave, which can push the oil beyond the spray swath.

Mounting the spray arms on the bow allows the vessel to travel faster and also allows for longer spray arms because the freeboard area is often greater at the bow. This combination allows for the optimization of the amount of oil that can be treated (increasing the encounter rate) with a limited dispersant payload.

Fire hoses or monitors are also used to apply diluted concentrate dispersants. However, optimum dilution of the dispersant is difficult to achieve because of the very high flow rates, and wastage of dispersant can be a common problem. The high-powered jet of water also makes it difficult to apply the dispersant as a uniform spray of droplets, causing them to pierce through the oil layer making it ineffective. Thus, fire monitors are generally not considered to be an effective application tool unless they are specially modified for the purpose.

Vessels offer certain advantages for dispersant spraying because they are usually readily available, easy to load and deploy, have cost advantages over aircraft, and can target dispersants to specific areas of a slick. They also have limitations, particularly for larger spills, because of the low treatment rate and the added difficulty of locating the heaviest concentrations of oil from the bridge of a vessel. An additional complication is when slicks become fragmented or form narrow windows, causing some dispersant to be sprayed onto clear sea. These problems can be partially overcome by controlling the operation from spotter aircraft.

3.2.1.2 *Aerial Spraying*

Spraying dispersant chemicals from an aircraft has the advantages of rapid response, good visibility, high treatment rates, and optimum dispersant use. In addition, aircraft allows treatment of spills at greater distances from the shore.

Two categories of aircraft are used: those designed for agricultural or pest control operations, which require minor modification for dispersant application, and those that have been specifically adapted for the application of dispersant. Various types of helicopters have also been adapted to spray dispersants, but the most common are configurations that are able to carry an underslung bucket spray system. The most suitable aircraft will be determined primarily by the size and location of the spill, but local availability can be a factor. The endurance, fuel consumption, turnaround time, payload, and ability to operate from short or improvised landing strips are important factors that should be considered as part of emergency response plans. The aircraft must be capable of operating at low altitudes and relatively low speeds (50–150 knots) and must be highly maneuverable.

Only Type 3 dispersants (described below) are suitable for aerial spraying since they require no mixing beyond that provided by the natural movement of the sea. The relatively low-dose rate required also makes the best use of available payload.

3.3 Types of Dispersants and Commercial Products

This section describes various commercially available dispersants. Where available, we have extracted information from MSDSs to summarize the properties and characteristics of the product.

There are other chemical products that have been developed for combating shorelines impacted by oil spills. These various chemical products can be used to wipe and clean hard surfaces that have been coated with oil. They are described in Chapter 4 along with adsorbents and booms.

Dispersants fall into three general categories based on product form and method of handling:

- Type 1 dispersants are based on hydrocarbon solvents with 15–25% surfactant. They are sprayed neat (i.e., undiluted) onto the oil (the product is already in a pre-dilution state when supplied by the provider). The chemical must be applied directly to the oil because seawater renders the product(s) ineffective. Typical dose rates are between 1:1 and 1:3 (dispersant:oil).
- Type 2 dispersants are dilutable, concentrated dispersants that are alcohol or glycol (i.e., oxygenated) solvent-based with a higher surfactant concentration. Dilution is normally 1:10 with seawater. The product COREXIT was given as an example in Chapter 2.
- Type 3 dispersants are also concentrated dispersants with a similar formulation to Type 2 products. However, they are designed to be used neat, and typical dose rates are between 1:5 and 1:30 (neat dispersant:oil).

Type 1 and 2 dispersants require thorough mixing with the oil after application to produce satisfactory dispersion. With Type 3 products, the natural movement of the sea is usually sufficient to achieve mixing. The lower application rates required with concentrates mean that Types 2 and 3 have largely superseded Type 1 dispersants for application at sea.

In severe sea conditions, breaking waves submerge the oil, which prevents direct contact between the dispersant and the oil. Dispersant sprayed onto water rather than oil will be ineffective.

One study reports field trials indicating that a wind speed between 4 to 12 m/s (about 5 to 25 knots) is optimum for application.[1]

Dispersants are toxic to marine life, animals, and sensitive ecosystems such as coral reefs and have generally low toxicity to humans. Some are more toxic than others. The aquatic toxicity of oil spill chemical dispersants is affected by many factors including exposure conditions, species, and life stage. A wide range of LC_{50} (Lethal Concentration values, i.e., the concentration at which 50% mortality of a test population is recorded) can be generated even for a single species, depending on exposure duration and temperature. Increasing exposure duration generally results in increased dispersant toxicity (i.e., lower LC_{50} values). It is generally reported that dispersants are less toxic when applied at sea than shown in laboratory testing due to the turbulence and dilution effects of waves. However, there are not steadfast guidelines. The following studies by Fiocco[2], et. al., the NRC[3], and Singer, et. al.[4] may be consulted for detailed information on toxicity studies.

Chemical dispersants can be applied to open water by a variety of methods. In general, workboats are more suitable for treating minor spills in harbors or confined waters. Large, multi-engine planes are best equipped for handling large offshore spills. Small, single-engine aircrafts and helicopters are suitable for treating smaller spills and near-shore areas. If dispersants are employed along shorelines to fight oil slicks, airborne exposure to aerosols is a concern. Appropriate warnings for safe work practices are mandatory, and use of respirators should be considered.

Factors that have an impact on dispersant effectiveness fall into two general areas: oil type and properties, and the environmental

---

[1]*Technical Information Paper – The Use of Chemical Dispersants to Treat Oil Spills*, The International Tanker Owners Pollution Federation Limited, London, United Kingdom.

[2]Fiocco, R. J., G. P. Canevari, J. B. Wilkinson, et.al., 1991, *Development of COREXIT 9580 – a chemical bleach cleaner*, Proceedings of the 1991 Oil Spill Conference, San Diego, California, pp. 395–400.

[3]NRC (National Research Council), 1989, *Using Oil Spill Dispersants on the Sea*, National Academy Press, Washington, D.C., 335 pp.

[4]Singer, M. M., D. L. Smalheer, R. S. Tjeerdema, and M. Martin, 1990, *Toxicity of an Oil Dispersant to the early life stages of four California marine species*, Environmental Toxicology and Chemistry, v. 9, pp. 1389–1397.

conditions under which the dispersants are applied. Specific factors of importance for oil type and properties include the following:

- Oil viscosity
- API gravity
- Wax content and Pour Point
- Emulsifiers

Environmental factors include the following:

- Water temperature
- Sea state (specifically the degree of mixing energy)
- Extent of weathering, i.e., the length of time the oil has been on the sea
- Water salinity

There are several techniques used to assess how effective dispersants are once they have been applied. These include a visual inspection. In other words, some dispersants under favorable conditions will show an almost spontaneous dissipation of the slick. More often, the effects of a dispersant take some time, and, hence, other methods such as remote sensing using infrared detectors can be employed to measure any changes in oil film thickness. Other methods include sampling the water column for oil and the application of fluorometry for continuous oil detection in the water column.

The following are commercially available products from which to choose. These products are not listed in any particular order. The authors make no recommendations or endorsements. The user needs to review the product literature and the relative toxicity information provided on MSDSs and contact the chemical supplier to discuss the application. The user must also review the chemicals with the onsite incident commander and ensure that the product can be legally applied. Some states and countries ban the use of certain dispersants. To supplement the product descriptions, tables have been prepared for the products reviewed. These tables are organized at the end of this chapter and, along with the information provided below, will assist the user in selecting a dispersant. The table descriptions are as follows:

- Table 3.1 provides physical properties of the chemical dispersants including the manufacturer's

Table 3.1 Physical properties of commercially available chemical dispersants.

Dispersant	Flammability	Flash Point (°F)	Pour Point (°F)	Specific Gravity	pH	Recommended storage Temperature (°F)			Shelf Life
						Max.	Min.	Optimum Range	
BIODISPERS (formerly PETROBIODISPERS)	Non-flammable	193	–26	0.965	7	>158	–13	ND	Unlimited
COREXIT® EC9500A	Non-flammable	181.4	<–71	0.95	6.2	170	–30	40 to 100	The shelf life of unopened drums is unlimited. Containers should always be capped when not in use to prevent contamination and evaporation of solvents.
COREXIT® EC9527A (formerly COREXIT 9527)	Non-flammable	163	<–40	0.98 – 1.02	6.1	170	–30	40 to 100	The shelf life of unopened drums is unlimited. Containers should always be capped when not in use to prevent contamination and evaporation of solvents.

Table 3.1 (cont.) Physical properties of commercially available chemical dispersants.

Dispersant	Flammability	Flash Point (°F)	Pour Point (°F)	Specific Gravity	pH	Recommended storage Temperature (°F)			Shelf Life
						Max.	Min.	Optimum Range	
DISPERSIT SPC 1000™" (aka, SEACARE E.P.A. (ECOSPERSE™ POLLUTION ABATEMENT))	Non-flammable	208	ND	0.995	10	180	–25	40 to 140	The shelf life of Dispersit SPC 1000 is unlimited in unopened containers. Containers must be kept closed when not in use to prevent contamination.
JD-109	Non-flammable	199	–2.2	1.02	9.45	120	–4	32 to 90	The shelf life of unopened drums is unlimited. Containers should be capped when not being used to prevent contamination and evaporation.
JD-2000™	Non-flammable	212	–36	0.99	7.54	120	–30	30 to 90	The shelf life of unopened drums is unlimited. Containers should be capped when not being used

									to prevent contamination and evaporation. Opened container should be used within 1 year for optimal performance. Poly containers are recommended for storage near wet environments, i.e., ships, harbors, ports, etc.
MARE CLEAN 200 (formerly MARE CLEAN 505)	Flammable; keep away from open flame.	212	10	0.95	7.7	122	21	32 to 86	10 years when stored indoors. Container will deteriorate before contents.
NEOS AB3000	Flammable; keep away from open flame.	212	Less than 32	0.924	8.0	158	32	50 to 140	5 years
NOKOMIS 3-AA	Non-flammable	None	25	1.031	9.0	160	32	32 to 140	In excess of 24 months or more if stored correctly in plastic drums.
NOKOMIS 3-F4	Non-flammable	None	28	1.0065	10.3	212	32	50	15 years or more if stored correctly in plastic drums.
SAF-RON GOLD" (aka, SF-GOLD DISPERSANT)	Non-flammable	>212	27.5	1.014	8.8	120	32	40–120	The shelf life is unlimited in unopened containers.

Table 3.1 (cont.) Physical properties of commercially available chemical dispersants.

Dispersant	Flammability	Flash Point (°F)	Pour Point (°F)	Specific Gravity	pH	Recommended storage Temperature (°F)			Shelf Life
						Max.	Min.	Optimum Range	
SEA BRAT #4	Non-flammable	200	4	0.994	9.45	120	35	ND	Indefinite when stored properly.
SEACARE ECOSPERSE 52 (FINASOL® OSR 52)	Keep well away from sources of ignition and heat. Avoid the accumulation of static electricity. Provide grounding.	230	–16.6	1.01	8.57	149	–4	41 – 95	The shelf life of unopened drums of FINASOL OSR 52 is three years from the production date when the product is stored indoors in optimum conditions between 5º and 35ºC. The shelf life can be extended after effectiveness testing.
ZI-400 (ZI-400 OIL SPILL DISPERSANT)	Non-flammable	>199	12.6	1.026	10.9	120° continuous, 140° up to 5 days	35	40 to 120	Unlimited in sealed polydrums of totes (as delivered).

recommended storage temperature and the product shelf life. Some of the products have flash points and, hence, are flammable materials. The reader should make note of these when handling and storing the products.

- Table 3.2 provides a summary of the manufacturer's reported values for toxic chemicals. The information provided in the table covers heavy metals, chlorinated solvents, cyanide, and other hazardous ingredients. Because many of the products do not disclose the surfactants and other ingredients used in them for proprietary reasons, we cannot state with certainty that these chemicals are safe to use. Each of the chemicals carries warning information and guidelines for safe handling. These safe work practices are included in the product descriptions below. All of the commercially available dispersants reviewed have low concentrations of toxic heavy metals, and some have cyanide and chlorinated solvents. The concentrations are reported to be trace levels, but they are still measurable and pose a risk to handlers. The published safety literature, such as Material Safety Data Sheets, for many of the products does not make full disclosures of the chemical ingredients, which is a concern. In terms of generalities, all of the products contain surfactants. Non-ionic surfactants are for all practical purposes non-toxic. These include sorbitan mono oleates, alkyl/phenoxy polyethoxy ethanols, and ethoxylated alkyl phenols. Anionic surfactants, if used, are slightly to moderately toxic and include alkyl sodium sulfates, sulfonates, and sulfonated petroleum oils. Cationic surfactants are moderately to very toxic ingredients. Additives that may be used in some of the products include alkyl and/or aryl substituted ammonium chloride, bromide or sulfate, as well as alkyl substituted "quats." Cationic surfactants, to our knowledge, are not used in dispersants. A number of the products employ solvents that are, generally, not widely published or disclosed on MSDSs. As an example, COREXIT contains over 60% surfactants plus ethylene glycol monobutyl ether (butoxy

Table 3.2 Toxic chemical ingredients reported in commercial dispersants. (Concentrations are ppm. ND means "No Data.")

Dispersant	Antimony	Arsenic	Barium	Cadmium	Chromium	Copper	Lead	Mercury	Nickel	Selenium	Silver	Zinc	Cyanide	Chlorinated Hydrocarbons
BIODISPERS (formerly PETROBIODISPERS)	ND	<2.50	ND	<0.75	<0.75	<0.50	<5.00	ND	<1.20	ND	ND	<0.50	3.9	<5.00
COREXIT® EC9500A	ND	0.16	ND	ND	0.03	0.1	ND	ND	ND	ND	ND	ND	ND	ND
COREXIT® EC9527A (formerly COREXIT 9527)	ND	<0.005	ND	<0.01	1.0	<0.2	<0.1	<0.003	<0.1	ND	ND	0.1	<0.01	<0.01
DISPERSIT SPC 1000™" (aka, SEACARE E.P.A. (ECOSPERSE™ POLLUTION ABATEMENT))	ND	<2.0	ND	<2.0	<2.0	<2.0	<2.0	<2.0	<2.0	ND	ND	<2.0	Not detected	Not detected
JD-109	ND	<10	ND	<10	<10	<10	<10	<1	<10	ND	ND	<10	<0.5	<1.4

JD-2000™	ND	<0.24	ND	<0.10	<0.10	<0.10	0.43	<0.10	<0.10	ND	ND	0.11	<0.20	<2.00
MARE CLEAN 200 (formerly MARE CLEAN 505)	ND	<0.50	ND	<0.100	<0.500	<0.250	<2.50	<0.0200	<0.250	ND	ND	0.611	<0.01	<0.10
NEOS AB3000	ND	<0.1	ND	<0.1	0.26	<0.05	0.21	<0.001	0.076	ND	ND	1.1	<0.05	<0.10
NOKOMIS 3-AA	ND	<0.12	ND	<0.25	<0.25	<0.25	<0.12	<0.0016	<0.25	ND	ND	<1.0	0.034	<0.10
NOKOMIS 3-F4	ND	0.3	ND	<5.00	<10.00	<10.00	<10.00	<0.05	<10.00	ND	ND	<10.00	<2.00	<1.00
SAF-RON GOLD" (aka, SF-GOLD DISPERSANT)	ND	<0.01	ND	<0.005	0.14	0.324	<0.005	<0.020	<0.005	ND	ND	0.0671	<0.20	<0.80
SEA BRAT #4	ND	<0.05	ND	<0.05	<0.05	<0.05	<0.05	<0.0002	<0.05	ND	ND	0.215	<0.05	<0.05
SEACARE ECOSPERSE 52 (FINASOL® OSR 52)	<10.0	<10.0	<10.0	<10.0	<10.0	<10.0	<10.0	<1.0	<10.0	<10.0	<10.0	<10.0	<0.4	<4.4
ZI-400 (ZI-400 OIL SPILL DISPERSANT)		<10	ND	<10	<10	<10	<10	<1	<10	ND	ND		<0.5	<1.0

ethanol). CONCAWE (Conservation of Clean Air and Water in Europe, the oil industry's European association for health, safety, and environment in refining and distribution) reports that oil spill dispersants are not hazardous to humans; however, they caution that products must be properly handled. They also state that there is "...no evidence that properly stored and properly used modern dispersants are harmful to man or the environment..." and that "risks to bystanders and terrestrial wildlife will be negligible providing dispersants are used correctly." In light of the recent disaster in the Gulf of Mexico, neither government nor industry assurances carry much credibility. We strongly recommend that any reader who is required to work with a chemical dispersant first insist upon specific product hazard communication/hazard control training. Such training should be extensive and not simply a walkthrough of the product MSDS. Handlers should minimize their contact with the chemicals by doing the following:

- Wear personal protective equipment such as nitrile gloves, Saranex™ suits, and chemical goggles.
- Position yourself upwind or sidewind of the application areas.
- Promptly wash any skin that has come into contact with the chemical dispersants.
- Treat work clothing as hazardous waste. The clothing either needs to be properly decontaminated or disposed of as a hazardous waste.

- Table 3.3 reports effectiveness of each product as claimed by the manufacturer. The data compiled in Table 3 are laboratory bench test results for the swirling flask dispersant effectiveness test for two different crude oils. Suppliers should be contacted for similar test results for diesel and fuel oils. The average of the two sets of data is included in Table 3.3, and for convenience the products have been ranked in order of most effective to least effective.
- Table 3.4 compiles marine toxicity lab data as reported by each supplier/manufacturer. The data provided in the table are LC50 (50% Lethal Concentration) in parts

Table 3.3 Comparisons of swirling flask dispersant effectiveness tests for dispersants.

Ranking*	Dispersant	Effectiveness (%)		
		Prudhoe Bay Crude	South Louisiana Crude	Average of Prudhoe Bay and South Louisiana Crudes
1	MARE CLEAN 200	63.97	84.14	74.06
2	SEACARE E.P.A. (DISPERSIT SPC 1000™)	40	100	73
3	ZI-400 OIL SPILL DISPERSANT	50.1	89.8	69.9
4	SAF-RON GOLD	84.8	53.8	69.3
5	JD-2000™	60.4	77.8	69.1
6	NOKOMIS 3-AA	63.2	65.7	64.5
7	NOKOMIS 3-F4	62.2	64.9	63.5
8	JD109	26	91	58.5
9	SEA BRAT #4	53.55	60.65	57.1
10	BIODIPSERS	51.0	63.0	57.0
11	NEOS AB3000	19.7	89.8	54.8
12	SEACARE ECOSPERSE 52 (FINASOL® OSR 52)	32.5	71.6	52.1
13	COREXIT® EC9527A	37.4	63.4	50.4
14	COREXIT® EC9500A	45.3	54.7	50.0

*Ranking most effective to least effective

per million (ppm). For convenience, the products have been ranked in order of least toxic to most toxic to marine life.

Oil Slick Dispersants Limited was established in 1998 to supply the Marine Emergency Services around the world with the dispersant

Table 3.4 Comparison of reported marine toxicity LC50 (ppm).

Ranking*	Product	Species		
		Menidia Beryllina	Mysidopsis Bahia	Menidia Bahia
1	MARE CLEAN 200	1996.00 96-hr	938.00 48-hr	ND
2	JD-2000™	407.00 96-hr	90.50 48-hr	ND
3	NEOS AB3000	91.1 96-hr	33.0 48-hr	ND
4	NOKOMIS 3-AA	34.22 96-hr	20.16 48-hr	ND
5	ZI-400 OIL SPILL DISPERSANT	31.76 96-hr	20.96 48-hr	ND
6	SEA BRAT #4	30.00 96-hr	14.00 48-hr	ND
7	NOKOMIS 3-F4	29.80 96-hr	32.20 48-hr	ND
8	SAF-RON GOLD	29.43 96-hr	63.00 48-hr	ND
9	COREXIT® EC9500A	25.20 96-hr	32.23 48-hr	ND
10	COREXIT® EC9527A	14.57 96-hr	24.14 48-hr	ND
11	BIODIPSERS	13.46 96-hr	78.90 48-hr	ND
12	SEACARE ECOSPERSE 52 (FINASOL® OSR 522	11.66 96-hr	9.37 48-hr	ND
13	SEACARE E.P.A. (DISPERSIT SPC 1000™)	3.5 96-hr	ND	16.6 48-hr
14	JD109	1.90 96-hr	1.18 48-hr	ND

*Ranking least toxic to most toxic

called **Superdispersant-25**. No specific information was available on the product at the time this publication was prepared.

- Recommended Application Procedure:
  - Application Method: Superdispersant-25 has been used in such recent incidents as the Sea Empress and Braer operations. As a Type 2/3 dispersant, it may be applied from the air as well as from

sea vessels. In the Isle of Wight sea trials MCA Project MSA 10/9/180, published in July 2004, Superdispersant-25 was found to be the most cost-effective dispersant tested.

- Concentration/Application Rate: The recommended rate of application of Superdispersant-25 is 1 part of dispersant to 20–30 parts of oil, but this depends on many factors, which include the type of oil, viscosity, degree of weathering, ambient temperature, and prevailing conditions.
- Conditions for Use: In general, the oil should be treated quickly because weathering increases the viscosity of the oil, and this necessitates a subsequent increase in dispersant requirement. As Type 2 (water dilution mode), Superdispersant-25 is used premixed at a ratio of 1 part of dispersant to 10 parts of seawater and sprayed onto the slick using surface breaker boards or other enhanced agitation techniques to give greater mixing of the dispersant/oil emulsion. As Type 3 (undiluted mode), Superdispersant-25 is used in its neat form, utilizing appropriate spraying methods from aircraft or vessels onto the slick. In all circumstances, the Type 3 mode is the preferred method as this gives far greater efficiency in the dispersal of the oil slick.

- Contact information: Beck Cottage, Main Street, Elvington, York, YO41 4AG, England; Phone: +44 (0)1904 607910; Fax: +44 (0)1904 607911: Email: info@oilslickdispersants.co.uk. http://www.oilslickdispersants.co.uk/index.htm.

Polychem Dispersit SPC 1000 is an EPA-reviewed, NCP-listed, water-based dispersant to be used on oil discharges or oil spills in accordance with the procedures set forth by the U.S. Coast Guard and other governmental agencies. Polychem Dispersit SPC 1000 has been included in the NCP Product Schedule for use as a dispersant for oil spills. Polychem Dispersit SPC 1000 is an aqueous composition with emulsifiers, dispersants, and a water-soluble coupling solvent. Polychem Dispersit SPC 1000 contains no petroleum solvents. Independent testing (J L Ross Env. Ltd.) confirmed that

Dispersit has effectiveness in fresh and brackish water. The supplier claims that Polychem Dispersit SPC 1000 has an overall effectiveness of 73%. Dispersit is particularly well suited for application via a vessel-based education system because it maintains effectiveness when diluted with water unlike petroleum-based products.

- Special Handling and Worker Precautions for Storage and Field Application: This product should be used in a well-ventilated area and preferably with a respirator. The user should wear appropriate skin protection such as a tyvek suit, neoprene rubber gloves, and goggles.
- Recommended Application Procedure:
  - Application Method: Spray the dispersant on the oil slick by any conventional method, e.g., spray boom with pressure nozzle or rotating atomizers mounted on helicopters or airplanes.
  - Concentration/Application Rate: For light duty dispersing, a dispersant to oil ratio of 1 part Polychem Dispersit SPC 1000 to 50 parts of crude oil or an application rate of 2–6 gallons (7.6 liters–8.1 liters) per ton of oil is recommended. For heavy duty dispersing, a dispersant to oil ratio of 1:10 to 1:20 or about 45–60 gallons (170–227 liters) dispersant per ton of oil is recommended. This rate is dependent on the type of spill, weathering, temperature, and thickness of the slick.
  - Conditions for Use: A timely application ensures the highest degree of performance. Polychem Dispersit SPC 1000 may be used for both fresh and seawater. The dispersant is also effective in minimizing the volatile emissions from the oil spill.

- Surface active agents: Confidential, not disclosed
- Solvents: Water
- Additives: None
- Distributor's Contact Information: US Polychemical Corporation, 584 Chestnut Ridge Rd, Chestnut Ridge, NY, 10977; Phone: (800) 431-2072, (845) 356-5530 x335.

COREXIT was developed early on by ExxonMobil. This is a family of products that have a long evolutionary history. The first product

was introduced in 1967 as a specially formulated dispersant for the marine environment. The original product was known as COREXIT 7664, which was a weak, water-based product that is no longer manufactured. In 1972, the first self-mix concentrate (COREXIT 9527) was introduced to allow for aircraft application. In 1992, the product became available for application on heavy, weathered, and emulsified crude oils (COREXIT 9500). The COREXIT products are the principal U.S. dispersants within the arsenal to combat oil spills. It is important to note that this product is not intended for direct application to shorelines. Instead, another product named CORTEX 9580 is a beach cleaner developed for that application. Additionally, CORTEX 9527 applications to weathered or heavy oils have not been highly successful. The current supplier is Nalco Energy Chemicals. The following provides information on the COREXIT dispersants currently marketed. Refer to the tables for chemical-specific information on properties, effectiveness, and toxicity.
COREXIT EC9527A has been highlighted by the U.S. EPA. The following information was obtained from the U.S. EPA web site.[5]

- Recommended Application Procedure:
  - Application Method: COREXIT® EC9527A is most effectively applied by aircraft. However, application with boat spray booms, boat fire monitors, and by handheld sprayers and backpacks has been successfully done on a number of spills and trials. Aircrafts provide the most rapid method of applying dispersants to an oil spill and a variety of aircraft can be used for spraying. For aerial spraying, COREXIT® EC9527A is applied undiluted. Typical application altitudes of 30 to 50 feet have been used, although higher altitudes may be effective under certain conditions. Actual effective altitudes will depend on the application equipment, weather, and aircraft. Careful selection of spray nozzles is critical to achieve desired dose levels, since the droplet size must be controlled. Many nozzles used for agricultural spraying are of low capacity and produce too

[5]http://www.epa.gov/oem/content/ncp/products/corex952.htm

fine a spray. A quarter-inch open pipe may be all that is necessary if the aircraft travels at 120 mph (104 knots) or more since the air shear at these speeds will be sufficient to break the dispersant into the proper sized droplets. COREXIT® EC9527A may also be applied by workboats equipped with spray booms mounted ahead of the bow wake or as far forward as possible. The preferred and most effective method of application from a workboat is to use a low-volume, low-pressure pump so the chemical can be applied undiluted. Spray equipment designed to provide a 5–10% diluted dispersant solution to the spray booms can also be used. COREXIT® EC9527A should be applied as droplets, not fogged or atomized. Natural wave or boat wake action usually provides adequate mixing energy to disperse the oil. Recent tests have indicated that a fire monitor modified with a screen cap for droplet size may also be useful for applying COREXIT® EC9527A. Due to the increased volume output and the greater reach of the fire monitor, significantly more area can be covered in a shorter period of time. Spray systems should be calibrated at temperatures anticipated to ensure successful application and dosage control.

- Concentration/Application Rate: A treatment rate of about 2 to 10 U.S. gallons per acre, or a dispersant to oil ratio of 1:50 to 1:10 is recommended. This rate varies depending on the type of oil, degree of weathering, temperature, and thickness of the slick.
- Conditions for Use: As with all dispersants, timely application ensures the highest degree of success. Early treatment with COREXIT® EC9527A, even at reduced treatment rates, can reduce the "mousse" forming tendencies of the spilled oil. COREXIT® EC9527A is useful on oil spills in salt water.

- Surface active agents: The product contains these, but the ingredients are not listed on the MSDS.
- Supplier's Contact Information: Nalco Energy Services, L.P., P.O. Box 87, Sugar Land, TX 77487-0087;

Customer Services: Phone: (800) 333-3714; Product Management: Office: (281) 263-7336, Mobile: (281) 202-8126; Email: kapreston@nalco.com (Ms. Kathryn Preston)

BIODISPERS (formerly PETROBIODISPERS) is a water-based dispersant. Refer to the tables for chemical-specific information on properties, effectiveness, and toxicity. The following information provides the manufacturer's recommendations on applications and special handling requirements.

- Special Handling and Worker Precautions for Storage and Field Application:
  - Ventilation: Use general ventilation to minimize exposure to vapor or mist.
  - Skin and eye contact, protective clothing, and treatment in case of contact: In the case of skin contact, immediately flush with copious amounts of water. Remove and wash contaminated clothing and shoes. If irritation persists, see a doctor. In the case of eye contact, it may cause eye irritation. Immediately flush eyes with plenty of water. If irritation persists, see a doctor immediately. Hygienic practices – wear safety glasses. Wear gloves in accordance with routine laboratory safety precautions.
  - Temperatures of phase separations and chemical changes: No phase or chemical separation observed.

- Recommended Application Procedure:
  - Application Method: Recommended application is by aircraft, fireboat monitors, or similar apparatus.
  - Concentration/Application Rate: Concentration varies with spilled material – solvents at 5%, light oils at 10%, medium oils at 10%, and heavy oils at 10%.
  - Conditions for Use: Data is for water temperature of 40°F to 65°F. There are no known application restrictions.

- Surface Active Agents: Confidential, not disclosed.
- Solvents: Water

- Additives: None
- Supplier's Contact Information: Petrobiotech LLC; P.O. Box 813 Newport, NH, 03773; Phone: (203) 966-4573; Fax: (561) 966-0920.

JD-109 is manufactured by GlobeMark Resources Ltd. Refer to tables below for additional information.

- Special Handling and Worker Precautions for Storage and Field Application:
  - Ventilation: Use with ventilation equal to unobstructed outdoors in a moderate breeze.
  - Skin and eye contact, protective clothing, and treatment in case of contact: Avoid skin and clothing contact. If skin contact occurs, immediately wash with large amounts of soap and water (if possible). Remove any contaminated clothing and shoes. Launder before reusing. If irritation persists, seek medical assistance. For areas where contact is likely, wear long sleeve shirt, chemical resistant gloves, and chemical resistant goggles.
  - Temperatures of phase separations and chemical changes: None

- Recommended Application Procedure:
  - Application Method: The most rapid and effective method of applying JD-109 to an oil spill is by spraying. Temperature ranges for application should not drop below 32°F or exceed 120°F. For aerial spraying, typical application altitudes of 30 to 50 feet are advised, although higher altitudes may be used if conditions warrant. Spray nozzle should be about ¼ inches in diameter flying at 104 knots or more to create enough water shear sufficient to break the dispersant into proper sized droplets. A ½ inch diameter nozzle may be needed for temperatures below 50°F. JD-109 can also be applied by workboats with spray booms mounted as far forward as possible of the bow or wake. The most effective application from a workboat is a low-volume, low-pressure pump.

  - Concentration/Application Rate: A treatment rate of about 2–10 US gallons (7.6 to 37.9 liters) per acre (4,840 square meters) or a dispersant to oil ratio of 1:50 to 1:10 is recommended. The rate may vary depending on the type of oil, degree of weathering, temperature, and thickness of slick.
  - Conditions for Use: As with any oil related spill, timely application of a dispersant will ensure the highest degree of success. Timely treatment with JD-109, even at low application rates, can counter the "mousse" forming effect of the spilled oil.

- Surface Active Agents: Anionic and nonionic, proprietary, surfactants
- Solvents: Proprietary, ester-based
- Additives: None
- Contact Information: GlobeMark Resources Ltd., 1205 Pine Heights Drive Atlanta, GA 30324, Mobile: (254) 231-2251; E-mail: joannie@globemarkresouces.com; http://www.globemarkresources.com.

JD-2000™ is another product made by GlobeMark Resources Ltd. (see JD-109 for contact information). Look-up tables supplement the data provided below.

- Special Handling and Worker Precautions for Storage and Field Application:
  - Ventilation: Use with ventilation equal to unobstructed outdoors in a moderate breeze.
  - Skin and eye contact, protective clothing, and treatment in case of contact: Avoid skin and clothing contact. If skin contact occurs, immediately wash with large amounts of soap and water (if possible). Remove any contaminated clothing and shoes. Launder before reusing. If irritation persists, seek medical assistance. For areas where contact is likely, wear long sleeve shirt, chemical resistant gloves, and chemical resistant goggles.
  - Temperatures of phase separations and chemical changes: None

- Recommended Application Procedure:
  - Application Method: JD-2000™ is free of phosphates, aromatic chlorinated solvents, branched ethoxylated alcohols, and hydrotreated distillates. Optimum temperature for application is above 30°F. For aerial spraying, typical application altitudes of 30 to 50 feet are advised, although higher altitudes may be used if conditions warrant. Spray nozzle should be about ¼ inch in diameter flying at 104 knots or more to create enough air shear sufficient to break the dispersant into proper sized droplets. A ½ inch diameter nozzle may be needed for temperatures from 30°F. JD-2000™ can also be applied by workboats with spray booms mounted as far forward as possible of the bow or wake. The most effective application from a workboat is a low-volume, low-pressure pump.
  - Concentration/Application Rate: JD-2000™ is an oil spill dispersant concentrate that may be diluted by 5–10% percent with water if needed. A treatment rate of about 2–10 U.S. gallons (7.6 to 37.9 liters) per acre (4,840 square meters) or a dispersant to oil ratio of 1:50 to 1:10 is recommended. The rate may vary depending on the type of oil, degree of weathering, temperature, and thickness of slick.
  - Conditions for Use: As with any oil related spill, timely (preferably within 48 hours) application of a dispersant will ensure the highest degree of success. Timely treatment with JD2000™, even at low application rates, can counter the "mousse" forming effect of the spilled oil.

- Surface Active Agents: Confidential, not disclosed
- Solvents: Listed as confidential.
- Additives: None

MARE CLEAN 200 (formerly MARE CLEAN 505) is a solvent-based dispersant, manufactured and supplied by Taiho Industries Co. See look-up tables for additional information beyond what is provided below.

- Special Handling and Worker Precautions for Storage and Field Application:
  - Ventilation: Ventilation is required. Use in closed room is not recommended.
  - Skin and eye contact, protective clothing, and treatment in case of contact: Use protective goggles to avoid eye contact. In case of eye contact, wash immediately with plenty of water and consult a physician.
  - Temperatures of phase separations and chemical changes: Phase separation does not relate to temperatures. Chemical changes may occur at temperatures above 194°F.

- Recommended Application Procedure:
  - Application Method: Sprinkle the dispersant on the oil spill, and then 5–10 minutes later stir the surface intensively. For convenience, MARE CLEAN 200 may be diluted with water if desired.
  - Concentration/Application Rate: Use 53–66 gallons of MARE CLEAN 200 per ton of oil.
  - Conditions for Use: The performance of MARE CLEAN 200 is not affected by water salinity. At temperatures below 40°F or in the case of a heavy crude oil spill, MARE CLEAN 200 should be used without dilution. MARE CLEAN 200 is an effective dispersant for any liquid hydrocarbon.

- Surface Active Agents: A mixture of sorbitan fatty acid esters, polysorbates, and polyoxyethylene fatty acid esters
- Solvents: Paraffinic hydrocarbons (CAS 74664-93-0)
- Additives: None
- Contact Information: Taiho Industries Co., Ltd. 21-44, 2-chome, Takanawa Minatoku, Tokyo, Japan Phone: (81) 33-445-8111 Fax: (81) 33-443-6333.

NEOS AB3000 is manufactured and supplied by NEOS Company Limited.

- Special Handling and Worker Precautions for Storage and Field Application:
  - Ventilation: Special ventilation is not required, but natural ventilation is recommended.
  - Skin and eye contact, protective clothing, and treatment in case of contact: Contact may cause skin and eye irritation. Goggles and rubber clothing are recommended during application. In case of contact with skin or eye, flush with copious amounts of freshwater. If severe, consult a doctor.
  - Temperatures of phase separations and chemical changes: Phase separation and chemical changes do not appear between the temperature range of 32°F to 158°F.

- Recommended Application Procedure:
  - Application Method: Spray neat concentrate on the oil slick in atomized form by means of a manual pump, or spray with a pump system incorporating an ejector system for drawing concentrate from the drum or stock tank. For aerial application, use a spray boom with pressure nozzles or rotating atomizers mounted on helicopters or airplanes.
  - Concentration/Application Rate: The application rate is 65 gallons of dispersant per ton of oil. 5 to 15 parts of dispersant to suctioned water is recommended for ejector systems. For aerial application, 75 to 125 gallons per ton of oil is recommended.
  - Conditions for Use: NEOS AB3000 can be used in salt water. It is effective with crude and residual heavy oil. The dispersant is also effective at controlling volatile emissions from the oil.

- Surface Active Agents: Nonionic and Cationic surfactants
- Solvents: Paraffins
- Additives: None
- Contact Information: NEOS Company Limited, Daisan Kendai Building 1-2, 3-chome Isobedori Chuo-ku, Kobe, 651-0084 Japan Phone: (81) 78-331-9384 Fax: (81) 78-272-4649.

NOKOMIS 3-AA is manufactured and supplied by Mar-Len Supply, Inc. Refer to the look-up tables for additional information.

- Special Handling and Worker Precautions for Storage and Field Application:
  - Ventilation: No special requirements
  - Skin and eye contact, protective clothing, and treatment in case of contact: In case of eye contact, flush with water. In case of skin contact, wash with water, and if swallowed, drink water to dilute and induce vomiting.
  - Temperatures of phase separations and chemical changes: None

- Recommended Application Procedure:
  - Application Method: For aerial spraying, spray from an aircraft flying with a controllable speed at altitudes of 30 to 50 feet. Spray nozzles should be sized to cover an area using 5 gallons per surface acre. NOKOMIS 3-AA can also be applied by ships or workboats with spray booms mounted as far forward as possible. NOKOMIS 3-AA will generally be marketed in 55-gallon plastic drums. Application of the products may be made directly from the drum. By using a direct connection from pump to drum, NOKOMIS 3-AA can be applied at full strength to the oil spill. The determination of whether to use NOKOMIS 3-AA at full strength or diluted with water must be determined by observations and evaluations made on-site at the oil spill. Portable engine-powered centrifugal water pumps are a satisfactory means of moving the dispersant from the container to the spill's surface. Diesel-powered larger capacity pumps may also be used. If the spill is confined to a small area, a hand-pump connected directly to the 55-gallon drum can be used. Hose diameters and lengths will relate to capacities required for the specific situation and distances from the pump to container and the spill. Fire hoses have been successfully used, and with hand pumps a small diameter hose is adequate. Fog nozzles on

the dispersing end of the hose provide a fine spray of NOKOMIS 3-AA full strength or diluted as appropriate. Any type of hose nozzle can be used, but it is preferable to use one with the ability to produce a spray rather than a coarse stream.

- Concentration/Application Rate: NOKOMIS 3-AA may be applied to oil spills at full strength or diluted with seawater. Where large areas of water may be covered with heavy concentrations of crude or Bunker C oil, it may be advantageous to apply NOKOMIS 3-AA at full strength. Where lighter fractions of petroleum are involved, it is possible that dilutions of up to 1 part NOKOMIS 3-AA to 30 parts water may be applicable. Approximately 5 gallons of NOKOMIS 3-AA can be used for one surface acre of oil spill.
- Conditions for Use: Once NOKOMIS 3-AA is applied to the water's surface, in order to obtain the most efficient emulsification of the oil it is necessary to agitate and mix dispersant, oil, and water thoroughly. In open unconfined areas, the use of ship propellers has been determined a practical way of accomplishing this purpose. By passing over the spill area on a grid system, the vessel's propellers will churn the water, causing the needed mixing. For a large spill, two or more vessels may be needed to apply the product and agitate the water. In and around piers and similar confined areas, it is necessary to apply the product from small boats, the shore, or the pier itself. Where it is impossible or impractical to supply agitation with propellers of a vessel, it is necessary to use pumps and a hose, applying salt water in the coarse stream and under sufficient pressure to cause surface turbulences and subsequent mixing.

- Surface Active Agents: Confidential, not disclosed
- Solvents: Propyleneglycol and water
- Additives: None
- Contact Information: Mar-Len Supply, Inc., 23159 Kidder Street Hayward, CA, 94545; Phone: (510) 782-3555; Fax: (510) 782-2032.

NOKOMIS 3-F4 is manufactured and supplied by Mar-Len Supply, Inc. Refer to the look-up tables for additional information.

- Special Handling and Worker Precautions for Storage and Field Application:
  - Ventilation: No special requirements
  - Skin and eye contact, protective clothing, and treatment in case of contact: In case of eye contact, flush with water. In case of skin contact, wash with water, and if swallowed, drink water to dilute and induce vomiting.
  - Temperatures of phase separations and chemical changes: None

- Recommended Application Procedure:
  - Application Method: Application of the product may be made directly from 33- and 35-gallon drums (marketing sizes) by fitting the drum with a "T" connection and pumping fresh or salt water across the junction, which will pull the dispersant from the drum into the water stream and onto the oil spill.
  - Concentration/Application Rate: For heavy concentrations of crude or Bunker C oil, apply the product undiluted. Where lighter fractions of petroleum are involved, the product can be diluted up to 1 part dispersant to 30 parts water.
  - Conditions for Use: It is effective in salt water and can be used on water of any temperature.

- Surface Active Agents: Confidential, not disclosed
- Solvents: None
- Additives: None
- Contact Information: Mar-Len Supply, Inc., 23159 Kidder Street Hayward, CA, 94545; Phone: (510) 782-3555; Fax: (510) 782-2032.

SAF-RON GOLD is manufactured by Sustainable Environmental Technologies Inc. The primary distributor of the product is TRK Enterprises, LLC.

- Special Handling and Worker Precautions for Storage and Field Application:
  - Ventilation: Special ventilation is not required.
  - Skin and eye contact, protective clothing, and treatment in case of contact: No special equipment is required; however, goggles are recommended. In the case of skin contact, if irritation occurs flush with water. Seek medical attention if irritation persists. In the case of eye contact, flush with plenty of water, and seek medical attention if irritation persists.

- Recommended Application Procedure:
  - Application Method: Sprayer, boat, aircraft
  - Concentration/Application Rate: A dispersant to oil ratio ranging from 1:50 to 1:10 is recommended. For heavy high-viscosity oils, 8 gallons of SAF-RON GOLD will disperse approximately 70 gallons of crude oil or Bunker C. Also, for use with heavy high-viscosity oils, SAF-RON GOLD can be used as a concentrate or diluted at a dispersant to water ratio of 1:10. For lighter oils, SAF-RON GOLD can be diluted at mixtures of 1:20 to 1:100. A recommended dispersant to water ratio of 1:40 can be used for most spills. Ratios are dependent upon the type of oil and weather conditions.
  - Conditions for Use: Any applicable condition determined by authorities for dispersant use

- Surface Active Agents: Proprietary blend of surface-active agents
  - Solvents: None
  - Additives: None
  - Manufacturer's Contact Information: Sustainable Environmental Technologies Inc., P.O. Box 30516, Mesa, AZ, 85275; Customer Services: Phone: (877) 853-2947, (800) 347-8950; Fax: (877) 853-2947, (480) 461-8798; 24-hour Emergency Number: (877) 853-2947; E-mail: info@sustainable-corp.com, bruce@sustainable-corp.com; www.sustainable-corp.com; www.saf-ron.com.

- Distributor's Contact Information: TRK Enterprises, LLC, 2000 Market St. 2530, Scottsville Road Suite 12, Bowling Green, KY, 42104; Phone: (270) 782-0882.

SEA BRAT #4 is manufactured by Alabaster Corp. The primary distributor is Garner Environmental Services.

- Special Handling and Worker Precautions for Storage and Field Application:
  - Ventilation: Normal
  - Skin and eye contact, protective clothing, and treatment in case of contact: Wear safety glasses or goggles, gloves, and rubber boots. Wash after each shift. Remove and wash contaminated clothing before reuse. No respiratory protection is required. Local exhaust is desirable. Mechanical exhaust is helpful in congested areas. In the case of skin contact, flush with water. Seek medical attention if irritation persists. In the case of eye contact, flush with water using eye cup or fountain for 15 minutes. Seek medical attention if irritation persists. If ingestion occurs, seek medical attention. No medical attention is required with inhalation.

- Recommended Application Procedure:
  - Application Method: Spray affected area with a high-pressure pump. Always apply in compliance with federal, state, and local laws.
  - Concentration/Application Rate: Dilution ratios of 1 part SEA BRAT #4 to 9 parts water for a 10% solution.
  - Conditions for Use: May be applied to the coastal waters of the U.S. It is designed for hydrocarbon spills on water of temperatures between 50°F and 90°F. It is best applied with nozzle pressure of between 80 psi and 100 psi with a direct hard spray and continuously moving the stream of water over the entire surface.

- Surface Active Agents: Proprietary blend of surface-active agents

- Solvents: None
- Additives: None
- Manufacturer's Contact Information: Alabaster Corp., 6921 Olson, Pasadena, TX, 77505; Phone: (281) 487-5482, (800) 609-2728; Fax: (281) 487-9014; E-mail: alabastercorp@aol.com.
- Distributor's Contact Information: Garner Environmental Services, 3197 Main Street, LaMarque, TX, 77568; Phone: (800) 935-0308; Fax: (409) 935-0678; E-mail: alabastercorp@aol.com.

SEACARE ECOSPERSE 52 (FINASOL® OSR 52) is manufactured by Total Fluides. The primary distributor of the product is Total Petrochemicals USA, Inc.

- Special Handling and Worker Precautions for Storage and Field Application:
  - Ventilation: Keep containers closed when not in use. Do not inhale vapors, fumes, or mists. In case of inhalation, take the person into fresh air. Ensure adequate ventilation is provided if there is any risk of vapors, mists, or aerosols forming. In the event of serious problems, call a doctor or summon medical assistance urgently.
  - Skin and eye contact, protective clothing, and treatment in case of contact: Wear appropriate long sleeve shirts, chemical-resistant gloves, and goggles. Avoid contact with skin and eyes. In case of contact, wash immediately with plenty of water. Avoid contact with clothing. In case of contact, wash immediately and remove contaminated clothing. Prolonged and repeated contact with the skin may cause skin disorders. In case of ingestion, do not induce vomiting to prevent aspiration into the respiratory tract.

- Recommended Application Procedure:
  - Application Method: FINASOL OSR 52 is applied effectively by aircraft or from boats. It can also be applied by handheld sprayers as well as with backpack spray units. Spraying must not give too fine or atomized droplets as in fog. It must be done as

moderately coarse droplets, with diameters in the range of 500–1000 µm, whether application is by surface vessel, aircraft, or land-based equipment. FINASOL OSR 52 is sprayed undiluted when applied from aircrafts, usually between 30 and 50 feet altitude, or at higher altitude in poor weather conditions. For boat spraying, FINASOL OSR 52 is sprayed undiluted using a low-pressure system. Spraying equipment providing 5–10% dilution of FINASOL OSR 52 can also be used.

  - Concentration/Application Rate: Use FINASOL OSR 52 from 1:5 to 1:50, according to the nature of the oil, the state of weathering, temperature, thickness of the slick, and other circumstances of the spill. A 1:10 ratio would correspond appreciatively to 26 gallons per ton of oil.
  - Conditions for Use: Diluted application in low salinity water is not recommended. Use only receptacles, joints, pipes, etc., that are resistant to hydrocarbons. Do not spray at high pressure (> 3 bar). FINASOL OSR 52 is designed to treat spills in salt water.

- Surface Active Agents: Confidential, not disclosed
- Solvents: Confidential, not disclosed
- Additives: Confidential, not disclosed
- Manufacturer's Contact Information: Total Fluides, 51 Esplanade du Général de Gaulle, 92907 Paris La Défense Cédex, France; Phone: +33-141-356-101; U.S.: (713) 483-5712; 24-hour Emergency Number: +33-1-41-35-65-00; E-mail: abdallah.bouhlassi@total.com; www.totalspecialfluids.com.
- Distributor's Contact Information: Total Petrochemicals USA, Inc., 1201 Louisiana St., Suite 1800, Houston, TX, 77002; Phone: (713) 483-5712; Fax: (713) 483-5700; www.totalspecialfluids.com.

ZI-400 is manufactured and distributed by Z.I. Chemicals.

- Handling and Worker Precautions for Storage and Field Application:
  - Ventilation: No special requirements

- Skin and eye contact, protective clothing, and treatment in case of contact: No special equipment or clothing required; however, goggles are recommended. If eye or skin irritation occurs, flush with plenty of water.

- Recommended Application Procedure:
  - Application Method: ZI-400 may be applied by the usual method of aerial or boat spraying. For smaller spills, a drum pump with sprayer may be used, mixing with water as required depending on the type and viscosity of oil being treated.
  - Concentration/Application Rate: On heavy oils, use ZI-400 directly on the spill or up to approximately a 1:10 dilution ratio (product:water). Lighter oils will require a 1:10 to 1:30 product to water dilution ratio. Warmer waters (greater than 78°F) and/or good agitation during application will require less of the product.
  - Conditions for Use: No limitations as to usage within optimum temperature parameters (application may be made at or above 35°F, with optimum above 48°F).

- Surface Active Agents: Proprietary
- Solvents: Proprietary
- Additives: None
- Contact Information: Z.I. Chemicals, 8605 Santa Monica Boulevard, #38201, Los Angeles, CA, 90069; Phone: (818) 827-1301; E-mail: sales@zichemicals.com; www.zichemicals.com.

4

Combating Spills at the Shoreline

4.1 Chemical Warfare

The arsenal of technologies and tools available for combating spills that impact shorelines are both manpower intensive and primitive. At best, this can be described as chemical and trench warfare.

Similar application of some dispersants as employed at sea can be used to remove oil from hard surfaces such as rocks, sea walls, and other manmade structures. However, it is important to remove the bulk of the stranded oil by other means first. Shores subjected to strong wave action are cleaned naturally to a certain extent, and they should not be sprayed unless the oil has to be removed immediately or is being washed up along shores as massive slicks.

Dispersants can be applied to the surface and scrubbed into the oil before flushing with seawater. The dispersed oil cannot be collected, and, for this reason, dispersant use on the shoreline should be restricted to areas of low environmental concern.

Shoreline cleaners may also be used but it is important to note that their mechanism of action is different from that of dispersants.

Cleaners are essentially detergents. Degreasers are often carried on board ships to deal with small spillages of oil on deck, but most are more toxic than dispersants and should not be used as a dispersant at sea or as a shoreline cleaner. Citrus-based chemical cleaners are favored as these usually have a lower environmental impact than traditional chemical cleaners. However, we have been unable to identify any such products.

Chemical products suitable for shoreline remediation of oil spills fall into the following categories:

- Surface-washing agents
- Bioremediation agents
- Miscellaneous oil spill control chemical agents such as oil solidifying agents

Some of the products act as detergents, i.e., they break up the oil and allow sea waves to carry the emulsion out to sea. Bioremediation agents use microbes to digest the oil. There are a few products that cause polymerization of the oil, thereby facilitating the manual collection of oily waste. Waste products in this form are considered hazardous and require special staging and disposal.

Many suppliers claim that their products are safe and biodegradable, yet there are few published studies and supplier field tests that support such claims. Many of the chemicals we surveyed show the presence of trace levels of toxic heavy metals and solvents. As in the case of dispersants, many suppliers do not disclose or are required to disclose the ingredients of their products due to the confidentiality exclusion under OSHA for MSDS documentation.

Another problem with these products, particularly bioremediation-based products, is that they must be applied under special conditions in order to achieve optimum results. Shoreline conditions under adverse weather conditions and rough terrain do not always permit optimum conditions of temperature, contact time, and mild agitation to be applied.

Many of the chemicals available are more toxic to marine life than dispersants based on reported laboratory scale testing performed by suppliers. While nearly all the product literature we examined highlights that marine toxicity data reported in the supplier's literature is for the concentrated product and not at dilution levels seen in application, the fact is this is the only relative toxicity assessment available and studies under actual conditions either don't exist or are very limited. In essence, there are no definitive assessments on marine and wildlife toxicity of the commercially available chemicals and certainly no extensive assessments that allow anyone to assure that the fate and transport of these chemicals when applied in large amounts will not have negative impacts on local environments and species.

The same level of uncertainty and concern is expressed for workers that use these chemicals. Because the chemical suppliers do not disclose the ingredients in their products and in many cases provide no guidelines for threshold limit values and/or time weighted averages, workers have little assurance that the products they apply in clean-up efforts are not harmful. None of the emergency responders involved in the post clean-up of the infamous 9/11 attack were ever advised of the respiratory effects from dusts and the toxins contained in them when they worked on clean-up operations. As we now painfully know, there are thousands of sick workers that were exposed to low-level toxins that were airborne. The same situation exists with many of the chemicals touted as the magic bullets for shoreline clean-up activities – these chemicals are applied as mists, sprayed on areas, and spread about as powders. They pose the same degree of potential inhalation hazards as those posed on workers involved in post-9/11 remediation activities. Regardless of how non-toxic suppliers claim their products are, workers should be cautious. Many product safety literature simply use terms such as "normal ventilation is adequate" or "use respirators when TLVs are exceeded," but no TLVs are cited for the products. In general, we have found that the safe recommended work practices are simply warning labels that, in many instances, offer little guidance.

We have compiled information on the commercially available products and organized the data into tables. The information contained in each of the tables is defined below. A few points of clarification are needed for the use of the data organized:

1. The data compiled in certain tables contains recommended handling and application practices. The reader should be advised that these recommendations

are the manufacturers' recommendations as stated in their respective product literature and are intended to be specific to their products. We do not agree with all of the recommendations, especially those that claim products are inherently safe. All of the products are chemical agents, and all of them may be considered hazardous to human health. Not enough data exists on chronic exposure to any of the products that are currently available. Therefore, our overall recommendation is that workers should be thoroughly versed with the MSDSs for each and every product that they are instructed to use before they apply it in the field and that appropriate skin and respiratory protection be used. When in doubt on the inherent safety of the product, Level B OSHA protection should be employed. Level C may be applied if the supervisor and worker have sufficient justification for a lower level of skin and respiratory protection. Levels of protection are explained in Chapter 6.

2. The abbreviation ND used in some tables means "non-detected." In previous chapters, the designation was used to denote "no data."
3. A dash numeric entry (–) indicates a zero value.
4. In table entries where there is no data, there simply is no data provided by the supplier, or the supplier's or manufacturer's literature reported data without any units, in which case we have ignored the entry.

The tables provided in this section are as follows:

- Table 4.1 – Product Identification Chart
- Table 4.2 – Product Stability and Storage Chart
- Table 4.3 – Product Physical Properties Chart
- Table 4.4 – Product Safe-Handling Chart
- Table 4.5 – Product Hazardous Ingredients Chart
- Table 4.6 – Product Marine Toxicity Chart – LC50 values in ppm (parts per million)
- Table 4.7 – Bioremediation Agent Effectiveness Chart
- Table 4.8 – Product Application Chart

Table 4.1 Product identification chart.

Product	Type Product / Application	Manufacturer/ Contact Information	Address	Phone and Fax	Email/Web Site
AQUACLEAN	Surface Washing Agent – No special applications identified	Madison Chemical Company, Inc.	3141 Clifty Drive Madison, IN 47250	Phone: (812) 273-6000; Fax: (812) 273-6002	Email: cara.cyrus@ madchem.com
BG-CLEAN™ 401	Surface Washing Agent – BG-CLEAN™ 401 is recommended as a cleaner for piers, seawalls, pilings, ship's hulls, and rocky beaches. It can be used on metal, wood, and plastics. BG-CLEAN™ 401 is also highly useful for cleaning contaminated solids such as soil. This includes a range from normal soil aggregate to 100% sand. Two conditions are required for maximum cleaning: thorough mixing of BG-CLEAN™ 401 with the oil and an adequate concentration of BG-CLEAN™ 401.	BioGenesis Enterprises, Inc.	610 W Rawson Avenue Oak Creek, WI 53154	Phone: (414) 571-6230; (414) 768-7100; Fax: (414) 571-6231; Fax: (414) 768-7106	Email: skelling@ aol.com

Table 4.1 (cont.) Product identification chart.

Product	Type Product / Application	Manufacturer/ Contact Information	Address	Phone and Fax	Email/Web Site
BIOSOLVE® HYDROCARBON MITIGATION™ AGENT	Surface Washing Agent – Used for Surface Washing Applications; BIOSOLVE® applied through power washers in light dilution is effective in attaining the removal of oils from rock, cobblestone, shorelines, and sea walls. In marsh or wetland applications, BIOSOLVE® prevents the oil from clinging to grasses and mangroves. May be used with salt or fresh water. Temperature is not relevant.	The Westford Chemical Corporation®	98 Concord Road Westford, MA 01886	Phone: (978) 392-0689 or (800) 225-3909 or (508) 878-5895 Fax: (978) 692-3487	Email: president@ biosolve.com/ Web site: http://www. biosolve.com
CN-110	Surface Washing Agent – Preferably fresh water, at temperature of 32°F.	Chemex, Inc.	107-B Balboa Drive Broussard, LA 70518	Phone: (337) 837-9148; Fax: (337) 837-2648	Email: chemex@ msn.com
COREXIT® EC7664A	Surface Washing Agent – COREXIT® EC7664A (1–3% in water) will protect beaches, marshes, rocks,	Nalco Energy Services, L.P.	P.O. Box 87 Sugar Land, TX 77487-0087	Customer Services: Phone: (800) 333-3714;+F4	Email: kapreston@ nalco.com

	and other shoreline structures from oil contamination if applied before the oil comes in contact with them.			Product Management: Office: (281) 263-7336; Mobile: (281) 202-8126	
CYTOSOL	Surface Washing Agent – The following shoreline types are appropriate for the use of CYTOSOL: Coarse sand beaches where petroleum has penetrated into sand; marsh areas and vegetated wetlands where oil has coated plants and become trapped; concrete bulkheads, rip rap, and piers that may have trapped oil; oiled pilings; gravel or cobble shorelines and rocky shores where oil has become trapped in pockets; and public beaches, fisheries, hatcheries, river banks, and other sensitive or high-impact sites. The CYTOSOL has been field tested successfully for removing oil from mussel beds and	CytoCulture International, Inc.	249 Tewksbury Avenue Point Richmond, CA 94801-3829	Phone (510) 233-0102; Fax: (510) 233-3777 Mobile: (561) 762-5440	Email: vwedel@aol.com; Email: rvw@cytoculture.com/ Web Site: http://www.cytoculture.com

Table 4.1 (cont.) Product identification chart.

Product	Type Product / Application	Manufacturer/ Contact Information	Address	Phone and Fax	Email/Web Site
	intertidal zones, pilings, and concrete rip rap. The CYTOSOL also proved effective in facilitating the removal of oil from the banks and vegetation along an oiled creek.				
DO-ALL #18	Surface Washing Agent – Recommended for cleaning petroleum fractions from beaches, rocks, piers, bilges, decks, waterline scum, rigs, platforms, tanks, barges, engine rooms, machinery, holding tanks, and grease traps.	Radcob Solutions, Inc.	4800 North State Road 7, Suite #105 Lauderdale Lakes, FL 33319	Phone: (954) 249-2178; Fax: (954) 894-6826	
ENVIROCLEAN (formerly ENVIRO CLEAN 165)	Surface Washing Agent – ENVIROCLEAN may be diluted with hard, soft, brackish, salt, or most waters of sufficient quality for operations.	Enviro Clean Services, LLC	P.O. Box 721090 Oklahoma City, OK 73172-1090	Phone: (405) 373-4545 Fax: (405) 373-4549	Email: info@envi-rocleanps.com

E-SAFE©	Surface Washing Agent – E-SAFE© works on all soil types and weather conditions that allow hydrocarbon penetration. E-SAFE© follows the same path, channel, or gradient as the contaminant. When visible detection reveals that E-SAFE© has been absorbed by, or has penetrated, the soil, water should be applied to the site. E-SAFE© is soluble in water and breaks the surface tension of the transporting water molecules.	PLUTUS Environmental Technologies, Inc.	P.O. Box 5104 Sevierville, TN 37864-5104	Phone: (865) 453-0060 Fax: (865) 908-6652	Email: CEO@plutusonline.com
F-500	Surface Washing Agent – No special applications identified	Hazard Control Technologies, Inc.	150 Walter Way, Fayetteville, GA 30214	Phone: (770) 719-5112; Fax: (770) 719-5117	
GOLD CREW SW	Surface Washing Agent – Equally effective with salt or fresh water.	Gold Crew Products & Services, LLC	P.O. Box 12032 Orange, CA 92869	Phone: (714) 288-8781 Fax: (714) 288-8730	Email: jfigueira@goldcrew.net
NALE-IT	Surface Washing Agent	SPL Control LLC	P.O. Box 627 Elemore City, OK 73433	Phone: (580) 788-2187	Email: splcontrol@aol.com

Table 4.1 (cont.) Product identification chart.

Product	Type Product / Application	Manufacturer/ Contact Information	Address	Phone and Fax	Email/Web Site
NATURE'S WAY HS	Surface Washing Agent – For cleaning heavy to lightly contaminated surfaces.	Integra Environmental, Ltd.	5825 Centralcrest Houston, TX 77092	Phone: (713) 680-1234 Fax: (713) 680-1608	Email: info@inte-graenvironmental.com
NOKOMIS 5-W	Surface Washing Agent – Fresh or salt water applications.	Mar-Len Supply, Inc	23159 Kidder Street, Hayward, CA 94545	Phone: (510) 782-3555; Fax: (510) 782-2032	
PETRO-CLEAN	Surface Washing Agent – Fresh or salt water applications.	Alabaster Corp.	6921 Olson Lane Pasadena, Texas 77505	Phone: (281) 487-5482 or (800) 609-2728; Fax: (281) 487-9014	Email: alabaster-corp@aol.com
PETRO-GREEN ADP-7	Surface Washing Agent – Applicable over wide range of salinities and temperatures and tested on several crudes.	Petro-Green, Inc.	P.O. Box 814665 Dallas, TX 75381	Phone/Fax: (972) 484-7336	Email: petrogreen@hotmail.com
PETROTECH 25	Surface Washing Agent – May be used with salt or fresh water. Temperature is not relevant.	Petrotech America Corporation	P.O. Box 46 Newport, NH 03773	Phone: (203) 966-4573 Fax: (561) 966-0920	

PREMIER 99	Surface Washing Agent	Gold Coast Chemical Products	2790 South Park Road Pembroke Park, FL 33009	Phone: (954) 893-0044; Fax: (954) 893-8884	Email: noslime@goldcoastchemical.com
PROCLEANS	Surface Washing Agent – May be used with fresh or salt water. Most effective if used on solid surfaces such as shoreline beaches and rocks contaminated with light and medium weight crude oils and refined petroleum products.	Eximco International, Inc.	5252 Gulfton, #2-B Houston, TX 77081	Phone: (713) 432-7889	Email: procleans@procleans.com
SC-1000™	Surface Washing Agent – May be used on any surface that is compatible with water.	GEMTEK® Products	3808 North 28th Avenue Phoenix, AZ 85017	Emergency Number: (602) 265-8586 Phone: (800) 331-7022 Fax: (602) 265-7241	Email: techsupport@infogemtek.com
SHEEN-MAGIC©	Surface Washing Agent	PLUTUS Environmental Technologies, Inc.	P.O. Box 5104 Sevierville, TN 37864-5104	Phone: (865) 453-0060 Fax: (865) 908-6652	Email: CEO@plutusonline.com
SIMPLE GREEN®	Surface Washing Agent – Equally effective in fresh water, estuarine, and marine environments at all temperatures.	Sunshine Makers, Inc.	15922 Pacific Coast Highway, Huntington Harbor, CA 92649	Phone: (800) 228-0709 / (562) 795-6000; Fax: (562) 592-3830	

Table 4.1 (cont.) Product identification chart.

Product	Type Product / Application	Manufacturer/ Contact Information	Address	Phone and Fax	Email/Web Site
SPILLCLEAN	Surface Washing Agent – Water Salinity: Water soluble, not applicable for use on water. Water Temperature: >32°F and <120°F. Types of Pollutants: Automobile motor oil, gasoline, diesel fuel, and transmission fluid. Ages of Pollutants: 0 to 20 years old.	Super Sat Ventures, Inc.	611 Riverview Drive Thiensville, WI 53092	Phone: (414) 840-9223	
SPLIT DECISION SC (formerly SPLIT DECISION)	Surface Washing Agent – Effective in fresh water, estuarine, and marine environments at normal climatic temperatures	Mantek Division of NCH Corporation	P.O. Box 660196 Dallas, TX 75266-0196	Phone: (972) 438-0202 / (800) 527-9919 ext. 2020	Email:gzimmerm@ nch.com
TOPSALL #30	Surface Washing Agent – Recommended for cleaning petroleum fractions from decks, platforms, bilges, rigs, and other seagoing equipment.	Superall Products LLP	22215 Tuwa Rd., Tomball, TX 77375	Phone: (281) 351-4800; Fax: (281) 351-4855	

TXCHEM HE-1000™	Surface Washing Agent – Pollutants should be of hydrocarbon in nature. Greater results occur at ambient temperatures greater than 50°F. Fresh or salt water can be used with no performance change.	Texas EnviroChem, Inc.	11659 Jones Road, PMB #348, Houston, TX 77070	Phone: (281) 728-3217	
JE1058BS	Bioremediation Agent (Nutrient Additive) – Product contains biosurfactant and has an ability to stimulate the biodegradation of oil by indigenous microorganisms. JE1058BS may be used in a wide range of environmental conditions of temperature and salinities where natural populations of oil-degrading microbes are active. The mechanical removal of as much oil as possible should be done before the product is introduced to the slick.	Japan Energy Corporation	Business Development Department, Bio Research Center 3-17-35 Niizo-Minami Toda-shi, Saitama 335-8502 Japan	Phone: (81) 48-433-2191 Fax: (81) 48-444-3223	Email: saeki@j-energy.co.jp
LAND AND SEA RESTORATION PRODUCT 001	Bioremediation Agent (Nutrient Additive) – Largely intended as an absorbent that will bind the	Land and Sea Restoration LLC	4147 Acorn Hill, San Antonio, TX 78217	Phone: (210) 650-5556, Fax: (210) 650-5567	

Table 4.1 (cont.) Product identification chart.

Product	Type Product / Application	Manufacturer/ Contact Information	Address	Phone and Fax	Email/Web Site
	oil on hard surfaces and can be swept up and removed as hazardous waste.				
MICRO-BLAZE®	Biological Additive – Intended for contaminated soils and sludge. Nutrients for microbes are included in the product. However, for longer-term bioremediatior projects, additional applications for Bio-Catalyst may be added to boost microbial activity.	Verde Environmental, Inc.	P.O. Box 8706 Houston, TX 77249-8706	Phone: (713) 691-6468, (800) 626-6598; Fax: (713) 691-2331	E-mail: bscogin@micro-blaze.com/ Web site: http://www:micro-blaze.com
OIL SPILL EATER II (OSE II)	Bioremediation Agent (Biological Enzyme Additive [previously listed as a Nutrient Additive]) – Product engulfs the hydrocarbons and prevents oil from attaching to the shoreline, rocks, or any equipment in its path. OSE II can remediate hydrocarbon-based material including	OSEI Corporation (Formerly Sky Blue Chems)	P.O. Box 515429, Dallas, TX 75251-5429	Phone: (972) 669-3390	E-mail: oseicorp@msn.com/ Web Site: www.osei.us

	chlorinated hydrocarbons, PCB's, dioxins, and some pesticides.				
OPPENHEIMER FORMULA (a.k.a, THE OPPENHEIMER FORMULA I, GENISIS WE-F, MIGHTY MIKE BPT, NATURAL ENVIRO 8000 BIOREMEDIATION, PETRO-TREAT)	Biological Additive – The formula is applicable to grass flats, marshes, and the open sea. It has also been proven effective in soil and fresh water oil clean-up and municipal waste treatment, septic tanks, and grease traps.	Oppenheimer Biotechnology, Inc.	P.O. Box 5919 Austin, TX 78763	Phone: (512) 474-1016 Fax: (512) 472-2909	E-mail: jen.neve@obio.com
S-200 a.k.a. S-200C, SHEENCLEAN, BILGE CLEAR	Bioremediation Agent (Nutrient Additive) – Intended application is to remediate oil sheens on water surfaces via microbial action. Bioremediation accelerator is used for the remediation of hydrocarbon spills or leaks.	International Environmental Products, LLC	Two Villanova Center 795 E. Lancaster Ave., Suite 280 Villanova, PA 19085	Phone: (610) 520-7665 Fax: (610) 520-7663	E-mail: info@iepusa.com
SPILLREMED (MARINE)®	Bioremediation Agent – Biological Additive: Microbiological Culture; Intended to enhance degradation of oil. SPILLREMED	Sarva Bio Remed, LLC	11 North Willow Street, Trenton, NJ 08608	Phone: (609) 695-4992; Fax: (419) 710-5831	E-mail: sales@sarvabioremed.com/ Web Site: www.sarvabioremed.com

Table 4.1 (cont.) Product identification chart.

Product	Type Product / Application	Manufacturer/ Contact Information	Address	Phone and Fax	Email/Web Site
	(MARINE)® is a marine salt formulation and is not effective in fresh water conditions. Optimal effectiveness is in the salinity range of 10 to 35 ppt.				
STEP ONE	Biological Additive – Intended to enhance biodegradation of oil.	B&S Research, Inc.	4345 Highway 21 Embarrass, MN 55732	Phone: (218) 984-3757 Fax: (218) 984-3212	E-mail: soilplus@2z.net
SYSTEM E.T. 20 (formerly MCW.B.20)	Biological Additive – An oleophilic, non-water soluble nutrient, which releases nitrogen and phosphorous enzymatically to enhance degradation rates of oil. Intended for soil remediation.	Environmental Restoration Services	9211 Lakewood Drive Windsor, CA 95492	Phone: (619) 253-0664; Phone: (760) 746-5145; Fax: (760) 746-2034	E-mail: ERS.BTI@gmail.com
VB591™, VB997™, and BINUTRIX® (formerly MYCOBAC TX-20)	Biological Additive (a nutrient additive that contains no microorganisms) – Product is used to stimulate natural oil-utilizing microbial populations to maximize	BioNutraTech Inc.	P.O. Box 2342 Branson, MO 65615	Phone: (417) 858-1150 Fax: (417) 858-1152	E-mail: shruza@bionutratech.com/ Web Site: http://www.bionutratech.com Mobile: (713) 301-0254

	biodegradation activity. VB591™, VB997™, and BINUTRIX® should be applied to spilled oil as soon as possible following spillage to stimulate natural oil-utilizing microbial populations to maximize biodegradation activity.				
WMI-2000	Biological Additive – Bioremediation of oil via microbial action. Intended for sludge applications.	WMI International, Inc.	4901 Milwee, Suite 109 Houston, TX 77092	Phone: (713) 956-4001 Evening: (713) 526-5829 Fax: (713) 956-7305	E-mail: wmi@wt.net E-mail: wmi.meor@yahoo.com
ALSOCUP	Miscellaneous Oil Spill Control Agent – Product is intended to collect petroleum-based products spilled on water or land. When the product comes into contact with a petroleum product, it chemically bonds and traps the petroleum. The oil/product mixture can be collected by skimming with pumps or mechanical devices.	REVCOM Associates	1550 Rimpau Ave. #53 Corona, CA 92881	Phone: (951) 737-0104 Fax: (951) 737-5500	E-mail: revcom@sbcglobal.net

Table 4.1 (cont.) Product identification chart.

Product	Type Product / Application	Manufacturer/ Contact Information	Address	Phone and Fax	Email/Web Site
AQUA N-CAP™ POLYMER	Miscellaneous Oil Spill Control Agent – Product is applied to the outer edges of spill area to prevent spill migration. Product may be broadcast onto the spill, whereby oils will clump into a soft cluster like material easily recoverable with a net or other means as appropriate.	RTA Systems, Inc.	8120 Mid America Boulevard, Suite 200A Oklahoma City, OK 73135	Phone: (405) 397-9108 Fax: (405) 610-2620	E-mail: bjohnson@ RTASys.com
CIAGENT (formerly CI AGENT, CHEAP INSURANCE, & PETRO-CAPTURE)	Miscellaneous Oil Spill Control Agent – Product encapsulates and solidifies oil.	CIAGENT Solutions, LLC	11760 Commonwealth Dr. Louisville, KY 40299	Phone: (502) 267-0101; (800) 255-6073 Fax: (502) 267-0181	E-mail: dan@ ciagent.com
ELASTOL	Miscellaneous Oil Spill Control Agent – Product can be applied to the surface of an oil/water separator tank or added to a waste stream before it enters a separation	Action Additives, Inc. 205 Industrial Road	P.O. Box 965 Ducktown, TN 37326	Phone: (423) 496-5000; (800) 496-5110	

	tank. Treated oil may increase the performance of containment and recovery equipment.				
PES-51	Miscellaneous Oil Spill Control Agent – Intended as a shoreline and surface treatment supplement and used for tank cleaning and equipment decontamination.	Practical Environmental Solutions (Formerly known as Petroleum Environmental Services)	P.O. Box 12563 San Antonio, Texas 78212	Phone: (210) 493-7172 Fax: (210) 493-7172	E-mail: simsbi@aol.com
RAPIDGRAB 2000™	Miscellaneous Oil Spill Control Agent – Intended to reduce and confine floating oils by the oleophitic synergistic effect of contraction and congealment into a physical state that simplifies clean-up. RAPIDGRAB 2000™ is designed to be used after oil has been contained by booms or other similar apparatus	GlobeMark Resources Ltd.	1205 Pine Heights Drive Atlanta, GA 30324	Mobile: (254) 231-2251	E-mail: joannie@globemarkresouces.com Web Site: http://www.globemarkresources.com

Table 4.1 (cont.) Product identification chart.

Product	Type Product / Application	Manufacturer/ Contact Information	Address	Phone and Fax	Email/Web Site
WASTE-SET #3200®	Miscellaneous Oil Spill Control Agent – Oil Spill Solidifying Agent – Product is intended to prevent spills from spreading. Surrounds the spill to prevent migration. Can be used to reinforce containment line with a boom.	Environmental & Fire Technology, LLC	3374 West River Dr. NW Grand Rapids, MI 49544	Phone: (616) 784-0770 Fax: (616) 784-5018	E-mail: eandft@comcast.com
ZYME-FLOW	Miscellaneous Oil Spill Control Agent – Product is intended to make heavy crudes more pumpable and break the adhesion between oils and soil, rock, or sand.	United Laboratories, Inc.	320 37th Avenue St. Charles, IL 60174	Phone: (630) 377-0900 (800) 323-2594 Fax: (630) 377-0960	E-mail: nsherrel@unitedlabsinc.com

Table 4.2 Product stability and storage chart.

Product	Flammability	Maximum Storage Temperature	Minimum Storage Temperature	Optimum Storage Temperature Range	Shelf Life
AQUACLEAN	Non-flammable	140°F	50°F	>50°F to <104°F	Shelf life is 18 months if stored between 50°F and 104°F and away from acids.
BG-CLEAN™ 401	Non-flammable	180°F	–25°F	68°F	The shelf life of unopened containers of BG-CLEAN™ 401 is unlimited. Containers should always be capped when not in use to prevent contamination and evaporation.
BIOSOLVE® HYDROCARBON MITIGATION™ AGENT	Non-flammable	120°F (50°C)	35°F (1.5°C)	60°F (15°C)	BIOSOLVE® has a 10+ year shelf life if unopened.
CN-110	Non-flammable	120°F	–40°F	70°F–80°F	CN-110 has a shelf life of 2 years.

Table 4.2 (cont.) Product stability and storage chart.

Product	Flammability	Maximum Storage Temperature	Minimum Storage Temperature	Optimum Storage Temperature Range	Shelf Life
COREXIT® EC7664A	The flash point of COREXIT® EC7664A is 116°F. It is, therefore, classified as flammable by IMO rules but not by DOT rules. Keep away from heat, sparks, and open flame.	130°F	10°F	40°F to 100°F	The shelf life of unopened drums of COREXIT® EC7664A is unlimited. Partially used containers should be kept tightly closed to retard evaporation of solvents.
CYTOSOL	Non-flammable. However, keep the product away from heat and avoid contact with strong oxidizing agents. Ensure proper disposal of product-	110°F	39°F	55°F	Closed container: 10 years in a dry environment. Open container: 1 year in a warm, humid environment. The product does not deteriorate appreciably over time, but it will grow bacteria if water condensation accumulates in the container.

	saturated absorbents, rags, and combustible materials to avoid the possibility of spontaneous combustion.				
DO-ALL #18	DO-ALL #18 is water-based, contains no solvents, and is non-flammable.	100°F	35°F	40°F to 100°F	DO-ALL #18 has a shelf life of at least 2 years.
ENVIROCLEAN (formerly ENVIRO CLEAN 165)	Non-flammable	120°F	28°F	30°F – 120°F	Unlimited if unopened.
E-SAFE©	Non-flammable until >170°F	160°F	–15°F	40°F – 110°F	Unlimited if left in unopened containers stored at 40°F–110°F and away from direct sunlight.
F-500	Non-flammable	130°F	17°F	35°F – 130°F	The shelf life is 15 years when stored between 35°F and 130°F in unopened containers.

Table 4.2 (cont.) Product stability and storage chart.

Product	Flammability	Maximum Storage Temperature	Minimum Storage Temperature	Optimum Storage Temperature Range	Shelf Life
GOLD CREW SW	Non-flammable	When above 120°F, keep container closed and stored in a cool dark place. Evaporation may change product's characteristics.	Product will freeze below 25°F. No phase separation will occur. If frozen, thaw, and stir well.	25°F to 120°F	20 years (unopened)
NALE-IT	Non-flammable – Equally effective with fresh or salt water	NA	>32°F	40°F to 200°F	Indefinite
NATURE'S WAY HS	Non-flammable	130°F	32°F	90°F	5 year shelf life
NOKOMIS 5-W	Non-flammable (water-based)	212°F	32°F	50°F	15 years or more if stored correctly in plastic drums

PETRO-CLEAN	Non-flammable	120°F	35°F	None recommended	Indefinite when stored properly
PETRO-GREEN ADP-7	Non-flammable	140°F	26°F	40°F to 110°F	Indefinite shelf life
PETROTECH 25	Non-flammable	45°C	–5°C (It can be stored below its freezing point without detectable loss of performance, provided that it is warmed to within liquid range prior to application.)	>–5°C and <+45°C	PETROTECH 25 has an unlimited shelf life when maintained in the factory-sealed containers and stored within the prescribed temperature limits. Climatic factors such as humidity have no effect on closed container storage, although the product is hygroscopic and will absorb water if left for long periods in open containers.
PREMIER 99	None	125°F	0°F	0°F – 75°F; Avoid freezing the product. If material freezes and separation is experienced,	Unopened material – at least 5 years is expected. Opened material – at least 2 years is expected.

Table 4.2 (cont.) Product stability and storage chart.

Product	Flammability	Maximum Storage Temperature	Minimum Storage Temperature	Optimum Storage Temperature Range	Shelf Life
				it should be warmed and mixed together.	
PROCLEANS	Non-flammable	130°F	35°F	50°F to 100°F	Approximately 2 years at recommended temperatures if unopened.
SC-1000™	Non-flammable	None	Room temperature	70°F to 90°F	Minimum of 5 years
SHEEN-MAGIC©	Non-flammable until >170°F	160°F	–8°F	40°F – 110°F	Unlimited if left in unopened containers stored at 40°F–110°F and away from direct sunlight
SIMPLE GREEN®	Non-flammable	140°F	34°F	>32°F and <140°F	SIMPLE GREEN® has an unlimited shelf life.
SPILLCLEAN	>210°F	<210°F	>32°F	50°F – 70°F	3 years; avoid extreme heat, and store in a dry, cool area.

SPLIT DECISION SC (formerly SPLIT DECISION)	Not flammable	140°F	27°F	2–140°F	The recommended shelf life is one year.
TOPSALL #30	Non-flammable	120°F	20°F	40°F to 100°F	Not less than two years
TXCHEM HE-1000™	Non-flammable	95°F	32°F	75°F	One year in a sealed container
JE1058BS	Non-flammable	105°F	NA	0°F – 68°F	The shelf life of the product is approximately 2 years if stored in a cool dry area.
LAND AND SEA RESTORATION PRODUCT 001	Flash point is 250°C (482°F)	65°C (150°F)	0°C (32°F)	0°C to 65°C (32 to 150°F)	The shelf life is more than 5 years in unopened, original shipping container. Store in a cool, dry place. Avoid moisture prior to use.
MICRO-BLAZE®	Non-flammable	120°F	35°F	NA	Minimum 10 years, with proper storage, in original containers. Freezing will not kill these microbes; however, extreme heat (over 180°F) for long periods of time will kill the microbes.

Table 4.2 (cont.) Product stability and storage chart.

Product	Flammability	Maximum Storage Temperature	Minimum Storage Temperature	Optimum Storage Temperature Range	Shelf Life
OIL SPILL EATER II (OSE II)	Water-based, non-flammable	120°F	None; OSE II can freeze and thaw without adverse effects	72°F	OSE II has a recommended shelf life of 5 years. After 5 years at optimum storage temperature, there is an approximate 10% decrease per year in product capability.
OPPENHEIMER FORMULA (a.k.a, THE OPPENHEIMER FORMULA I, GENISIS WE-F, MIGHTY MIKE BPT, NATURAL ENVIRO 8000 BIOREMEDIATION, PETRO-TREAT)	Non-flammable	130°F	32°F	82°F	The shelf life of the product is approximately 5 years.

S-200, a.k.a. S-200C, SHEENCLEAN, BILGE CLEAR	Non-flammable	132°F	32°F	55°F – 85°F	The recommended shelf life of S-200 is 1½ years if kept in the unopened original container.
SPILLREMED (MARINE)®	Non-flammable	100°F	32°F	Ambient	The recommended shelf life is one year.
STEP ONE	Non-flammable	90°F	-30°F	32°F to 70°F	Over 3 years
SYSTEM E.T. 20 (formerly MCW.B.20)	Non-flammable			4°C – 20°C (39°F – 68°F). Place in a dry area, shaded from sunlight.	2 years if maintained at 4°C (39°F)
VB591™, VB997™, BINUTRIX® (formerly MYCOBAC TX-20)	Non-flammable	>140°F	NA	50°F to 80°F	The shelf life is approximately three years if kept dry. Store in dry location and avoid contact with moisture. Prolonged storage may result in the formation of soft clumps, which are easily broken by mechanical disruption.

Table 4.2 (cont.) Product stability and storage chart.

Product	Flammability	Maximum Storage Temperature	Minimum Storage Temperature	Optimum Storage Temperature Range	Shelf Life
WMI-2000	Non-flammable	100°F	35°F	45°F – 90°F	Store in a cool, dry location. Keep partially used containers tightly closed. Prolonged exposure to high temperature and humidity may lower activity of the product. The shelf life of WMI-2000 is 2 years if stored at ambient temperatures.
ALSOCUP	NA	163°F	0°F	0°F to 163°F	5 years
AQUA N-CAP™ POLYMER	Non-flammable	185°F	NA	185°F	Not limited
CIAGENT (formerly CI AGENT, CHEAP INSURANCE, & PETRO-CAPTURE)	Flammable solid at >210°F	190°F	None	50°F – 80°F	> 5 years if stored in cool dry area, away from direct sunlight.

ELASTOL	Autoignition temperature for the product is 392°F. Product is combustible in liquid form and can form combustible mixtures and possibly explosive mixtures above the flash point. Use water spray to cool fire-exposed surfaces and to protect personnel. Isolate "fuel" supple from fire. Use foam, dry chemical, or water spray to extinguish fire. Avoid spraying water directly into storage containers due	~160°F	–100°F	40–70°F	At least 5 years when stored at temperatures below 150°F.

Table 4.2 (cont.) Product stability and storage chart.

Product	Flammability	Maximum Storage Temperature	Minimum Storage Temperature	Optimum Storage Temperature Range	Shelf Life
	to danger of boil over. Product is volatile and gives off invisible vapors. Either the liquid or vapor may settle in low areas or travel some distance along the ground or surface to the ignition source where they may ignite.				
PES-51	124°F	NA	NA	NA	6 years (unopened drum); 1 year (opened drum)
RAPIDGRAB 2000™	Non-flammable	120°F	0°F	32°F – 90°F	The shelf life of unopened drums of RAPIDGRAB 2000™ is unlimited.

					Containers should be capped when not being used to prevent contamination and evaporation.
WASTE-SET #3200®	Product will not burn unless pre-heated. Avoid elevated temperatures (>200°F) for extended periods (>5 days).	200°F	None	<200°F	Unlimited
ZYME-FLOW	Non-flammable	120°F	0°F	0°F – 120°F	1 year minimum

Table 4.3 Product physical properties chart.

Product	Flash Point	Pour Point	Viscosity	Specific Gravity	pH	Surface Active Agents	Solvents	Additives	Solubility
AQUACLEAN	None detected	–20°F	16.5 cP at 25°C (77°F)	1.06	11.8	Anionic and nonionic synthetic surfactants	Confidential	Confidential	Completely soluble
BG-CLEAN™ 401	>76.6°C (>170°F)	–4°C	24 SUS at 100°F	1.006	8.227	Proprietary	Proprietary	Proprietary	Miscible in water
BIOSOLVE® HYDROCARBON MITIGATION™ AGENT	NA, Water-based >200°F (93.3°C)	32.9°F (0.5°C)	77.5 Centistrokes (concentrate), 490 centipoise (concentrate), 15 centipoise at 6%, at 60.08°F or 15.6°C	1.025	9.37 +/– .5	Confidential	Confidential	Confidential	Complete-true solution formed with water
CN-110	None	30°F	45.7 cst at 78°F	1.025	11.4	Trace amounts of a sulfonated compound	None	Complex silicate solution	NA
COREXIT® EC7664A	116°F	7°F	25 cst at 100°F, 34 cst at 60°F	1.02	6.14	Confidential	Isopropanol, Water	None	NA
CYTOSOL	360°F	10°F	4.15 CST @ 104°F	0.8877	Neutral	None	No Petroleum Distillates	Confidential	14 ppm in fresh water, 7 ppm in sea water

DO-ALL #18	>200°F	32°F		1.07	13.1	Confidential	None	Confidential	Soluble with water
ENVIROCLEAN (formerly ENVIRO CLEAN 165)	>200°F	28.8°F	9 Centipose	1.028	8.63	Confidential	None	Confidential	Complete in water
E-SAFE©	>170°F	–27ºC	11SFS at 100°F	1.0118	8.04	Confidential	Confidential	Confidential	Soluble
F-500	>200°F	–17°F	54-56 centist-okes at 72°F	0.99	7	Confidential	None	None	Complete
GOLD CREW SW	>200°F	25°F	33.87 CST	1.035	9.76 +/– 0.01	Confidential	None	Confidential	Complete
NALE-IT	>212°F	30°F		1.02	6.8 – 7.2	Confidential	None	Confidential	Soluble in fresh and salt water
NATURE'S WAY HS	NA	Flowable at all tempera-tures above freezing	<100 CPS	1.01	8 – 9.5	Confidential	Confidential	Confidential	NA
NOKOMIS 5-W	None	28°F	31.00 cps	1.0065	10.4	Confidential	None	None	Completely water soluble
PETRO-CLEAN	> 200°F	–17°F		0.99	8.05 (10% solution, s.u.)	Confidential	None	Confidential	100%

Table 4.3 (cont.) Product physical properties chart.

Product	Flash Point	Pour Point	Viscosity	Specific Gravity	pH	Surface Active Agents	Solvents	Additives	Solubility
PETRO-GREEN ADP-7	>212°F	22°F		1.035	10.5	Confidential	Confidential	Confidential	NA
PETROTECH 25	None	0°C	700 CP	1.03	7.5	Confidential	Confidential	Confidential	100%
PREMIER 99	160°F	NA	15 cps	1.01	10 – 11.5	Active 8000* – Purity Chemical <10%	<5%	<5%	Complete in Water
PROCLEANS	>186°F	34.6°F	2.41 cST at 104°F	1.01	6.8	Anionic and nonionic	None	None	NA
SC-1000™	>212°F	25°F	<10 cps at 25°C	1.009	10.2 – 10.5	Confidential	None	Confidential	Soluble in water
SHEEN-MAGIC©	>170°F	–22ºC	11SFS at 100°F	1.0127	8.1	Confidential	Confidential	Confidential	Soluble
SIMPLE GREEN®	>200°F	None	2.0 Centistokes at 78°F	1.0257	9.5	Confidential	Confidential	Confidential	Infinitely miscible (Increasing salt con-centrations in marine ecosystems will lead to complexes with the

									product that may become visible at ratios above one part product to 99 parts seawater.)
SPILLCLEAN	>210°F	27.0°F	6400cPs at 25°C	1.03	7.2	Confidential	Confidential	Confidential	Water-soluble
SPLIT DECISION SC (formerly SPLIT DECISION)	None	NA	20 cps – Brookfield #1 spindle at 20 rpm at 73°F	1.075	7	Confidential	Water	Confidential	Complete
TOPSALL #30	>212°F	20°F	30.53 SUS at 100°F	1.06	12.6	Confidential	None	Confidential	Miscible with water
TXCHEM HE-1000™	>210°F	23.0°F	1.99 cSt	0.999	7.2	Confidential	NA	Confidential	NA
JE1058BS	No data	No data	No data	No data	No data	No data	No data	No data	No data
LAND AND SEA RESTORATION PRODUCT 001	250°C (482°F)	NA	NA	NA	NA	NA	NA	NA	NA
MICRO-BLAZE®	NA	NA	NA	NA	NA				

Table 4.3 (cont.) Product physical properties chart.

Product	Flash Point	Pour Point	Viscosity	Specific Gravity	pH	Surface Active Agents	Solvents	Additives	Solubility
S-200 a.k.a. S-200C, SHEENCLEAN, BILGE CLEAR	>212°F	40°F	100cp at 70°F	0.996	4.5	NA	NA	NA	Dispersible
SPILLREMED (MARINE)®									
STEP ONE	>212°F	–7°F	2.8 cst at 50°F	1.03	7.2 (1:2.5:80 BC101: MSE2.5: water mixture)	Noedal	None	Phosphoric acid as P205 3% emulsi-fier <.5% by weight	Infinite
AQUA N-CAP™ POLYMER	230ºF	NA	NA	0.91	4.08	NA	NA	None	Negligible
CIAGENT (formerly CI AGENT, CHEAP INSURANCE, & PETRO-CAPTURE)	No flash observed at 210.F	Not appli-cable; solid samples	Not appli-cable; solid samples	0.94	7.81	None	None	None	Negligible
ELASTOL	176ºF	–71ºF	3.46 cst at 77ºF	0.83 at 60ºF	6.8 – 7.0	Confidential	Confidential	Confidential	Insoluble

PES-51	124°F	–50°F at 30 min.	30 cst at 28°C	0.840 at 25°C	6.7	Confidential	Confidential	Confidential	NA
RAPIDGRAB 2000™	110°C	30°F		0.84	6.95	Oleophitic surfactant	Proprietary formulation oleophitic surfactants	None	Miscible in oil and solvents
WASTE-SET #3200®	No flash observed at 310°F	NA	NA	0.94	NA	NA	None	None	Insoluble
ZYME-FLOW	None	32°F	<10 cps	0.99	7.0 – 8.0	Confidential	NA	Confidential	Soluble

Table 4.4 Product safe-handling chart.

Product	Flammability	Ventilation	Skin and Eye Contact, Protective Clothing, and Treatment in Case of Contact
AQUACLEAN	Non-flammable	Handle in a well-ventilated space. Local exhaust is recommended if TLV's are exceeded.	Avoid eye and skin contact. In case of such contact, immediately flush with large amounts of cool water for at least 15 minutes, and call a physician. Wear protective eye goggles, impermeable protective gloves, and protective clothing when handling the product. Wash contaminated protective equipment (including clothing, shoes, eye goggles, gloves, etc.) thoroughly in soap and water, rinse repeatedly in clean water, and dry before reuse. Avoid ingestion and breathing dusts, mists, or fumes. In case of ingestion, induce vomiting, give water, and call a physician. In case of inhalation, move the affected person to fresh air, and call a physician.
BG-CLEAN™ 401	Non-flammable	Avoid prolonged breathing of vapors.	BG-CLEAN™ 401 is an organic compound with a slightly alkaline pH. No extraordinary precautions are required during handling. Avoid unnecessary contact with skin or eyes. No special protective clothing is required. Body areas exposed to BG-CLEAN™ 401 may be flushed with water to clean off the chemical.

BIOSOLVE® HYDROCARBON MITIGATION™ AGENT	Non-flammable	Normal	Flush contaminated eyes thoroughly with water for 15 minutes, and get medical attention. Remove contaminated clothing, wash exposed area with soap and water, and wash clothing before reuse. Get medical attention if irritation develops. Get medical attention for ingestion. No medical attention is necessary with inhalation. There are no special storage requirements or special handling precautions. Use good normal hygiene.
CN-110	Non-flammable	None needed	As in handling any industrial chemical, the standard precautions of wearing chemical-resistant gloves and eye protection are recommended.
COREXIT® EC7664A	The flash point of COREXIT® EC7664A is 116°F. It is, therefore, classified as flammable by IMO rules but not by DOT rules. Keep away from heat, sparks, and open flame.	Avoid prolonged breathing of vapors. Use with ventilation equal to unobstructed outdoors in moderate breeze.	Avoid contact with skin or eyes and breathing mist or vapors. The use of gloves and goggles is recommended. In case of contact, flush exposed area with water.

Table 4.4 (cont.) Product safe-handling chart.

Product	Flammability	Ventilation	Skin and Eye Contact, Protective Clothing, and Treatment in Case of Contact
CYTOSOL	Non-flammable. However, keep the product away from heat and avoid contact with strong oxidizing agents. Ensure proper disposal of product-saturated absorbents, rags, and combustible materials to avoid the possibility of spontaneous combustion.	Product is not volatile. However, in the event of aerosol inhalation, immediately move victim to fresh air. If victim has stopped breathing, give artificial respiration, preferably by mouth to mouth. Get medical attention immediately.	CYTOSOL contains no volatile hydrocarbons or petroleum constituents. However, as a precautionary measure, wear gloves and safety glasses meeting the specifications of ANSI Standard Z87.1. Avoid breathing aerosols. Avoid prolonged contact with skin.
DO-ALL #18	DO-ALL #18 is water-based, contains no solvents, and is non-flammable.	Normal room ventilation	Avoid contact with skin, eyes, and clothing. Wear safety glasses or goggles and protective gloves when handling. In case of eye contact, flush immediately with water for at least 15 minutes. If persistent irritation occurs, call a physician. For skin contact, wash thoroughly with soap and water.

ENVIROCLEAN (formerly ENVIRO CLEAN 165)	Non-flammable	Normal	Use good normal hygiene and avoid contact with skin and eyes. Gloves and goggles are recommended for field application and product handling. In case of contact with eyes, flush with water for 15 minutes and seek medical attention. For contact with skin, wash area with soap and water.
E-SAFE©	Non-flammable until >170°F	Handle in a well-ventilated space. Local exhaust is recommended if TLV's are exceeded.	Avoid eye and (sensitive) skin contact. In case of contact, immediately flush with large amounts of cool water for at least 5 minutes. Wear protective eye goggles when using any chemicals. Impermeable protective gloves are recommended for sensitive skin types. Protective clothing is not required. Rinse contaminated clothing, shoes, goggles, and gloves in simple tap water to remove any chemical residue. Avoid ingestion and breathing dusts, mists, or fumes. In case of ingestion, drink several glasses of water. Do not induce vomiting. In case of inhalation, move affected person to fresh air.
F-500	Non-flammable	Local exhaust – no special requirements	Eye contact may cause eye irritation. Flush with water. Wear splash-proof protection goggles or full-face shield. If irritation persists, contact a physician. Skin contact may cause mild irritation. Flush with water. Wear

Table 4.4 (cont.) Product safe-handling chart.

Product	Flammability	Ventilation	Skin and Eye Contact, Protective Clothing, and Treatment in Case of Contact
			rubber and vinyl gloves and apron. Product has been tested and found to be non-skin sensitizing per OECD 406. Ingestion may cause gastrointestinal irritation. Drink plenty of water to dilute. Do not induce vomiting. If irritation persists, contact a physician. In case of inhalation, elevated temperature vapors may cause irritation to respiratory tract. Remove to fresh air.
GOLD CREW SW	Non-flammable	Normal	Extensive testing indicates that GOLD CREW SW is non-hazardous and non-toxic to humans; however, good hygiene practices should always be followed as outlined below. In case of contact with eyes, flush with water and get medical attention if required. In case of skin contact, remove contaminated clothing, wash exposed area, and wash clothing before use. If irritation develops, get medical attention. In case of ingestion, get medical attention if required. No precautions are considered necessary in the case of inhalation.

NALE-IT	Non-flammable	Workers should be in well-ventilated area. If in a confined area, use a respirator.	Workers should wear protective goggles or safety glasses. Prolonged contact with skin may result in dryness.
NATURE'S WAY HS	Non-flammable	No special precautions are required, but in case of inhalation, move person to fresh air.	Flush contaminated skin with plenty of water for at least 15 minutes. Consult a physician if irritation develops. Protective gloves are recommended for extended or prolonged contact, (e.g., immersing hands). Tight fitting safety goggles are recommended for handling the product in concentrated form, especially if contacts are worn.
NOKOMIS 5-W	Non-flammable (water-based)	No special requirements	In case of eye contact, flush with water. In case of skin contact, wash with water. If swallowed, drink water to dilute and induce vomiting.
PETRO-CLEAN	Non-flammable	Normal	In case of contact with eyes, flush with water using eye cup or fountain for 15 minutes. Seek medical attention if irritation persists. Wash contaminated clothing and footwear before reuse. In case of ingestion, seek medical attention. No medical attention is required for inhalation.
PETRO-GREEN ADP-7	Non-flammable	Use normal household type ventilation when storing and handling.	Avoid contact with skin and eyes. In the event of contact, flush with clean water. Prolonged contact with skin can cause drying. Treat skin with common hand lotion. The use of gloves and goggles is recommended.

Table 4.4 (cont.) Product safe-handling chart.

Product	Flammability	Ventilation	Skin and Eye Contact, Protective Clothing, and Treatment in Case of Contact
PETROTECH 25	Non-flammable	Normal	No special handling is required beyond that prescribed under the general OSHA rules for nonhazardous, non-flammable, and noncorrosive liquids. This is not to suggest that contact with eyes and skin is totally without risk for allergy. Contact with eyes and skin is to be avoided, as suggested under OSHA procedures.
PREMIER 99	None	Adequate ventilation is needed if in a closed environment.	In case of skin and eye contact, detergents will defat skin and eyes. May cause irritation. Use protective gloves and splash goggles for manual cleaning. If splashed on skin, wash with copious amounts of water. If ingested, drink diluted vinegar or lemon juice. Get medical attention.
PROCLEANS	Non-flammable	Normal	Flush contaminated eyes thoroughly with water for 15 minutes and get medical attention. Remove contaminated clothing, wash exposed area with soap and water, and wash clothing before use. Get medical attention if irritation develops. Get medical attention if ingested. No medical attention is necessary if inhaled.

SC-1000™	Non-flammable	General room ventilation is satisfactory.	Non-irritating, no first aid needed. May be an eye irritant. Do not spray into eyes; safety glasses are recommended. If irritation does occur, rinse thoroughly with water.
SHEEN-MAGIC©	Non-flammable until >170°F	Handle in a well-ventilated space. Local exhaust is recommended if TLV's are exceeded.	Avoid eye and (sensitive) skin contact. In case of contact, immediately flush with large amounts of cool water for at least 5 minutes. Wear protective eye goggles when using any chemicals. Impermeable protective gloves are recommended for sensitive skin types. Protective clothing is not required. Rinse contaminated clothing, shoes, goggles, and gloves in simple tap water to remove any chemical residue. Avoid ingestion and breathing dusts, mists, or fumes. In case of ingestion, drink several glasses of water. Do not induce vomiting. In case of inhalation, move affected person to fresh air.
SIMPLE GREEN®	Non-flammable	Not required	SIMPLE GREEN® is safe for use on skin and will not cause irritation in the majority of users. Avoid contact with eyes; irritation may result. Wear splash goggles or full face-shield and have eye washing equipment available in areas where potential is high for eye contact. No special precautions or additional protective equipment are required

Table 4.4 (cont.) Product safe-handling chart.

Product	Flammability	Ventilation	Skin and Eye Contact, Protective Clothing, and Treatment in Case of Contact
			during handling or use. SIMPLE GREEN® is provided with a Material Safety Data Sheet (No. 1002).
SPILLCLEAN	>210°F	General room ventilation is expected to be satisfactory.	Product may cause mild skin irritation. Wear impervious gloves, safety goggles, and full cover clothes. Eye wash and safety shower should be accessible. In case of eye contact, flush immediately with water for at least 15 minutes. In case of skin contact, wash with soap and water. If swallowed, give two glasses of water (do not induce vomiting). Seek medical attention for eye and skin contact and ingestion.
SPLIT DECISION SC (formerly SPLIT DECISION)	Not flammable	Not required	SPLIT DECISION SC is not aggressive to skin and will not cause irritation in the majority of users. Avoid contact with eyes; irritation may result. Wear safety glasses with side

			shields if the method of use presents the likelihood of eye contact. No special precautions or additional protective equipment are required during handling or use.
TOPSALL #30	Non-flammable	Normal type household ventilation is adequate for handling and storage.	In case of contact with eyes, wash thoroughly with large amounts of water. If irritation persists, seek medical attention.
TXCHEM HE-1000™	Non-flammable	Recommended, respirator in confined area	Wear rubber gloves and eye protection. In case of eye contact, wash eyes thoroughly for 15 minutes, including upper and lowers lids, and seek medical attention. For skin contact, irritation is possible. Wash with soap and water for 15 minutes. If irritation persists, seek medical attention.
JE1058BS	Non-flammable	Use adequate ventilation. Treat product as a hygroscopic powder.	Avoid contact with eyes and do not take internally. Upon contact with eyes, flush with plenty of fresh water. Avoid breathing dust. Normal protective equipment for handling of powder, such as a dust mask and eye shield, can be used.
LAND AND SEA RESTORATION PRODUCT 001	Flash point is 250°C (482°F)	Normal ventilation	Protective clothing and glasses. In case of eye contact, flush eyes with water or eyewash and refrain from rubbing. For skin contact, wash with mild soap and apply cream if itching or redness occurs. Avoid inhalation.

Table 4.4 (cont.) Product safe-handling chart.

Product	Flammability	Ventilation	Skin and Eye Contact, Protective Clothing, and Treatment in Case of Contact
			In case of inhalation, seek fresh air. Wear protective clothing and eye goggles. Workers should wear single-use nuisance dust masks appropriate for fine particulate dust.
MICRO-BLAZE®	Non-flammable	Normal room ventilation	Avoid eye contact. Wear protective gloves, and wash hands with soap and water after handling the product. Wash contaminated clothing and footwear before reuse.
OIL SPILL EATER II (OSE II)	Water-based, non-flammable	Needs no ventilation; aqueous-based product; does not emit hazardous vapors.	OSE II is not a primary dermal irritant. Avoid eye contact, and wear goggles if possible for the spray to come in direct contact with eyes. Facilities for quick and copious eye flushing should be provided and prompt medical attention should be sought if exposure and irritation persists. Protective rubber gloves are suggested during handling. Before mixing, the product has a smell of fermentation. The product does not give off any harmful vapors.

OPPENHEIMER FORMULA (aka, THE OPPENHEIMER FORMULA I, GENISIS WE-F, MIGHTY MIKE BPT, NATURAL ENVIRO 8000 BIOREMEDIATION, PETRO-TREAT)	Non-flammable	Special ventilation is not required. Treat product as a hygroscopic powder. In closed spaces, use dust protective measures.	The application of the dry powder requires the usual precautions of a dust irritant to membranes. The material is easily removed by washing or flushing, and a protective mask and eye glasses are recommended in confined areas or use.
S-200, a.k.a. S-200C, SHEENCLEAN, BILGE CLEAR	Non-flammable	Ventilation of the work place is not necessary.	In case of contact with skin or eyes, wash with plenty of water.
SPILLREMED (MARINE)®	Non-flammable	No special ventilation is required.	On contact with product, wash with plenty of water, and for eye contact consult your physician. Use of gloves is recommended during field application to protect eyes from spray.
STEP ONE	Non-flammable	Normal ventilation is adequate.	Avoid contact with skin, especially when there are open wounds. In case of skin contact, wash with soapy water. The use of protective gloves is recommended. Avoid contact with eyes. In case of eye contact, immediately flush eyes with plenty of water

Table 4.4 (cont.) Product safe-handling chart.

Product	Flammability	Ventilation	Skin and Eye Contact, Protective Clothing, and Treatment in Case of Contact
			continuously for at least 15 minutes. Consult a physician. The use of protective goggles is recommended. Avoid inhalation and ingestion. It is recommended that workers wear dust masks and not eat or smoke while handling the product(s).
SYSTEM E.T. 20 (formerly MCW.B.20)	Non-flammable	No special ventilation is required under normal use.	In its primary form, no special handling or storage is required. Avoid excessive inhalation, and protect the nose and mouth with a dust protection mask. Wear protective gloves, and wash hands with soap and water after handling the product.
VB591™, VB997™, BINUTRIX® (formerly MYCOBAC TX-20)	Non-flammable	Normal ventilation is adequate.	Use normal precautions and protective equipment for handling any type of powder, such as dust protectors and eye shield. Avoid contact with eyes and do not take internally. Upon contact with eyes, flush with water immediately for a minimum of 15 minutes. If redness or irritation continues, contact a physician. Avoid breathing dust.

WMI-2000	Non-flammable	Avoid breathing dust or aerosol due to possibility of sensitization. Use adequate ventilation.	Minimize contact with eyes, skin, and clothing. Wash hands thoroughly after handling culture.
ALSOCUP	NA	NA	ALSOCUP is practically non-irritating to the eyes and is non-irritating to the skin. Safety glasses and protective clothing should be worn when applying the material. The product is not expected to cause irritation to the nose, throat, or respiratory tract.
AQUA N-CAP™ POLYMER	Non-flammable	Work in a well-ventilated area. Avoid excessive inhalation of dust.	No recommendation is made specifying the need for personal protective clothing to prevent skin contact. Person(s) should wash daily at the end of each work shift. Exposed areas should be thoroughly washed with water. Product may cause irritation as any normal dust. Wear appropriate eye protection, compliant with ANSI Standard Z87.1, to prevent eye contact. If eye irritation or redness is experienced, promptly wash eyes with large amounts of water for at least 15 minutes, occasionally lifting the lower and upper lids. Should eye irritation persist, seek medical attention immediately. Contact lenses should not be worn when working with this product. Wear appropriate respiratory

Table 4.4 (cont.) Product safe-handling chart.

Product	Flammability	Ventilation	Skin and Eye Contact, Protective Clothing, and Treatment in Case of Contact
			protection, compliant with OSHA standard 29 CFR 1910.134, to prevent inhalation. If a respiratory disorder is observed, remove person(s) from the work area immediately. Should respiratory disorder persist, seek medical attention immediately. In case of ingestion, rinse mouth with water and seek medical attention immediately.
CIAGENT (formerly CI AGENT, CHEAP INSURANCE, & PETRO-CAPTURE)	Flammable solid at >210°F	Normal ventilation is adequate.	Product may cause slight eye irritation if in contact with eyes. Use protective eye goggles when handling the product. In case of eye contact, flush eyes with water. In an enclosed environment, workers should wear a dust mask for personal comfort. In case of a confined space fire, do not enter without full bunker gear and positive pressure NIOSH approved self-contained breathing apparatus. Maintain fire watch at 450°F.

ELASTOL	Autoignition temperature for product is 392°F. Product is combustible in liquid form and can form combustible mixtures and possibly explosive mixtures above the flash point. Use water spray to cool fire-exposed surfaces and to protect personnel. Isolate "fuel" supple from fire. Use foam, dry chemical, or water spray to extinguish fire. Avoid spraying water directly into storage containers due to danger of boil over. Product is	Product is moderately volatile. Ensure proper ventilation. Avoid formation and inhalation of spray mist and vapors. Do not wear contaminated clothing. Keep away from fire and heated objects. Avoid formation of aerosols. Avoid squirting and use dosage pump when transferring.	Protection depends on use and the condition of the work environment. Ensure good personal hygiene. Wash hands regularly with soap and water. Avoid contact with skin and eyes and inhalation of fumes/aerosols. All new protective gear must be of CE standards. There must be an eye rinsing kit available at the work place. Where there is danger of inhalation, use mask with brown filter. If the product is sprayed and there is a risk of aerosol formation, use mask with combination filter A2-P2. When used regularly, measuring routines for solvent measurement should be implemented. Level of contamination should be kept as low as possible.

Table 4.4 (cont.) Product safe-handling chart.

Product	Flammability	Ventilation	Skin and Eye Contact, Protective Clothing, and Treatment in Case of Contact
	volatile and gives off invisible vapors. Either the liquid or vapor may settle in low areas or travel some distance along the ground or surface to the ignition source where they may ignite.		
PES-51	124°F	Handle product in a normal, well-ventilated place.	Although PES-51 is expected not to pose any specific health hazard, the following precautions are recommended due to possible irritation from the biological by-products contained in the product: Avoid contact with skin, eyes, and clothing. Avoid prolonged or repeated contact with skin, avoid breathing mist, and do not take internally. Keep product away from heat, sparks, and flames, and store in a cool, dry, well-ventilated place, away from incompatible materials.

			Vent container in warm weather to relieve pressure. Do not cut, grind, weld, or drill on or near product containers. Handle empty containers just as would the full ones.
RAPIDGRAB 2000™	Non-flammable	Use with ventilation equal to unobstructed outdoors in a moderate breeze.	Avoid skin and clothing contact. If skin contact occurs, immediately wash with large amounts of water and soap (if possible). Remove any contaminated clothing or shoes. Launder before reusing. If irritation persists, seek medical assistance. For areas where contact is likely, wear long sleeve shirt, chemical resistant gloves, and chemical resistant goggles.
WASTE-SET #3200®	Product will not burn unless preheated. Avoid elevated	Avoid generating dust clouds.	May cause mild mechanical irritation to eyes, skin, and respiratory tract. Dust may cause coughing and watery eyes. Pre-existing eye, skin, and respiratory disorders

Table 4.4 (cont.) Product safe-handling chart.

Product	Flammability	Ventilation	Skin and Eye Contact, Protective Clothing, and Treatment in Case of Contact
	temperatures (>200°F) for extended periods (>5 days).		may be aggravated by this product. Wear NIOSH-approved respirator to prevent overexposure. Refer to transitional occupational exposure limits established by OSHA in 29CFR 1910.1000. Flush eyes with water, wash skin with soap and water, and if molten, treat as for burns. Remove person(s) to fresh air if excessive amounts of dust have been inhaled. Protective gloves and safety glasses should be worn. If material is released or spilled, sweep up or vacuum and place in an approved container.
ZYME-FLOW	Non-flammable	No mechanical ventilation is required.	Avoid eye contact due to possible mild eye irritation. In case of eye contact, flush eyes with water.

Table 4.5 Product hazardous ingredients chart.

Product	Concentrations Measured in Parts Per Million (PPM)										
	Arsenic	**Cadmium**	**Chromium**	**Copper**	**Lead**	**Mercury**	**Nickel**	**Silver**	**Zinc**	**Cyanide**	**Chlorinated Hydro-carbons**
AQUACLEAN	1.1	0.16	0.2	0.66	1.83	<0.01	0.8		0.29	0.06	4.2
BG-CLEAN™ 401	<0.118	<0.020	<0.125	0.049	0.166	<0.02	<0.125		0.28	0.002	ND
BIOSOLVE® HYDROCARBON MITIGATION™ AGENT	ND<0.2	ND<0.18	ND<0.5	ND<0.6	<0.2	ND<0.07	ND<0.6	ND<0.4	0.51	ND<0.01	ND<0.5
CN-110	<0.22	<0.088	0.109/0.131	0.238/ 0.234	<0.275	<0.0001	<0.11		<0.374/ <0.369	0.25	ND
COREXIT® EC7664A	0.2	<0.05	1	<0.2	0.1	0.003	0.4		0.15	<0.01	<0.01
CYTOSOL		ND	ND	ND	ND	ND	ND	ND	ND	ND	ND
DO-ALL #18	ND (<0.02)	ND (<0.02)	ND (<0.02)	0.33	ND (<0.02)	ND (<0.00)	ND (<0.10)		0.28	0.26	ND (<0.30)
ENVIROCLEAN (formerly ENVIRO CLEAN 165)	<0.002	<0.001	<0.005	<0.002	<0.005	<0.0002	<0.005		<0.002	<0.1	<0.05
E-SAFE©	<0.005	<0.0008	<0.025	0.061	0.082	<0.007	<0.003		0.214	0.2	ND
F-500	<0.03	<0.01	<0.01	<0.05	<0.025	<0.005	<0.02		<0.07	<0.02	<0.10
GOLD CREW SW	<1.0	<0.10	<1.0	<1.0	<0.5	<0.02	<1.0		0.44	ND	ND

Table 4.5 (cont.) Product hazardous ingredients chart.

Product	Concentrations Measured in Parts Per Million (PPM)										
	Arsenic	Cadmium	Chromium	Copper	Lead	Mercury	Nickel	Silver	Zinc	Cyanide	Chlorinated Hydro-carbons
NALE-IT	0.82	ND	ND	0.173	ND	ND	ND		0.18	ND	ND
NATURE'S WAY HS	<0.003	<0.005	<0.010	0.02	<0.002	<0.0002	<0.030		0.26	<0.020	ND
NOKOMIS 5-W	0.037	<0.050	<0.050	<0.050	<0.025	<0.0064	<0.050		1.7	0.042	<0.500
PETRO-CLEAN	<0.10	<0.10	<0.01	<0.01	<0.002	<0.01	<0.01		<0.01	<2.0	<1.0
PETRO-GREEN ADP-7	ND	0.12	ND	0.4	0.6	0.004	0.75		1.32	0.074	<1.0
PETROTECH 25	<0.10	<0.08	<0.10	<0.26	0.047	<0.0005	<0.40		0.256	0.7	ND
PREMIER 99	<0.01	<0.0005	<0.1	<0.1	<0.005	<0.001	<0.5		<0.1	<0.005	ND
PROCLEANS	0.69	ND	ND	ND	ND	ND	ND	ND	0.738	ND	ND
SC-1000™	1.33	ND	ND	0.1	ND	ND	ND		0.2	ND	ND
SHEEN-MAGIC©	<0.005	<0.0008	0.026	0.023	0.085	<0.007	<0.003		0.231	0.2	ND
SIMPLE GREEN®	<5.0000	<0.0233	0.115	<0.7500	0.0776	<0.0125	<2.3000		<4.4000	<1.0000	<1.0
SPILLCLEAN	0.02	<0.007	<0.01	<0.01	<0.02	<0.00003	<0.01		<0.01	0.137	<0.01
SPLIT DECISION SC (formerly SPLIT DECISION)	<0.05	<0.03	<0.03	0.2	<0.05	<0.03	<0.03		0.08	<0.1	<0.1
TOPSALL #30	0.05	<0.002	0.031	0.25	0.02	<0.01	<0.01		0.25	<0.1	<0.1

TXCHEM HE-1000™	ND	ND	0.0366	0.0402	0.0189	ND	0.0125		0.29	<0.05	ND
STEP ONE	<0.0200	<0.0030	0.0859	0.0177	0.009	0.0002	<0.0080		0.0602	0.283	<0.0050
ALSOCUP	0.0211	0.09	0.456	0.445	0.454	0.00492	0.474		0.433	ND	NA
AQUA N-CAP™ POLYMER	<1.0	<0.05	<0.325	17.25	<0.08	0.02	<0.225		<0.45	<0.1	ND
CIAGENT (formerly CI AGENT, CHEAP INSURANCE, & PETRO-CAPTURE)	<1.0	<2.5	<2.0	<2.5	<5.0	<0.050	<5.0		<2.5	<0.50	ND
ELASTOL	ND	ND	ND	ND	ND	ND	ND		1.4	ND	ND
PES-51	<0.005	<0.01	<0.05	<0.05	<0.05	<0.005	<0.01		<0.05	<1.00	<0.01
RAPIDGRAB 2000™	ND	ND	ND	ND	ND	<0.1	ND		ND	2.8	ND
WASTE-SET #3200®	ND	ND	0.44	0.98	0.08	ND	0.38		3.1	ND	ND
ZYME-FLOW	0.005	0.002	0.002	0.02	0.007	0.0004	0.07		0.04	0.3	<1.00

Table 4.6 Product marine toxicity chart – LC50 values in ppm (parts per million).

Product	Type Product	Species		
		Menidia beryllina at 96-hr	Mysidopsis bahia at 48-hr	Average Test Results
CN-110	Surface Washing Agents	52,233.00	12,262.00	32,247.50
BG-CLEAN™ 401	Surface Washing Agents	787.00	477.00	632.00
PETROTECH 25	Surface Washing Agents	601.00	350.00	475.50
CYTOSOL	Surface Washing Agents	738.00	124.00	431.00
COREXIT® EC7664A	Surface Washing Agents	87.06	583.91	335.49
PREMIER 99	Surface Washing Agents	565.70	94.70	330.20
E-SAFE©	Surface Washing Agents	329.00	257.00	293.00
DO-ALL #18	Surface Washing Agents	66.10	288.00	177.05
NATURE'S WAY HS	Surface Washing Agents	152.14	193.00	172.57
SHEEN-MAGIC©	Surface Washing Agents	183.00	161.00	172.00
NALE-IT	Surface Washing Agents	273.30	69.00	171.15
TOPSALL #30	Surface Washing Agents	156.60	115.80	136.20
PETRO-CLEAN	Surface Washing Agents	100.00	110.00	105.00
PROCLEANS	Surface Washing Agents	83.73	83.98	83.86
TXCHEM HE-1000™	Surface Washing Agents	91.33	65.52	78.43
SIMPLE GREEN®	Surface Washing Agents	27.90	77.60	52.75
AQUACLEAN	Surface Washing Agents	70.70	32.70	51.70

ENVIROCLEAN (formerly ENVIRO CLEAN 165)	Surface Washing Agents	27.80	22.60	25.20
F-500	Surface Washing Agents	10.00	32.00	21.00
SC-1000™	Surface Washing Agents	26.40	15.20	20.80
SPILLCLEAN	Surface Washing Agents	24.30	10.00	17.15
GOLD CREW SW	Surface Washing Agents	13.80	20.40	17.10
NOKOMIS 5-W	Surface Washing Agents	10.46	21.52	15.99
PETRO-GREEN ADP-7	Surface Washing Agents	11.61	10.56	11.09
SPLIT DECISION SC (formerly SPLIT DECISION)	Surface Washing Agents	8.27	8.20	8.24
BIOSOLVE®	Surface Washing Agents	6.40	3.60	5.00
SPILLREMED (MARINE)®	Bioremediation Agents	392.30	343.80	368.05
JE1058BS	Bioremediation Agents	91.70	72.09	81.90
S-200	Bioremediation Agents	39.69	21.33	30.51
AQUA N-CAP™ POLYMER	Miscellaneous Oil Spill Control Agent	484,000.00	104,000.00	294,000.00
WASTE-SET 3200®	Miscellaneous Oil Spill Control Agent – Oil Spill Solidifying Agent	10,000.00	5,431.00	7,715.50

Table 4.6 (cont.) Product marine toxicity chart – LC50 values in ppm (parts per million).

Product	Type Product	Species		
		Menidia beryllina at 96-hr	Mysidopsis bahia at 48-hr	Average Test Results
CIAGENT (formerly CI AGENT, CHEAP INSURANCE, & PETRO-CAPTURE)	Miscellaneous Oil Spill Control Agent	2,227.00	2,617.00	2,422.00
ELASTOL	Miscellaneous Oil Spill Control Agent	1,333.00	94.00	713.50
ALSOCUP	Misc. Oil Spill Control Agents	100.00	100.00	100.00
PES-51	Miscellaneous Oil Spill Control Agent	137	54.00	95.50
ZYME-FLOW	Miscellaneous Oil Spill Control Agent	35.00	26.00	30.50
RAPIDGRAB 2000™	Miscellaneous Oil Spill Control Agent	5.13	2.29	3.71

Table 4.7 Bioremediation agent effectiveness in percentage (%). (Effectiveness is defined as the % decrease in weight of oil on day 28 of test.)

Product	Control (%)	Nutrient (%)	Product (%)
OIL SPILL EATER II (OSE II)	16.50	52.00	85.40
SPILLREMED (MARINE)®	6.00	76.00	85.00
JE1058BS	5.90	60.00	81.80
STEP ONE	–	19.00	51.00
WMI-2000	–	25.00	44.00
S-200	(18.57)		27.82
LAND AND SEA RESTORATION PRODUCT 001	–		25.18
SYSTEM E.T. 20 (formerly MCW.B.20)	–	19.00	18.00
VB591™, VB997™, BINUTRIX® (formerly MYCOBAC TX-20)	<1	17.60	18.00
MICRO-BLAZE®	<1	17.60	12.00
OPPENHEIMER FORMULA	(6.40)	50.90	10.40

4.2 Booms and Barriers

Oil containment booms are temporary floating barriers composed of plastic, fabric, and other buoyant materials. They are meant to prevent oil spills from washing up on beaches and marshes and to concentrate the spill enough so that it can be skimmed or vacuumed from the surface by other equipment. Most booms have built-in stainless steel connectors that link booms together to cover large areas of water. Booms may be encased with a lint-free poly sock that prevents shedding. Their primary uses are for deflection, collection, and protection. When materials like naphtha or gasoline are spilled (i.e., hydrocarbons with low flash points), booms may be

Table 4.8 Product application chart.

Product	Application Method	Concentration/Application Rate
AQUACLEAN	AQUACLEAN may be introduced with a pressure spray to cover the affected area after the appropriate dilution or concentration is prepared. After application, agitate the spill area with water using a solid stream flow. For manual cleaning with AQUACLEAN, follow instructions in the product data sheet provided by the manufacturer.	For small spills, dilute 3 to 1 with water and apply as above. For large spills, prepare a 33% solution of AQUACLEAN and apply through a foam eductor at 6% setting with ~90 gallons per minute flow at the nozzle. For spills on shorelines and beaches, dilute product 50% with fresh water and apply using a pressure spray to cover the entire contaminated area. Then rinse with fresh water. Water temperature should be above 41°F.
BG-CLEAN™ 401	BG-CLEAN™ 401 is applied by one of two methods: 1) For situations where clean-up starts shortly after contamination and the oil has not yet weathered, use a high pressure spray. BGCLEAN™ 401 is proportioned into a stream of water and sprayed directly on the surface to be cleaned at a rate of 60–70 gpm; 2) For difficult cleaning situations such as weathered crude oil on a rocky beach, a three-step procedure should be used. First presoak the contaminated area by proportioning BG-CLEAN™ 401 into a low-pressure stream of water at a rate of about 20 gpm. Allow the spray to soak for 15–30 minutes. Then	Dilute BG-CLEAN™ 401 with water. The amount of spray required varies with the amount of oil. Concentrations ranging from 5:1 to 100:1 have been found useful, depending on hydrocarbon concentrations and viscosity. During application, adjust the concentration of BG-CLEAN™ 401 to suit the cleaning effect desired.

	proportion BG-CLEAN™ 401 into a high-pressure water stream and apply at a rate of 60–70 gpm. Selection of the best procedure depends on site conditions. BG-CLEAN™ 401 effectiveness is increased when applied with water heated to between 100°F and 180°F. As ambient temperature decreases, the benefits of applying heated spray increase. The efficiency of BG-CLEAN™ 401 depends directly on the mixing of BG-CLEAN™ 401 molecules with the hydrocarbon molecules. Given good mixing, BG-CLEAN™ 401 can be used on all hydrocarbons except asphalt.	
BIOSOLVE® HYDROCARBON MITIGATION™ AGENT	Dilute or use eductors to specified rate and apply through fire hose, power washers, steam powered units, or chemical boom sprayers with nozzles that produce a shearing action. Special nozzles to apply the solution as droplets are not necessary. For shoreline clean-up involving heavy or weathered crude, presoak to a 6% solution may be necessary. Rigs and Platforms: BIOSOLVE® is used to inert undersea pipelines before plugging and abandonment, de-gas tanks and platforms during workover operations, and wash drill cuttings to remove oils and prevent sheens on surface waters.	BIOSOLVE® is a highly concentrated product and must be diluted with water before use. Dilution ratios vary depending on site-specific conditions. Dilution ratio's normally run at 6%, 3%, or 1%. For heavy, mousse, or weathered oil, a 3% to 6% solution should be applied. For light or refined products, apply at 2% to 3%. For sheens, apply at .5 to 1%. Since testing shows that BIOSOLVE® quickly emulsifies weathered crude, it is not critical to apply immediately after a spill occurs; impact considerations can be fully considered prior to action taken.

Table 4.8 (cont.) Product application chart.

Product	Application Method	Concentration/Application Rate
CN-110	There are no application restrictions. The preferred application method is by spraying and/or applying the product over the pollutant (primarily oil of any viscosity) on the stained area. Immediately, the product begins to work on breaking the physical bond between the oil and the object desired to be cleaned. The full effect of CN-110 on hard, porous and most non-porous materials usually takes a minimum of 30 minutes with a maximum retention time of 60 minutes. Then, the freed oil is sprayed off with available water. The oil is then safely and effectively skimmed without dispersed oil remaining in the affected waters. If "herding" or "corralling" a slick is desired, CN-110 can also be applied without the increased toxicities by dispersing minute particles of oil into the affected water stream.	Application concentration for cleaning shorelines is approximately 1 gallon per 75–100 square feet of affected area. Depending on depths of shorelines, this amount will give a residual of much less than 1 ppm of CN-110 in the affected waters. This residual should be lower than the toxicity of the spilled oil. If "herding" is desired, the application concentration should be in its original state while either spraying CN-110 around the slick or dropping small amounts from a plane. The resulting residual concentration in this case should be the same as shoreline cleaning or less. If it is desired that CN-110 be added to a high-pressure cleaning system or steam, the dilution rate should be no lower than 10%. This will lower the CN-110 residual in the affected waters to the parts per billion range.
COREXIT® EC7664A	Spray from boats, aircraft, fire education systems on boats, helicopter buckets, handheld or backpack sprayers, or from hoses attached to small pumps. COREXIT® EC7664A should be applied in the form of small droplets and never as a fog or mist.	COREXIT® EC7664A diluted 1–3% in water sprayed on the shoreline before the oil comes ashore will protect the beaches, marshes, rocks, etc., from oil contamination. Full strength COREXIT® EC7664A can be sprayed onto the surf, which washes the shoreline. The dilute

		solution is sprayed on the shoreline area. COREXIT® EC7664A is also useful in washing oil from beaches, sea walls, docks, boats, and spill clean-up equipment. Use a 1–3% solution of COREXIT® EC7664A in water and apply from portable sprayers. If oil deposits are heavy or weathered, a short presoak with a hydrocarbon-based shoreline cleaner like COREXIT® EC9580A is recommended.
CYTOSOL	CYTOSOL is applied to oiled shorelines to extract and recover weathered petroleum by flotation with passive water deluges from header pipes or manual spraying. Remaining residual hydrocarbons are biodegraded, either passively by intrinsic bioremediation or aggressively by enhancing the process with controlled amounts of nutrients and/or acclimated cultures of bacteria cultured from the site when approved by local, state, and federal agencies. The CYTOSOL process is most suitable for the treatment of heavily oiled shorelines that do not respond well to conventional treatments or that are considered too sensitive for mechanical/pressure wash strategies. Prior to the application of CYTOSOL, collection booms, oil skimmers,	CYTOSOL may be applied with a variety of spraying or washing equipment, depending on the scale and type of shoreline to be cleaned. The product is to be used only neat and undiluted for direct application to spilled oil. For large beach areas, CYTOSOL can be sprayed from water trucks or workboats equipped with pumps, hoses, and nozzles to deliver the product as an aerial spray. In smaller applications, CYTOSOL may be applied with hand sprayers or portable pumps to spray the product directly onto oiled surfaces. Dose rates will vary with the type and amount of petroleum spilled, the extent of weathering, and other site-specific conditions, including temperature, porosity of shoreline, and residence time available to let the product contact the oil. In general, the ratio of

Table 4.8 (cont.) Product application chart.

Product	Application Method	Concentration/Application Rate
	sorbent pads, or other appropriate containment and collection mechanisms must be deployed and operational.	applied CYTOSOL to crude oil is between 0.5:1 and 1:1. The quantity of CYTOSOL applied should be approximately equivalent to the quantity of petroleum accumulated on the shoreline or as required to dissolve and remove weathered oil. After application, the product should be allowed to penetrate and dissolve the weathered petroleum for at least one hour, preferably longer. Cold weather applications will require more contact time before initiating recovery. In tidal areas, it is advisable to apply the CYTOSOL as the tide is ebbing (receding) to maximize contact time. Trapped oil may continue to be released for several days, requiring that containment devices be left in place.
DO-ALL #18	May be diluted with fresh or salt water. For general use, apply by spray, mop, etc. Agitate severe spots, and rinse thoroughly for residue-free surface.	To clean oil from beaches, rocks, and piers as a shoreline cleaner, first use pre-treatment or soaking, using 1 part DO-ALL #18 diluted with 3 parts of water. Allow 30 to 45 minutes to soften viscous oil deposits (soak time may vary with temperature, oil density, and degree of weathering). After the soaking period, dilute 1 part of DO-ALL #18 with 15 parts of water and

		apply with a high-pressure water hose (50–100 PSI). Oil released can then be removed by vacuuming, skimmers, or absorbents. For normal cleaning of shop machinery, bilges, decks, waterline scum, rigs, and platforms, dilute 1 part DO-ALL #18 with up to 20 parts of water. For heavy cleaning and degreasing of tanks, barges, engine rooms and soot, oil-stained concrete, and petroleum-based drilling muds, dilute 1 part DO-ALL #18 with up to 15 parts of water. For severe cleaning of holding tanks, grease traps, black magic, crude oil, and Bunker C, dilute 1 part DO-ALL #18 with up to 5 parts of water. For steam cleaning, dilute 1 part of DO-ALL #18 with up to 50 parts of water. For pressure washing, dilute 1 part DO-ALL #18 with up to 30 parts of water. Do not use undiluted on composition floors, water-based painted surfaces, or aluminum.
ENVIROCLEAN (formerly ENVIRO CLEAN 165)	Dilute product with water prior to use. Apply product via manual pressure applicator (i.e., hand pump) or other delivery equipment to impacted surfaces. Allow product to soak prior to rinse/recovery if heavily saturated with oil. On hard surfaces, product may be applied and agitated with a brush or other	ENVIROCLEAN is a concentrated product and must be diluted prior to use. Concentrations vary based upon type of oil and degree of surface saturation. For light ends (i.e., gas condensate/gasoline), a 1% – 2% solution may be used to flush area to containment. (A higher concentration solution of 6% should be used if

Table 4.8 (cont.) Product application chart.

Product	Application Method	Concentration/Application Rate
	physical means prior to recovery of effluent. Product may also be pre-mixed and used in a batch tank or through an eduction/metering system for flushing and recovery operations.	flammable vapor presents an explosion hazard.) For lighter oils (i.e., No. 2 fuel oil), a 3% solution may be used to flush impacted areas to containment. For heavy ends (i.e., crude oil/No. 6 fuel oil), a 6% solution of ENVIROCLEAN should be applied and allowed to soak into containment/substrate for up to an hour to allow for the release of oil. The surface should then be flushed to containment with a 1% – 3% solution of ENVIROCLEAN. Product may be used as noted above through any pressure or steam equipment. The use of a "hot" unit is recommended when oils are at low temperature or have been weathered.
E-SAFE©	E-SAFE© may be introduced with a pressure spray to cover the affected area. The contaminated area should be thoroughly moistened. Following the spray application, the treated area should be soaked with water to facilitate penetration. Heavy soil such as clay will be cleaned by ESAFE©, but tilling or aerating the soil will rapidly shorten penetration time. Sand or loam may require heavy dosage with E-SAFE© because hydrocarbon migration	E-SAFE© should be applied full strength. Beginning treatment for hydrocarbon contaminated surface is one gallon of E-SAFE© per 100 square foot of surface area. This dosage is recommended when the ambient temperature is 72°F and humidity is moderate. Higher temperature or lower humidity will increase the need for repeated applications or a higher volume of E-SAFE© per application.

	is so rapid in these soil types. Treatment of fouled beach areas should have E-SAFE© sprayed on all contaminated surfaces. Any remaining hydrocarbons may then be effectively vacuumed or wiped away from the treated surfaces. Incidental wave action or rainfall will enhance coverage and penetration by E-SAFE©.	
F-500	Use standard fire apparatus spray nozzle with agitation onto and into the spill.	1 part F-500:8 parts hydrocarbon:32 parts water (fresh or salt). Water temperature: 33°F – 211°F. Effective on both polar and non-polar hydrocarbons.
GOLD CREW SW	Apply through hand pump sprayer and allow to soak.	As a presoak, dilute 20 parts water to 1 part SW. For crude oil, allow about 1 hour. For medium distillates, allow 30 minutes. For light distillates, allow 15 minutes. Time may vary depending on weather conditions. After allowing the solution to presoak, wash the area in the following manner: apply through a power washer or through a steam powered unit at 1 percent, 3 percent, or 5 percent, depending on oil viscosity and temperature.
NALE-IT	NALE-IT may be applied using a pressure sprayer.	Use for pit closures, surface hydrocarbon spills, compressor stations, pipeline and flow line leaks, well head and tank farm leaks, and

Table 4.8 (cont.) Product application chart.

Product	Application Method	Concentration/Application Rate
		highway spills. Mix 1 part NALE-IT with 20 parts water (petroleum products).
NATURE'S WAY HS	NATURE'S WAY HS may be applied by spraying, pouring, dispensing equipment, or manual or automatic scrubber machines. Do not mix the product with any other cleaner, and do not allow any other cleaners to contact surfaces where HS is being used. Hot water should never be used.	Apply NATURE'S WAY HS full strength (undiluted) to surface area with sprayer, concentrating on areas with heaviest contamination first. To speed the clean-up of shorelines and beaches, pressure spray into cracks and crevices prior to scrubbing. Agitate and scrub well using power brushes, hand brushes, brooms, or mops. If possible, allow the scrubbed solution to remain on the surface for at least 15 minutes. If allowed to soak overnight, the next morning the treated surface must be re-wet with water, re-agitated, and rinsed. Additional product will not be necessary at that time. For heavily contaminated surfaces, NATURE'S WAY HS should always be used full strength. For moderate accumulations, HS may be diluted to as little as 2 oz. per gallon of cool water for cleaning light contamination. For average contamination, a dilution of 12 oz. per gallon is recommended (1 part NATURE'S WAY HS to 10 parts water). For light use, 4–6 oz. of product per gallon of water will be sufficient.

NOKOMIS 5-W	NOKOMIS 5-W is a concentrated liquid, non-hazardous cleaner. Can be used full strength or diluted with fresh or salt water. NOKOMIS 5-W is marketed in plastic containers.	Dilution ratios are determined by the individual cleaning application. For washing equipment contaminated with soluble crude and/or bunker fuel oil, use a 5:1 cleaning solution. Spray or brush onto surface, agitate with a stiff bristle brush, and rinse with fresh or salt water. For pressure washing, pre-spray surface with concentrate and allow to dwell on surface one to ten minutes (depending on age of oil). For shoreline areas and beaches with rocky breakwaters, contain the area of beach sand and gravel by creating a sand bather around a small, contaminated area. Spray contaminated area with a 10:1 solution. Add salt or fresh water, and agitate surface to help release oil from sand. Allow time for oil to surface and recover oil. To clean larger tidal rock surfaces, apply 5:1 or 20 percent mixture of NOKOMIS 5-W. Spray solution onto rocks with a garden type pump sprayer. Agitate with a stiff bristle brush. Allow to dwell on surface until surface looks soluble. Contain with a boom or use oil absorbent pads to recover oil.
PETRO-CLEAN	PETRO-CLEAN is applied through power washers or even garden type sprayers in light dilution and is very effective in removing petrochemical hydrocarbons from rocks,	Dilute or use eductors to specified rate and apply through fire hose, power washers, or sprayers to contaminated area. PETRO-CLEAN is a highly concentrated product and must be

Table 4.8 (cont.) Product application chart.

Product	Application Method	Concentration/Application Rate
	shorelines, sea walls, bridges, and highways. In wetland applications, PETRO-CLEAN prevents hydrocarbons from attaching to grasses, trees, rocks, etc.	diluted before use. Dilution ratios vary depending on the specific conditions of the contaminated site. Normal recommended dilutions are from 0.5% to 6%. On heavy or weathered crude, presoaking with 6% may be necessary. For light or refined products, apply as 3% to 6% solution. For sheens on water, apply a 0.5% to 1.0% solution.
PETRO-GREEN ADP-7	For spills on soils, sandy, or rocky areas, ADP-7 can be pre-mixed into water or educted into a water flow to be sprayed directly onto the oil spill.	For spills on land, typical dilution is one gallon of ADP-7 diluted in one barrel (42 gallons) of water (about 2.4%), which is enough to wash one barrel of oil on the surface. After loose oil is vacuumed off soil surfaces, the application rate is 100 barrels of solution per acre. Supplier notes that the conditions for each spill dictate proper choice of surface washing agent. PETRO-GREEN ADP-7 has been tested in both tropical and arctic environments as well as with fresh water and a variety of produced water salinities. It has been found to be effective over these ranges. PETRO-GREEN ADP-7 has also been tested on a wide variety of fresh crude oil and product spills. There are no specific limitations to use, however, and an aged spill is likely

		to be nonresponsive as it would. be to any environmentally sound surface-washing agent. For aged or asphaltic spills, the solution may be heated to 150°F in a hot water washing rig or a common oil field "hot oiler" truck to increase effectiveness
PETROTECH 25	It is recommended that the surface washing agent be applied to contaminated surfaces in one of three ways: (1) a mixture of ten parts PETROTECH 25 to ninety parts water, either fresh or salt water, should be discharged over the surface by a commercial pressure washer, hot or cold; (2) the same mixture is discharged on the surface using standard fire fighting apparatus preferably at a nozzle pressure of 100 PSI plus; or (3) PETROTECH 25 in concentrated form can be applied to the surface and manually or mechanically brushed and then water applied.	Generally, PETROTECH 25 is applied neat in its factory supplied concentrate form. However, for lighter oils where educting or proportioning equipment is available, an aqueous solution of PETROTECH 25 between 3 and 10 percent may be used. Broadly speaking, PETROTECH 25 has no solvent action. For any given oil spill of a specific nature, the application rate is unaffected by neat or diluted application as long as the concentrate to oil ratio remains the same. Applications of diluted PETROTECH 25 in the 6 – 10 percent range via fire fighting equipment can be used on automobile and aviation gasoline's together with the lighter oils such as diesel and all jet fuels. The dosage of PETROTECH 25 to be used depends upon two factors: (1) the nature of the oil, and (2) the energy used in its application to the contaminated substrate.

Table 4.8 (cont.) Product application chart.

Product	Application Method	Concentration/Application Rate
PREMIER 99	Apply by spray, mop, or pressure washer, etc. Agitate severe spots. Rinse thoroughly for residue free surface.	Normal cleaning – 1 part PREMIER 99 to up to 20 parts water Heavy cleaning – 1 part PREMIER 99 to up to 15 parts water Severe cleaning – 1part PREMIER 99 to up to 5 parts water Steam cleaning – 1 part PREMIER 99 to up to 50 parts water Pressure wash – 1 part PREMIER 99 to up to 30 parts water
PROCLEANS	Apply specified dilute solution using fire hoses or heated pressure washers onto the contaminated solid surface of the spill.	Use ten parts water to one part product. Dilution rates may be adjusted to suit different job conditions. Apply 10 to 15 gallons of diluted PROCLEANS to one cubic yard of contamination. May be used with fresh or salt water. Warmer temperatures may improve results. It is most effective if used on solid surfaces such as shoreline beaches and rocks contaminated with light and medium weight crude oils and refined petroleum products.
SC-1000™	SC-1000™ may be used on any surface that is compatible with water. The product may tarnish some soft aluminum surfaces if not adequately diluted and rinsed with water.	For beach sands or light rocky soils, burning or creating a temporary shallow wash tank is most successful. Fill one-half of the tank with contaminated beach sand, then add a solution

	SC-1000™ is a highly concentrated cleaning compound capable of a multitude of cleaning applications with dilutions ranging from full strength to far in excess of 1:350. Depending on the specific factors in the cleaning environment, the desired cleaning speed with the least amount of SC-1000™ can be determined. When diluting, it is recommended that SC-1000™ be added to pre-measured water to minimize foaming in the solution.	of SC-1000™ and clean water (ocean or fresh) to cover the sand. Gentle agitation will release the oil to the surface of the tank where it can be collected. For rocky surfaces, spray with a 20 percent solution of SC-1000™ using a horizontal eductor sprayer, spraying side-to-side, allowing the soil to dwell for several minutes before spraying top-to-bottom with clean ocean or fresh water to rinse oil into perimeter oil booms, blankets, or impermeable sheeting. For washing marine vegetation, use a non-pressure/impact spraying equipment to dispense a 0.01 percent SC-1000™ solution and allow to stand for 5–10 minutes before a final rinsing with fresh or ocean water. For washing marine equipment, use SC-1000™ at 20 percent solution for wet oils and bunker crude (preferably warmer than 80°F) and spray or wipe. Apply directly to equipment and allow to dwell for 1-2 minutes. Then, spray rinse with fresh or ocean water. For hardened oils, fuels, and viscous lubricants, apply SC-1000™ blended with SC-Supersolve™ (a non-toxic, low aromatic, water miscible solvent) at the ratio of 80/20 then dilute with water to a 50 percent solution, spray or wipe onto surface, and let stand for 1–2 minutes before rinsing. Heavy Cleaning

Table 4.8 (cont.) Product application chart.

Product	Application Method	Concentration/Application Rate
		Examples – Dilution full strength 1:5; diesel engines, auto parts, baked on oil or lube grease, dried oil/enamel, latex paints, thick food syrups, insect smears, dried animal or vegetable fats, hard resins, thick dust-laden oily dirt, asphalt and grass or plant stains. Average Cleaning Examples – Dilution 1:5 up to 1:20; automotive work counters and tools, food and beverage processing equipment, oily or food-laden floors, manufacturing work areas, vehicle maintenance, shipping containers, utility equipment, and parts washers. General Maintenance Examples – 1:20 up to 1:100; vehicle washing, general janitorial for offices/schools/hospitals/recreation and related equipment, pressure sprayers, food preparation and storage, painted/plastic laminate surfaces, sports equipment, general cleaning, immersion tanks, and ultrasonics.
SHEEN-MAGIC©	SHEEN-MAGIC© recommended treatment is light misting of contaminated surface. Heavier hydrocarbons such as crude oil should be boomed or vacuumed away as much as possible before application of the	One ounce per 50 square yards of surface is the normal treatment for diesel spills. Heavier hydrocarbon sheens may require additional treatment to remove the contaminant. SHEEN-MAGIC© works in all weather conditions that

	SHEENMAGIC©. Treatment of fouled beach areas should have SHEENMAGIC© sprayed on all contaminated surfaces. Incidental wave action or rainfall will enhance coverage and penetration by SHEEN-MAGIC©. Any remaining hydrocarbons may then be effectively vacuumed or wiped away from the treated surfaces. Aerial application should be delivered at height and air speed that will create air shear necessary to deliver droplets as per nozzle manufacturer's recommendations. Mist spray from surface vehicles may be delivered from power fogger units. Hand sprayers are generally used for small treatment areas.	allow hydrocarbon penetration. SHEEN-MAGIC© should be used as soon as the temperature and weather conditions allow delivery to the spill area. Heavier concentrations may be required when the temperature/humidity will cause rapid evaporation of the spray.
SIMPLE GREEN®	Spray on oily surface. Product is equally effective in fresh water, estuarine, and marine environments at all temperatures. SIMPLE GREEN® contains no known EPA Priority Pollutants.	For open water, spray concentrated product directly on the surface of oil at a ratio of 4 parts of oil to 1 part of SIMPLE GREEN®. Site conditions may warrant alternative procedures to maintain effectiveness.
SPILLCLEAN	SPILLCLEAN comes in two forms: "Concentrate" and "Original Formula." SPILLCLEAN "Concentrate" is 30% active ingredients and 70% water, while SPILLCLEAN "Original Formula" is 10% active ingredients and 90% water.	For surface and land application only: SPILLCLEAN "Concentrate," 30% active ingredients and 70% water – For spills less than 2 gallons, squeeze a one-half inch ribbon of straight product around the perimeter of the oil/gasoline spill, followed by an "X" pattern

Table 4.8 (cont.) Product application chart.

Product	Application Method	Concentration/Application Rate
	SPILLCLEAN "Concentrate" and SPILLCLEAN "Original Formula" are for on-land and surface use only. They should not be used around or near bodies of water when there is a reasonable chance of mixing in with water. Once the contaminant/SPILLCLEAN "Concentrate" solution mixes with the body of water, "control" of the contaminant will be lost since SPILLCLEAN is water-soluble. SPILLCLEAN is a formula designed to solubilize and emulsify oil and oil-type products such as motor oil, gasoline, diesel fuel, and transmission fluid. When such products are accidentally spilled and need to be cleaned up from hard surfaces such as streets and driveways, SPILLCLEAN is an effective material to use. Being a surfactant-type material, SPILLCLEAN allows the spilled contaminant to be emulsified and then lifted and separated from the hard surface. The resulting spill/contaminant can then be picked up, vacuumed, or flushed with water, all in accordance with federal, state, and local regulations. For SPILLCLEAN "Concentrate," squeeze	over the body of the spill. For spills greater than 2 gallons, squeeze a one-half inch ribbon of straight product around the perimeter of the oil/gasoline spill. The general purpose of SPILLCLEAN "Concentrate" is to "contain" the contaminant from migrating, and therefore should not be diluted with water during application. SPILLCLEAN "Original Formula," 10% active ingredients and 90% water – Concentration rate is based on equipment used and disposal method. Typical usage will be 0 to 100 parts water plus 1 part SPILLCLEAN "Original Formula" all applied to 1 part contaminant (oil/gasoline spill). The resulting mixture of SPILLCLEAN and oil must be disposed of in accordance with EPA, federal, state, and local agencies that regulate disposal of contaminant. The resulting SPILLCLEAN contaminant mixture should be wet-vacuumed up, placed in a proper container, and disposed.

	product directly around outside perimeter of oil/gasoline/antifreeze spill using squeeze bottle container. This process is designed to "contain" the contaminant and keep if from migrating down a hard surface like a city street, parking lot, or garage floor. SPILLCLEAN "Concentrate" is more viscous than oil/gasoline and will "hold" the contaminant in place. Once the contaminant is contained, the user can apply SPILLCLEAN "Original Formula" over the "body" of the contaminant if the spill is large (over 2 gallons). If the spill is small (less than 2 gallons), the user can apply either SPILLCLEAN "Concentrate" or SPILLCLEAN "Original Formula" to the "body" of the contaminant spill. Using a broom or broom-like applicator, sweep SPILLCLEAN in with the contaminant until thoroughly blended in place. The original bright red color of SPILLCLEAN "Concentrate" and SPILLCLEAN "Original Formula" will dissipate when it is thoroughly mixed with the contaminant. For SPILLCLEAN "Original Formula," apply over the "body" of the contaminant if the spill is large (over 2 gallons). If the spill is small (less than 2 gallons), the user can apply	

Table 4.8 (cont.) Product application chart.

Product	Application Method	Concentration/Application Rate
	SPILLCLEAN "Concentrate" or SPILLCLEAN "Original Formula" to the "body" of the contaminant spill. SPILLCLEAN "Original Formula" is designed to "attach" to a contaminant and remove it from a hard surface like a city street, parking lot, or garage floor. It is recommended to be used on any size contaminant spill, large or small. This product is designed to penetrate into the surface pores and molecularly "attach" with the contaminant, thus making the contaminant "water soluble." The contaminant can be removed from the pores (of the asphalt, concrete, epoxy surface, etc.) with a 60 psi water supply. The user can then dispose of the resulting mixture according to regulations. This product can be applied using a vast number of different applicators, especially with regard to fire-fighting equipment. Using a broom or broom-like applicator, sweep SPILLCLEAN in with the contaminant until thoroughly blended in place. When SPILLCLEAN'S bright red color has dissipated (upon mixing) with the	

	contaminant, it can be wiped clean with a rag or wet-vacuumed, collected, and disposed of according to EPA, federal, state, and local regulations. Additionally, a dry sorbent may be applied to the resulting SPILLCLEAN/contaminant/water mixture, shoveled up, and disposed of according to applicable regulations.	
SPLIT DECISION SC (formerly SPLIT DECISION)	Spray on oily surface. SPLIT DECISION SC forms a loose emulsion with oil that separates within seconds. Oil displaced from surface can be skimmed from the rinse water or absorbed with an oil absorbent.	For heavy oil, apply at a concentration of 1 part SPLIT DECISION SC to 3 parts water. For lighter oil on non-porous substrates, use up to 1 part SPLIT DECISION SC to 30 parts water. For application by hot water pressure washers or steam cleaner, SPLIT DECISION SC may be diluted with as much as 50 parts water.
TOPSALL #30	For general use, apply by spray, mop, etc. Agitate severe spots and rinse thoroughly for a residue-free surface. Do not use undiluted on composition floors, water-based painted surfaces, or aluminum.	For normal cleaning (rigs and platforms, bilges, decks, waterline scum, shop machinery and equipment) use 1 part TOPSALL #30 (SUPERALL #38) to up to 20 parts water. For heavy cleaning (degassing tanks and barges, engine rooms and soot, oil stained concrete, petroleum based drilling muds) use 1 part TOPSALL #30 (SUPERALL #38) to up to 15 parts water. For severe cleaning (Black magic, Bunker C, crude oil, holding tanks, grease traps) use 1 part TOPSALL #30 (SUPERALL

Table 4.8 (cont.) Product application chart.

Product	Application Method	Concentration/Application Rate
		#38) to up to 5 parts water. For steam cleaning use 1 part TOPSALL #30 (SUPERALL #38) to up to 50 parts of water. For a pressure wash use 1 part TOPSALL #30 (SUPERALL #38) to up to 30 parts water.
TXCHEM HE-1000™	TXCHEM HE-1000™ is applied by pressure sprayer, heated pressure washer, fire hoses, or mixing on or into hydrocarbon contaminated media.	TXCHEM HE-1000™ should be diluted to a one (1) part concentrated chemical to ten (10) parts water solution. Apply ten to fifteen gallons of diluted solution to one cubic yard of contaminated media. Agitate to slurry. The dilution rate may change due to level of hydrocarbon contamination. Please see manufacturer's instructions for additional application dilutions. Excess washing solution should be disposed of according to local, state, and federal regulations.
JE1058BS	JE1058BS contains biosurfactant and has an ability to stimulate the biodedgradation of oil by indigenous microorganisms. JE1058BS, which is a powder, can be applied using conventional spraying equipment, e.g., a powder mist duster attached with boom type multi-hole head.	JE1058BS is applied in a product/oil ratio of 1:10. No pre-mixing or dilution by water is required.

LAND AND SEA RESTORATION PRODUCT 001	On water, spread over contaminated area at 1 to 3 ratios. On soil, blend to depth equivalent to contamination level. On hard surfaces, spread over contaminated area. Make sure the spill is completely absorbed. For best results agitate with broom. (It will not damage cured asphalt if used as an absorbent for spills.) Remove and dispose of in accordance with all state and federal laws after absorption is complete.	Use 1 part 001 to 3 parts hydrocarbon.
MICRO-BLAZE®	Use normal spray equipment, fire or response equipment, eductor setups, water trucks, etc., as methods of application. Mix MICRO-BLAZE®/water mixture with contaminated soils and liquids thoroughly for maximum contact. Pick up treated contamination after volatile hazard has been negated per local regulatory parameters. For general bioremediation: For in-situ soils, MICRO-BLAZE®/water mixture can be tilled into the contaminated area. For shallower contamination, areas can be over-sprayed with normal spray equipment, eductor setups, water trucks, etc. For deeper contamination, application can be applied through underground setups using perforated piping per regulatory	For more viscous or less hazardous contamination, apply MICRO-BLAZE® at a 3% solution mixed with water (3 parts MICRO-BLAZE®, 97 parts water). When bioremediating soils, generally, per every 10 cubic yards of contaminated soils, use one gallon of concentrate, diluted with water according to contamination type. Water Salinity: Can be mixed with any fresh, brackish, or brine. However, brine reduces the effectiveness by 10%. Water Temperature: 35°F – 180°F; pH: 4 to 11.5. Temperature: 32°F – 120°F. Nutrient Requirements: Nutrients for microbes are included in product. However, for longer-term bioremediation projects, additional applications for Bio-Catalyst may be added to boost microbial activity. Type and Ages of Pollutants:

Table 4.8 (cont.) Product application chart.

Product	Application Method	Concentration/Application Rate
	recommendations. For wastewater and other operational by-product sludges and soils extracted from contaminated areas, a bioslurry or bioreactor can be setup using MICRO-BLAZE® and water in the process.	For use on organics and hydrocarbon-based materials. These strains of bacteria provide the capability of biodegrading various straight chained, branched chained, aromatic, and poly-nuclear aromatic hydrocarbons found in diesel and other fuels. Age of contamination is not a factor as much as its density. Tar-like substances may need to be cut for timely remediation.
OIL SPILL EATER II (OSE II)	A. Use surface spray apparatus, such as small hand held tanks, back pack, large mixing tanks with mechanical pumping devices, vessels with booms for spraying wide paths, or spray devices on airplanes or helicopters. B. OSE II can be applied by eductor systems from vessels, fire trucks, etc. Set the eductor system to 2% and apply 1 gallon of mixed OSE II to each spilled gallon of hydrocarbon. OSE II can remediate hydrocarbon-based material including chlorinated hydrocarbons, PCB's, dioxins, and some pesticides. • As the age of spilled hydrocarbons increases, the time necessary for bioremediation increases. In general, fresh crude gasoline	General – OSE II generally takes 3 to 30 minutes to penetrate the molecular walls of hydrocarbons. However, once you spray OSE II on the hydrocarbons, OSE II attaches itself and will eventually engulf the hydrocarbons regardless of where the hydrocarbons may spread on the surface of salt or fresh water. Additionally, once you spray OSE II, the hydrocarbons cannot attach itself to the shoreline, rocks, or any equipment in its path. OSE II breaks down the adhesion properties of hydrocarbons and causes hydrocarbons to float, thereby eliminating secondary contamination of the water column or any other areas and holding the contaminated area to the waters surface,

	of BTEX takes from 72 hours to 30 days to completely bioremediate. • Variations of seawater salinity should have no effect, but as long as microbial life can exist, OSE II will be effective. • OSE II bioremediation slows somewhat at temperatures below 40°F. OSE II, however, will continue to work at any liquid water temperature that will sustain microbial life. Optimum Conditions: a. pH: 7.0 b. Temperature: 72°F c. Salinity Ranges: Fresh water to salt water d. Maximum and Minimum pH: 3.5 – 8.0 e. Maximum and Minimum Temperature: 28°F – 128°F f. Maximum and Minimum Salinity Levels – Salinity level above that will support microbial activity and will adversely affect OSE II's performance g. Enzyme Shelf Life: Up to 5 years when properly stored h. Enzyme Optimal Storage Conditions: 72°F is optimal. Enzyme range is freezing to 120°F. Never leave OSE II in direct sunlight for more than a couple of hours	the original contaminated area. • If OSE II is to be used on ocean spills or on intertidal zones, OSE II should be mixed with ocean water. • If OSE II is to be used on lakes, rivers, streams, ponds, or on land, mix the product with water from a lake, stream, or pond. • If you are performing a clean-up, make sure the water used to mix with OSE II and the water used to keep the area saturated is the type of water normally associated with that area. • If you use fresh water in an area normally contacted with salt water or vice versa, the different types of bacteria and competition could occur, not to mention the problems with salinity for fresh water organisms. (Note: Do not mix tap water with OSE II if possible; chlorine in tap water slows bacterial enhancement.) **Spills on Water**: Dilute each gallon of OSE II with 50 gallons of fresh, brackish, or salt water, depending on the water associated with the area that has been impacted by the spill. Apply OSE II at a ratio of 1 gallon mixed OSE II to each gallon of hydrocarbon spilled. Apply using hand-held sprayers, tank sprayers, and

Table 4.8 (cont.) Product application chart.

Product	Application Method	Concentration/Application Rate
		booms from vessels, helicopters, or airplanes by spraying the perimeter first then working toward the middle of the spilled area. Next, spray the entire surface of the spill. If the spill is very heavy (more than 2 inches thick), it is recommended that OSE II be applied every day until you have met a 1:1 ratio of OSE II and water mixture to spilled oil/hydrocarbons. • Use 1 gallon OSE II for every 50 gallons of hydrocarbons. • Use 1 drum of OSE II for every 2,750 gallons of hydrocarbons. • If you know gallons of hydrocarbons spilled, multiply gallons of hydrocarbons by 0.02 to get amount of OSE II needed (gallons of hydrocarbons x 0.02 = gallons of OSE II). • If you know barrels of crude oil spilled, multiply barrels of crude oil by 0.015 to get drums of OSE II needed (barrels of crude oil x 0.015 = drums of OSE II). • If you do not know gallons of hydrocarbons or barrels of crude oil, multiply size of spill by

		0.0023 to get drums of OSE II needed or by 0.12 to get gallons of OSE II needed [(yards long x yards wide x inches thick) x 0.0.0023 = drums of OSE II or (yards long x yards wide x inches thick) x 0.015 = gallons of OSE II]. Intertidal Zone: Mix each 55 gallon drum of OSE II with 2,750 gallons of fresh, brackish, or salt water. The water used is determined by the type of water associated with the site. OSE II should be applied as the tide recedes (if there is a tide), and once the tide comes in the application should cease until the tide recedes again. Additional applications should only be warranted if spill has been allowed time to percolate into the depths of the soil. If there is no tide, but waves have pushed the spill into the intertidal zone, then there will be direct access to the spill at all times. If possible, use string or stakes to grid off the beach or intertidal zone area, and then you can calculate how much premixed OSE II to apply to a given area. If unable to grid off an area, then calculate how much OSE II to apply, and then determine how much premixed OSE II will flow through a nozzle (gallons per minute). Then, let an application technician know how many gallons to apply in a given area, and this can be determined by

Table 4.8 (cont.) Product application chart.

Product	Application Method	Concentration/Application Rate
		applying product for a certain time period to get the correct amount of OSE II applied to gain the 1:1 ratio. Note: If the intertidal zone is associated with the sea, then mix OSE II with salt water. If the spill area is in an area of brackish water, then mix OSE II with brackish water. If the intertidal zone is associated with fresh water such as lakes, rivers, streams, ponds, creeks, aquifers, or drinking water wells, then use fresh water to mix OSE II.
OPPENHEIMER FORMULA (a.k.a, THE OPPENHEIMER FORMULA I, GENISIS WE-F, MIGHTY MIKE BPT, NATURAL ENVIRO 8000 BIOREMEDIATION PETRO-TREAT)	Optimal application is by powder seeding directly at the source of oil contamination. Application over larger areas of water surface can be accomplished by aerial powder dusting, spraying a water mixture of nutrients/formula over the oil area, or dusting with mechanical powder pumps. It may be used in a wide range of environmental conditions of temperature and salinities, ranging from fresh to 100 ppt salts. The formula is especially valuable in grass flats, marshes, and the open sea. It has also proved effective in soil and fresh water oil clean-up and municipal waste treatment, septic tanks, and grease traps.	The application rate at the source oil slick is 1 pound per cubic yard of oil. In the open sea slick, the basic application rate is 10 pounds per acre, but this may vary in different situations related to speed of clean-up. In estuaries, a larger biomass is recommended and may be at a rate of up to 100 pounds per acre, depending on the oil type and concentration. Heavier oils may require a higher biomass than light oils.

S-200 a.k.a. S-200C, SHEENCLEAN, BILGE CLEAR	S-200 is a light amber liquid, and as such it can be sprayed with traditional liquid spraying equipment, such as pressurized sprayers or backpack sprayers. S-200 is a bioremediation accelerator used for the remediation of hydrocarbon spills or leaks. It is a non-intrusive, cost effective remedy for the clean-up of these hydrocarbons. These hydrocarbons include but are not limited to gasoline, No. 2 up to No. 6 diesel fuel, jet fuels, kerosene, lubricating oils, hydraulic oils, and crude oils.	The recommended application rate is approximately 10% of the S-200 as compared to the hydrocarbon being remediated. In other words, use 1 pound of S-200 for every 10 pounds of hydrocarbon you are remediating. For applications on soil contaminated below 1 foot, the soil is turned and the material is applied. This allows air to circulate and, thereby, maintains an aerobic condition. On a surface application, the amount to use is approximately 0.1 to 1 pound of S-200 for every square yard of surface area. The amount sprayed is dependent on the amount of hydrocarbon on the surface you are spraying. To remediate a sheen on water, the same application ratio for a surface application is used.
SPILLREMED (MARINE)®	SPILLREMED (MARINE)® is a ready-to-use liquid product. It can be sprayed undiluted over the spill in open water conditions.	SPILLREMED (MARINE)® is applied in a product:oil ratio of 1:10. Rate of application is one gallon per 10 minutes.
STEP ONE	Components of the product STEP ONE (BC101) and STEP ONE (MSE2.5) are only sold together. B&S Research will make application recommendations based on contamination and ppm of contamination at time of purchase order and employ specified application methods (spray, plowing, agitation, etc.) appropriate for a particular situation.	Nominally, the STEP ONE water mixture can be used on most hydrocarbons, including crude and refined petroleum products, pesticides, etc., on land, fresh water, or ocean water. Laboratory tests also indicate effectiveness in destruction of PCB's. In certain circumstances, B&S Research, Inc. may require laboratory soil and water contamination tests before the product may be used.

Table 4.8 (cont.) Product application chart.

Product	Application Method	Concentration/Application Rate
SYSTEM E.T. 20 (formerly MCW.B.20)	SYSTEM E. T. 20 can be used on a broad range of hydrocarbon compounds (including PAHs) found in open and closed water systems (salt or fresh) or in soil (beach, sands, or inland soil types). SYSTEM E.T. 20 protocols can be used with indigenous and introduced bacteria. SYSTEM E.T. 20 is an oleophilic, non-water soluble nutrient, which releases nitrogen and phosphorous enzymatically, thus allowing SYSTEM E.T. 20 to be used in water and soil applications, including: conditions requiring fast degradation rates due to regulations, limited space, or cost; under conditions where weathered or heavy oils, sludge, or PAHs exist that require consistently high populations of hydrocarbon degrading bacteria to metabolize these long or more complex carbon chains; tropical or sub-tropical conditions that support large, indigenous, non-beneficial bacterial populations that would otherwise dilute the effectiveness of standard bioaugmentation approaches; aqueous conditions in which soluble or time release nutrient compounds would be washed away or diluted,	Concentration and rate of application will vary from site to site, depending on the type of contaminant and the area contaminated. Before applying SYSTEM E. T. 20, QET recommends that the site be evaluated for its physical and chemical soil characteristics (i.e., pH, salinity, conductivity, physical parameters, moisture content, hydrocarbon type and concentration, nitrogen content, etc.) and toxicity of both the soil and the contaminant by biological screening tests. Basic biological requirements of SYSTEM E. T. 20, site conditions, and SYSTEM E. T. 20 capabilities dictate the appropriate procedures for application of SYSTEM E. T. 20. Biological requirements for SYSTEM E. T. 20 bacteria are nutrients, adequate oxygen supply, hydrocarbon food source, neutral pH, and a 15% minimum moisture level for soil remediation applications.

	i.e., beaches, frequent rains, floods, streams, stream banks, estuaries, and areas affected by tides; sensitive environmental conditions that demand low or no toxicity; toxicity in soil or water during bioremediation may be stimulated by adding broadly available nutrients, which may activate disease carrying bacilli or pathogens; controlled nutrient release to SYSTEM E. T. 20 bacteria prevents over-nitrification; conditions requiring the elimination of oil discoloration caused by the presence of polar fractions; and conditions where salt water conditions exist or where only salt water is available for bacteria application. 1. Application Methods include biopile, land farm, pump and treat, bioreactor, bio slurry, in situ, windrow, topical application to discolored rocks or soil, and topical application to beaches.	
VB591™, VB997™, BINUTRIX® (formerly MYCOBAC TX-20)	Application for localized spills can be done using hand-held pressurized dust blowers. For inland waterways, ship channels, marinas, or coastal wetlands, large dust blowers mounted on barges or ships can be employed to apply the product. Application by aircraft using conventional dust spraying systems is recommended for treatment of large per acre	Recommended initial application rate is 5 to 15 pounds of VB591™, VB997™, BINUTRIX® per barrel of spilled oil. Follow-up applications at 48 to 72 hour intervals should be adjusted to allow for reduction in oil due to clean-up activities and natural loss by evaporation, droplet formation, and dispersion and microbial activity. Applications should not exceed 250 pounds

Table 4.8 (cont.) Product application chart.

Product	Application Method	Concentration/Application Rate
	uncontained spreading spills or spills in open waters or at sea.	per application. VB591™, VB997™, BINUTRIX® should be applied to spilled oil as soon as possible following spillage to stimulate natural oil utilizing microbial populations to maximize biodegradation activity. Application of VB591™, VB997™, BINUTRIX® to spilled oil does not significantly alter the physical consistency of the spilled oil, and it will not adversely affect conventional clean-up activities, nor will conventional and removal activities adversely affect the activity of VB591™, VB997™, BINUTRIX®. Dispose of waste in accordance with local, state, and federal regulations. Information regarding use of VB591™, VB997™, BINUTRIX® in conjunction with chemical dispersing agents is at present not available.
WMI-2000	The culture should be activated for 2 hours in water, when applicable. Conditions for use vary depending on substrate concentration and type, pH, temperature, availability of nutrients, oxygen content, and liquid medium (static or flowing).	Inoculation concentration is 5–9 billion spores per gram. Temperature requirements are between 35°F and 100°F. The optimum dissolved oxygen content is 2 ppm at the sludge-to-water interface. Nitrogen and phosphorous are key nutrients that promote cultural growth and should be maintained at concentrations

		of 15–20 ppm and 2.5–5ppm, respectively. The optimum pH range is between 7.0 and 8.0.
ALSOCUP	ALSOCUP is used to collect petroleum-based products spilled on water or land. When ALSOCUP comes into contact with a petroleum product, it chemically bonds and traps the petroleum. Once captured, spilled product cannot be released from ALSOCUP. ALSOCUP does not absorb water and will not sink. The product can be collected by skimming with pumps or mechanical devices.	ALSOCUP is used by applying one pound of product to each 10 pounds of spilled petroleum products. Water temperature does not appear to affect the ability of ALSOCUP to collect petroleum products. Heavy seas may prevent containment of an oil spill, and the product would lose efficiency if the oil product dispersed. Booms and dams that contain the petroleum product into an area are most effective when using ALSOCUP. In water, the motion of the sea will mix the product with oil. On land, a mechanical means is necessary to combine the oil with ALSOCUP. ALSOCUP will collect petroleum products at any stage; however, it is most effective early in the spill before dispersion.
AQUA N-CAP™ POLYMER	Apply AQUA N-CAP™ along the outer edges of the spill area to prevent spill migration. Proceed by applying product onto remaining spill area until a light white excess of product is observed. AQUA N-CAP™ may be broadcast onto the spill by hand or with a mechanical blower. Lighter hydrocarbons and synthetic oils will clump into a soft cluster like material easily recoverable with a net or	Apply AQUA N-CAP™ at a rate of 0.5 to 2 lbs per gallon of spilled hydrocarbon. Depending on the type of hydrocarbon and recovery operations employed, more or less product may be used to achieve recoverable solid mat/residue. Allow approximately 30–60 minutes of contact time. Product will form an elastic or rubber-like mat material. Mat-like material may be removed whole or segmented as appropriate.

Table 4.8 (cont.) Product application chart.

Product	Application Method	Concentration/Application Rate
	other means as appropriate. AQUA N-CAP™ differs from traditional sorbents in that it initially behaves like a synthetic sorbent then as a solidifier as the molecular microencapsulating process occurs. Once encapsulated, the spilled oil is captured within the polymer matrix. AQUA N-CAP™ is designed to clean-up a wide variety of crude oil, refined hydrocarbon products (i.e., fuels, oils, hydraulic fluids), and synthetic fluids spilled on water and surfaces.	
CIAGENT (formerly CI AGENT, CHEAP INSURANCE, & PETRO-CAPTURE)	For small scale spills on water (salt or fresh), CIAGENT can be broadcast directly onto the spill, spreading a thin layer from the outer edge into the middle of the spill. Agitation is not necessary. The product is also available in oil absorbent, water repellant booms and pillows. For large spills on water (fresh or salt), CIAGENT may be deployed with an air or water stream directed at the leading edge of the spill. Because of its fine particulate nature, static buildup may occur if the product is applied in dry form at a high rate of delivery.	In general, a 10% to 30% by weight application is required to solidify light, medium, and heavy oils. Solidification may occur faster if additional CIAGENT is applied. CIAGENT is equally effective in fresh or salt water and under any weather conditions; however, colder temperatures may slow the solidification process. The product is most effective on water of temperatures between 32°F and 120°F. Depending on the age and/or viscosity of the material, varying amounts of CIAGENT may be required to obtain complete solidification. The recovered

	Dispensing material should be properly grounded to prevent this. The product is relatively non-abrasive and should not harm machinery or pumping systems. Agitation is not necessary. The polymer has a specific gravity of less than one and will tend to separate and move to the surface when mixed with water. Therefore, in the case of water-carrying media, the product should be introduced "just in time" as the media is sprayed, or agitation will be needed to keep the product suspended. The product will remove oils and other hazardous hydrocarbon-based materials from fresh or salt water. CIAGENT may be used in a flow-through filter to remediate contaminated water. CIAGENT encapsulates and solidifies the oil while continuing to float even after saturation. The solidified oil or hazardous material may be removed from water using a vacuum pump or fish net. The material should be put into appropriate containers and disposed of in accordance with federal, state, and local regulations. For spills on land, the product would be applied in the same manner as in the water-based spill situation. Agitation is not necessary. To recover solidified oil or hazardous materials from	solidified oil or hazardous materials may be landfilled, incinerated, used as a secondary fuel, or otherwise disposed of according to federal, state, and local regulations.

Table 4.8 (cont.) Product application chart.

Product	Application Method	Concentration/Application Rate
	spills on land, the spent material may be collected and swept up using an industrial vacuum cleaner, broom, or shovel. The material should be put into appropriate containers and disposed of in accordance with federal, state and local regulations.	
ELASTOL	ELASTOL can be applied to the surface of an oil/water separator tank or added to a waste stream before it enters a separation tank. ELASTOL floats on water, dissolving when it comes in contact with petroleum products (e.g., gasoline, diesel, jet fuel, fuel oil, crude, etc.). It remains effective for long periods but naturally degrades when exposed to the atmosphere over longer periods. ELASTOL-treated oil may increase the performance of containment and recovery equipment: • Containment booms – may have higher current and tow speed capability • Mechanical skimming equipment – may be improved 2 to 10 times its normal recovery rates	ELASTOL can be used at very low concentrations from 100–1,500 ppm (0.01% to 0.15%). The amount of ELASTOL to be used is dependent on the viscosity of the oil being removed. For example, one gallon of ELASTOL will remove: • 13 gallons of gasoline, • 34 gallons of diesel, • 84 gallons of medium oil, or • 150 gallons of heavy oil. ELASTOL may be used on fresh or salt water (under warm or cold temperatures). Prompt application may maximize effectiveness. The older the pollutants are, the few light oil elements remain in the oil, which may reduce the effectiveness of recovery.

	• Drum skimmers – may recover little or no water with the oil. When applied early to a spill, ELASTOL may reduce emulsification and dispersion of oil. In addition, it may reduce the penetration of oil into porous soils and sandy beaches.	
PES-51	The following PES-51 application methods are applicable for the full range of PES-51 industrial uses, including shoreline and surface treatment, tank cleaning, and equipment decontamination. The selection of the method(s) will be dependent on the level and extent of hydrocarbon contamination, type of oil, its degree of weathering/emulsion, and the nature and type of surface to be treated or cleaned. Equipment availability, logistics, and manpower requirements should also be considered. Application methods may be combined, if necessary. In addition, for shoreline cleaning, the treatment area will be boomed and contained prior to PES-51 usage. For equipment decontamination, the use of portable de-con pools or secondary containment liners are recommended. ***Portable Equipment*** – Hand-Held Spraying – Spray PES-51 on the contaminated area using a Chapin Steel	The product comes already mixed and ready for use. Water temperature and salinity do not effect the product performance. PES-51 is effective against hydrocarbons only, and the age of the hydrocarbon is not relevant.

Table 4.8 (cont.) Product application chart.

Product	Application Method	Concentration/Application Rate
	Sprayer #1729 (or equivalent), 2.5 gallons capacity, or the AU 8000 MicroNair sprayer. After application, allow 3 to 5 minutes for soaking without allowing evaporation of PES-51 (weather dependent). When saturation is attained, hydrocarbon will be seen running off the impacted surface. Rinse the treated surface with available water (fresh or sea water) from the pump until no hydrocarbon remains. The water should be used at ambient temperature. Depending on the level and extent of contamination, a pressure washer may also be used for rinsing (ambient may be used). Collect the effluent hydrocarbons with absorbent booms and pads, and squeeze off the oil contaminants from the booms and pads for reuse as process oil. Airless Sprayer – Depending on the level and extent of the hydrocarbon contamination and the nature of the impacted surface, an airless type sprayer may be used for direct product application. Common types of airless sprayers are Airlessco, Graeco, or equivalents. These airless sprayers can have single or	

	multi-hose attachments and can include wand extensions as required. Application rate and pressure will vary depending on the equipment type and site-specific conditions. After spraying with PES-51, allow to soak for 3 to 5 minutes (weather dependent) avoiding evaporation, rinse/flush surfaces with pumps, fire hoses, deluge headers, or pressure washers (ambient). Pressure Washer with Syphon Feed System – Depending on the level and extent of the hydrocarbon contamination and the nature of the impacted surface proposed for treatment, a pressure washer may be used for direct product application. In most applications, hot water (greater than 120°F) is not necessary. Common types of pressure washers are Hotsy, Lambda, or equivalents. These pressure washers have a variable rate "detergent syphon feed" system for PES-51 application and can have single or multi-hose attachments that can include wand extensions. Application rate and pressure will vary depending on equipment type and site specific conditions. After spraying with PES-51, allow to soak for 3 to 5 minutes (weather dependent) avoiding evaporation, rinse/flush surfaces with pumps, fire hoses,	

Table 4.8 (cont.) Product application chart.

Product	Application Method	Concentration/Application Rate
	deluge headers, or pressure washers (ambient). Air Knife (Modified for PES-51 Application) – PES has developed a patent-pending modified air knife system for product application. This method was developed primarily for rocky, cobble, bedrock type shorelines with both surface and subsurface oil. The modified air knife delivers the PES-51 in both a liquid stream (125 psi) or as an aerosol. Compressed air is used to dilate subsurface sediments and allow for distribution of the PES-51. The air knife method is also applicable for surface treatment of impacted rocks, bulkheads, seawalls, rip-rap jetties, etc. After spraying with PES-51, allow to soak for 3 to 5 minutes (weather dependent) avoiding evaporation, rinse/flush surfaces with pumps, fire hoses, deluge headers, or pressure washers (ambient). For subsurface treatment, continue flushing with large quantities of low-pressure seawater at ambient temperatures. ***Mobile Equipment*** – Boat Spraying – The recommended application rate is 1 to 5 gallons	

	per 200 sq. ft. from a boat with speed of 1 to 3 knots, depending on the sea conditions and oil film thickness on the rocks. For a boat with a mounted AU-8110 MicroNair sprayer (or equivalent sprayer) and a spray swath of about 20 feet, traveling at approximately two knots, 25 acres/hr will be treated. After spraying, rinse PES-51 off the rocks with a hard, coarse spray of sea water. Standard size pumps with fire hoses or deluge headers may be used. Higher pressure rinses may be required if oil is thick and weathered. The shoreline may also be sprayed from the beach side, which will force the oil into the containment boom. Helicopter Deployed Spraying – Aerial spraying can be utilized for shore treatments and pretreatment with the AU 5000 atomizer (MicroNair) or equivalent sprayer. The recommended aerial application of PES-51 is 14 to 23 liters/minute. The AU 5000 (or equivalent) can be used with fixed-wing aircraft and helicopters operating at speeds of 90 MPH (145 km/hr) and more. The smaller AU 7000 sprayer (or equivalent) is recommended for use at airspeeds below 90 MPH. After spraying, the hydrocarbons can be rinsed off the shore rocks as described	

Table 4.8 (cont.) Product application chart.

Product	Application Method	Concentration/Application Rate
	above with hand-held pumps, deluge headers, or boat spraying. Vehicular Spraying – The recommended vehicular spraying is 50 to 150 ft2/gallon depending on climatic conditions. A MicroNair vehicle-mounted sprayer is recommended. This unit is a self-contained sprayer kit that combines the AU 8000 spray-head (or equivalent) with a powerful 4-stroke engine and a 60-liter chemical tank to give complete product coverage. After spraying, the hydrocarbons can then be rinsed off the shore rocks as described above with hand held pumps, deluge headers, or boat spraying.	
RAPIDGRAB 2000™	RAPIDGRAB 2000™ is a non-ionic liquid formulation that may be effectively applied by mist spraying onto floating oil slicks and sheens that are often found in harbors and at sea as the result of oil leakage and spills from ships. It instantly reduces and confines floating oils by the oleophitic synergistic effect of contraction and congealment into a physical state that greatly simplifies clean-up operations. Oil slicks and sheens may be reduced	RAPIDGRAB 2000™ should be mist sprayed full strength on surface oil until sufficient herding results are achieved. Results will vary depending on oil type, temperature, wave action, and viscosity. RAPIDGRAB 2000™ is designed to be used after oil has been contained by booms or other similar apparatus. Timely treatment with RAPIDGRAB 2000™, even at low application rates, can help to contain the sheen/spreading effect of the oil slick.

	by up to 1,000 percent or more. The broad temperature range for RAPIDGRAB 2000™ sprays are from 32°F to 120°F. Aerial Spraying – Typical application altitudes of 30 to 50 feet are advised, although higher altitudes may be used if conditions warrant. Spray nozzle should be about ¼ inches in diameter at 104 knots or more to create enough air shear sufficient to break the RAPIDGRAB 2000™ into proper sized droplets. A ½-inch diameter nozzle may be needed for temperatures below 50°F. Boat Spraying – Mist sprays may be applied from shipboard by power "foggers."	
WASTE-SET #3200®	Spills on Water – For spills on water, surround the spill with WASTE-SET® 3200 to prevent migration. Reinforce containment line with a boom or sweep if necessary. WASTE-SET® and the resulting reacted material will float. If the interior of the diked spill is not to be reclaimed, fill in the remaining surface area of the spill by broadcasting additional WASTE-SET® 3200 evenly over the spill until the WASTESET® no longer reacts with the spill. Alternatively, the product may be introduced under the surface of the spill to rise and react	Use one pound WASTE-SET® per five pounds of oil (may vary with viscosity and temperature).

Table 4.8 (cont.) Product application chart.

Product	Application Method	Concentration/Application Rate
	with oils in the suspension layer. WASTE-SET® can then be swept or netted from the surface of the water. Spills on Hard Surfaces – For spills on hard surfaces, apply WASTE-SET 3200® heavily to perimeter of spill to prevent migration. After surrounding the spill, reinforcement of the containment line thus formed with a boom or sweep may be advisable depending on the volume and flow rate. If the interior of the diked spill is not to be reclaimed, fill in the remaining surface of the spilled material by evenly distributing additional WASTE-SET 3200® until the WASTE-SET® no longer reacts with the spill and remains white and on the surface. Disposal – Dispose of reacted material in accordance with local, state, and federal regulation. Under the EPA Toxicity Characteristic Leaching Procedures (TCLP), WASTE-SET encapsulated material may be eligible for disposal in landfills. Incineration results in extremely low ash content. Dependent on the nature of the encapsulated material, it may	

	also be disposed of by incorporation into synthetic surfaces such as asphalt. WASTE-SET® will effectively "encapsulate" crude oil and petroleum-based products, such as gasoline, kerosene, and diesel fuel, in addition to the various following liquid and vapor-phase contaminants: aliphatic hydrocarbons, esters, aromatic hydrocarbons, ethers, chlorinated hydrocarbons, ketones, alcohols, and other hydrocarbons.	
ZYME-FLOW	Product is pumpable and can break the adhesion between oils and soil, rock, or sand. ZYME-FLOW can be applied where the contaminated soil or sand is temporarily removed and placed into a device that can provide mechanical agitation (e.g., a cement mixer or a fractionating tank). Unagitated tanks may also be used, but more time will be required to achieve the goal. Add ZYME-FLOW solution to the tank. As agitation continues, oils with a specific gravity of less than one will float, while oils with a specific gravity of greater than one will settle. The floating or settled oil can then be removed from the soil or sand. As ZYME-FLOW will not emulsify oils, the recovered oils are almost	The exact dilution of ZYME-FLOW will vary between 50:1 and 200:1 depending on temperature, viscosity, and type of petroleum product. ZYME-FLOW is effective in all non-frozen waters with greater efficacy as the temperature increases. The product efficiency is not affected by salinity, but it will be affected by tar sand.

Table 4.8 (cont.) Product application chart.

Product	Application Method	Concentration/Application Rate
	anhydrous, and the water can be collected and reused. Another method is to physically isolate a small area that is to be cleaned with damming collars, sand bags, or other means on the sides and a none permeable layer on the bottom (if possible). Then, wash the contaminated area with ZYME-FLOW solution using a pressurized water stream. Keep the area super-saturated with water for several hours to allow oil residues to float to the surface for collection. ZYME-FLOW can be used to wash the inside or outside of a ship or barge that has been contaminated with oil using a high-pressure sprayer. Bilges of a ship or barge can also be cleaned using a 1% ZYME-FLOW solution and allowing the natural movement of the ship to provide the agitation.	

Figure 4.1 Booms laid out to protect a coastal area. Courtesy of Wikimedia Commons[1].

employed to deflect spills in order to avoid the risk of explosion or to protect sensitive areas while the spill material evaporates. Booms that are deployed near coasts only work well in calm seas. In rough seas they tend to break up, or waves simply spill over them. For this reason, they must be tended to make sure they do not drift away from the areas they are supposed to protect. They also cannot be placed out too early. If they are, they can break up or become ineffective before oil actually arrives. Figure 4.1 shows booms being laid out to protect a coastline area impacted from the BP oil spill near Louisiana.

There are no standard design dimensions for booms. Booms may be improvised depending on spill conditions and available resources. Commercially available booms are manufactured with variable dimensions, which is out of necessity since they are subjected to the marine currents, the direction and velocity of the wind, and the height and frequency of the waves as well as different

[1]http://commons.wikimedia.org/wiki/File:Anti_oil_spill_booms_around_Breton_National_Wildlife_Refuge.jpg

depths. This means that the Incident Commander should carefully assess the areas to be protected by booms and select those that are most appropriate based on sea conditions. Booms are best thought of as the first response equipment to arrive at an oil spill, and efficient deployment is the first job to be done even before the skimmers and tanks are deployed. Emergency response plans should ideally address the type, the availability, and the estimated numbers or lengths required based on a worst case scenario of a spill and not wait until the spill occurs.

Booms come in many shapes and sizes and are made from different types of materials according to their use. The efficiency of a well-designed boom is much better, but the first response should be to try to stop the oil from spreading by using anything that may be at hand. If, at the moment of an oil spill, the crew does not have resources available for the response, it is worthwhile to put together planks, foliage, screens, and bags of sand or gravel. First responders who are responsible for the selection and application of the booms should first define their functions and know exactly where each type of boom is likely to be most effective. They should also take into consideration the prevailing weather conditions, durability, forms of storage, transportation, and the capacity for containment. Operations with booms also require knowledge of boat handling, use of anchors, support buoys, ancillary materials, and support equipment. Some systems have hydraulic-driven reels so that there are hoses and ropes on the ground and on vessels, which are potential hazards. These components should be stowed correctly so the job can be done safely. For example, all ropes should be coiled or laid in a manner such that personnel do not get entangled in them while deploying booms. There should be a team effort with safety being a prime concern. But again, a well thought out emergency response plan will take these factors into consideration so that, in the event of a release, emergency response teams are ready to be deployed with the proper equipment.

The effective capacity for response to oil spills is a combination of several factors, the primary ones being:

1. speed and efficiency of mustering the service crew;
2. quantity, availability, and applicability of the equipment for the response action; and
3. quantity and availability of personnel trained to do the work.

The Canadian Shipping Act establishes guidelines for planning equipment requirements for response. The containment boom formula that is relied upon by our Canadian neighbors is

$$B = 1.25 \times H,$$

where B = the amount of boom in meters required to contain the free floating oil, and H = the amount of oil spilled in m^3.

While we do not have a full account of the BP Deepwater Horizon disaster in the Gulf, best estimates are that the oil released could top 100 million gallons. Using the Canadian rule of thumb, approximately 473,573 m of boom will need to be deployed to protect coastlines impacted by the Gulf spill. This is equivalent to 1,553,319 ft of boom. But, as of July 2010, the British Petroleum Web site has noted that more than 2.7 million feet of booms have already been deployed or will be deployed shortly. This shows that, at best, the general rules of thumb provide gross under estimates of the resources that are required for some releases. Records maintained by the Coast Guard indicate that there are about 3.1 million feet of coastal protection boom available in the U.S. and about 430,000 feet of heavier-duty boom intended for use in open oceans. It is clear that resources currently being deployed to address the Gulf disaster are already stretched to dangerous limits.

Offshore boom is usually available in 250 m or 300 m lengths, so this is 5 booms, at most. When used in U or J configurations, the vessels will be a maximum of 100 meters apart. Therefore, these 5 systems will have 500 m of boom encountering the oil. Definitions of terms can be found in Table 4.12 at the end of the chapter. Key functional components of a boom are illustrated in Figure 4.2.

The float (which keeps the boom afloat) can be air, foam, bubble wrap, cork, or other material. Its function is to maintain the correct orientation of the equipment on the water, establishing freeboard in order to minimize the loss of oil over the boom. It is designed to float the boom under load and tension. If booms of the inflated type are employed, then it is important that caps be replaced and secured because trash and other foreign matter in the water can get into valves, which will release air and reduce the effectiveness of the boom.

The term "freeboard" refers to the height measured from the water line to the top of the float section. Booms chosen for an application must be based on an estimate of the minimum required freeboard based on sea conditions and wave heights.

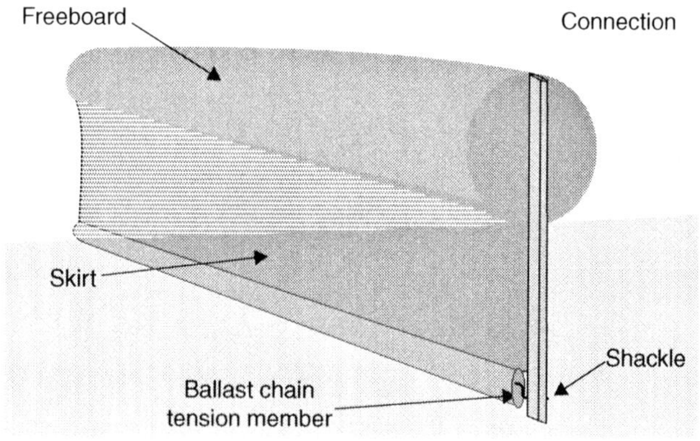

Figure 4.2 Key features of a boom.

The skirt is the continuous portion below the float that has the purpose of containing the oil that tends to pass under the boom. The distance measured from the water line to the lower limit of the skirt is called the draft. In shallow water applications, the skirt should not take up more than 20% of the water depth as it can cause an increase in current velocity, which will drag the boom under the water, thereby causing oil to flow over the top. In applications close to shorelines, too long of a skirt can cause the oil to flow under the boom.

The ballast is weight added to the bottom of the skirt. Its function is to maintain the barrier in a position perpendicular to the surface of the water and, in some cases, to provide the tension member for the boom. Stresses introduce by sea currents and waves are transferred away from the boom material to the ballast chain or wire rope. In no case should the tension member be lengthened because the stress will be transferred to the material usually causing the material to part. This will result in a failure of the boom's operation. Some manufacturers use steel or lead weights for the ballast, but water may also be used.

There are published guidelines for the construction of booms, the weight of the ballast, and the strength of materials required. These guidelines vary depending on the manufacturer. These design factors take into consideration the conditions of the sea. There are three general sea conditions that are used as a general guide to boom design, as shown in Table 4.9.

Table 4.9 Water body classifications for oil spill control strategies.

Water Body Class	Wave Height (m)	Condition Description
Calm	0–0.3	Small, short non-breaking waves
Protected	0–1	Small waves, some white caps
Open	0–2	Moderate waves, frequent white caps

Table 4.10 Boom specification criteria.

Boom property	Calm	Calm – current	Protected	Open
Min reserve buoyancy to weight ratio mm	150–600	200–600	450–1,100	900–2,300
Min total tensile strength N	2:1	3:01	3:1	7:1
Min skirt fabric tensile strength N	6,800	23,000	23,000	45,000
Min skirt fabric tensile strength N/50 mm	2,600	2,600	3,500	3,500
Min skirt tear strength N	450	450	450	450

Using the water body classifications in Table 4.9, the user can use the chart in Table 4.10 as a basis to select the type of boom most appropriate for the application.

Figure 4.3 shows the weight of the ballast as a function of boom size.

Other important features of booms that should be considered in selection are connections, anchors, anchoring components such as rope, and buoys.

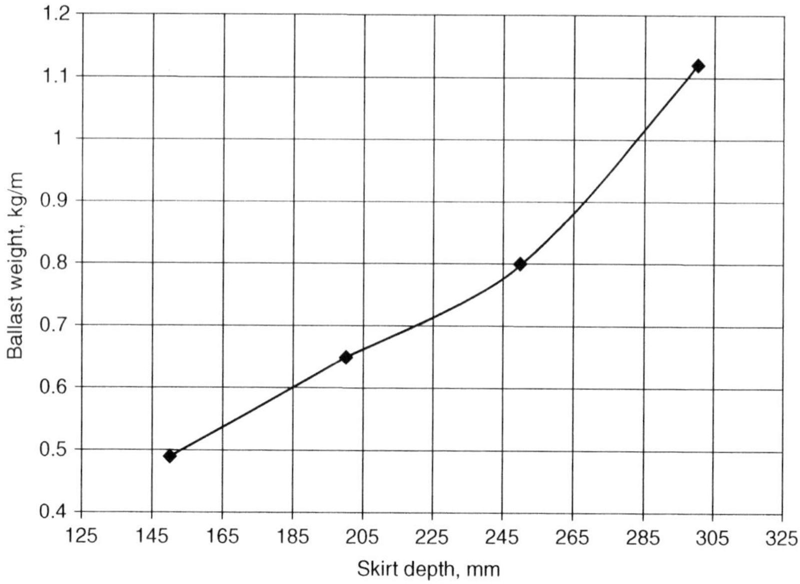

Figure 4.3 Ballast-boom size graph for 0.7 knot current.

- Connections are quick-connect devices that permit the coupling together of as many lengths of booms as are necessary. These are six standard types of connectors used by different manufactures, and they generally are not interchangeable. Connector types are ASTM, Universal Type 1, Universal Type 2, U.S. Navy, Hinge and Pin, and NOFI. When procuring booms, it is important to ensure that all connections are of the same type in order to reduce problems during applications in the field. Additionally, some connections come with plastic thumb screws that can be brittle during winter months or soft in hot climates. Others have spring loaded toggle pins. These types of connectors are generally fine for conducting training exercises, but during actual operations where booms may be in service for long periods of time, the use of bolts with wing nuts generally tend to be more reliable and offer fewer problems.
- Anchors maintain the booms in a stationary position. There are many variations of anchors but common ones

used during oil spill operations are the Bruce (mud) and Danforth (sand) types. The length of anchor cable should be approximately 5 times the local depth. This is attached to a chain to keep the rope from lifting the anchor in swell or wave conditions. Anchor points are the locations used for tying anchor ropes and serve to maintain a given configuration of the boom in the water. With many types of boom, anchors can only be fitted at the connections. Fastening rope around the boom to make an anchor point usually cuts the boom material with sea movement. The faster the current the more anchors will be needed, so shorter lengths of boom must be used to gain the extra connections where the anchors can be tied safely. In addition to giving consideration to anchor points, the selection of appropriate anchoring ropes is important. Polypropylene rope is preferred because it is light and floats. The rope should be of high quality and strength, and cost should not be a consideration. The anchoring rope employed is the link to the boom and should always be of a lesser breaking strain than that of the boom itself simply because rope is inexpensive compared to booms and the job they are required to perform.
- Tripping buoys serve an important function of protecting both the anchors and the boom. A tripping buoy will show the location of anchoring points and prevent anchors from being dragged out of the mud or sand by small boats and machinery. A thin rope attached to the head of the anchor leading to the buoy will assist with removing the anchor from the mud or sand. A length of rope equal to 1.5 times the local depth is generally acceptable.

Figure 4.2 has illustrated the general features of a boom. However, it is important to note that there are several types of booms that can be employed. These include fence type, curtain, fence curtain, shore-sealing, figure-of-eight, and self-inflating.

- Fence booms are made of rigid or semi-rigid material that is designed to retain the floating oil. They are more able to resist abrasion and easy to handle, clean, and store, in addition to being reliable. They are used in calm, quiet waters. The disadvantages are they have low stability in strong winds and currents, slight flexibility for towing, low efficiency in waves, and need reinforcement to prevent wear or breaking.

- Curtain booms are of the air-filled type and solid flotation devices. Air filled curtain booms have the advantage of being relatively compact when empty. They are efficient and easily stored and handled. The float elements are chambers filled with air. They are good for towing and installing and generally perform well in rolling seas. They have the disadvantage of low resistance to abrasion and are easily punctured if mishandled. They have a flexible skirt maintained by a ballast of either chain or wire rope, which is also their tension member. Solid flotation curtain booms are made of chambers filled with polyurethane, polystyrene, bubble rap, or cork. They display high reliability and efficiency when towed and are easily handled. The disadvantage is that they have a large volume from a logistics point of view. They also have little horizontal flexibility in rolling seas. Solid flotation booms are normally manufactured from oil resistant fabric coated with neoprene, polyurethane, or PVC.
- A fence curtain boom can be employed in calm waters. It can be inflated to become a curtain boom for rougher waters. The permanent flotation along both sides of the boom should provide more resistance in areas where abrasion is a problem.
- A shore-sealing boom is a device that consists of an air inflated flotation tube on top and two lower tubes that are filled with water to make up the skirt of the boom and its ballast. This type of boom is best suited for the intertidal section of the shoreline, from the high tide mark to the low tide mark where a curtain type boom is connected to the area below the low water mark. At low tide, the lower tubes settle on the substrate, maintaining a seal and preventing the passage of oil.

- Figure-of-eight booms have excellent wave following characteristics. They are constructed from two independent chambers, the upper chamber of air and the below chamber of water. These booms require a constant flow of air and water while being deployed. An air fan and water pump must be used to work continuously. In the event of a problem with the flow of air, the boom will sink. When deploying, if there is insufficient weight of water in the boom, they have been known to fly in windy conditions. This is an old design that is not common anymore due to its disadvantages outweighing its advantages.
- Self-inflating booms come in two varieties. The so-called yellow version has a spring down the length of each float section, which lies flat when the boom is rolled up on the reel. When the boom is deployed, the spring opens up and sucks air in through the valve. This type of boom needs to be deployed at least twice per year and stored in a dry location. One of the main problems is that the steel spring gets rusty and sticks to the material if left for a long time or stored in a humid environment. The second type has rigid plastic diamond structures inside the float section, which open and suck air in through a series of valves when deployed. The valves have to be opened as the boom is deployed and closed when the boom is recovered. The boom is stored vertically on a carousel, which can be difficult when recovering it as one or two workers are needed to stand on the carousel to manually stow the boom vertically. Workers have been known to get caught between the boom and carousel as it is recovered. The boom can be picked up by helicopter then taken to the spill site and dropped in the water where a boat picks up the tow end, and the boom unrolls as it moves away. If the helicopter is not an option, then it can be recovered onto a standard horizontal boom reel.

Table 4.11 provides a partial listing of distributors and manufacturers of different boom and barrier systems.

Table 4.12 provides definitions of important terms.

Table 4.11 Partial list of suppliers.

Distributor	Company Profile	Product Description	Contact Information
EnviroMet – Vancouver, WA	Offers a complete line of spill clean-up and pollution prevention products. Oil and chemical pads, rolls, booms, socks, pillows, granulars, spill kits, secondary containment, and drain mats. Brand Names: Enpac, Moltan Company, Spilltech, Zimmerlite	Supplies containment boom that is available in 100 ft sections. Product is designed for areas with high current, from streams to the ocean. Touted as having higher tension for extra strength. Boon has double-wall design with foam center. The flotation material is a 72″ (183 cm) rolled, closed cell polyethylene foam in a double layer chamber for extra protection. The fabric is PVC coated polyester, safety yellow. Dimensions: width - 18″, diameter (float) - 6″, skirtv- 12″.	1-800-590-2436
Chemtex, Inc. – Cumberland, RI	Distributor of oil spill containment booms. Containment booms are available in 3–6 in. float and 5–12 in. skirt. Containment booms are used to surround and contain sewage spills and retain debris. Applications include holding ponds, marinas, and spill response teams.	Products include the following: (1) Booms – Linkable booms are used to contain and absorb spills on or off water. The built-in stainless steel connectors link booms together to cover large areas of water. Booms are ncased with a lint-free poly sock that prevents shedding. (2) Containment Booms – Use containment booms eto surround	Local: (401) 312-3962 Toll Free: (888) 380-8604

		and contain oil spills and retain debris. They can be used in fresh or salt water and are ideal for holding ponds, marinas, or spill response teams. (3) Permanent Containment Booms – These are used to provide continuous facility protection and containment. This permanent boom system features easy deployment, high visibility, and abrasion resistance. Permanent containment booms are designed for deployment at permanent sites such as waterfront oil handling facilities, around ships in repair, or storage. This containment boom is more economical, lighter weight per foot, and has a higher buoyancy to weight ratio than other permanent booms. (4) Bilge Booms (Mini Booms) – Use in boat bilges of all sizes. Use the 8″ x 18″ in commercial size boats and the 5″ x 15″ in smaller pleasure crafts. Like the smaller bilge sock, these highly absorbent mini booms will help prevent pollution of our marine waters. Booms will absorb oils and fuels and will float on water.	

Table 4.11 (cont.) Partial list of suppliers.

Distributor	Company Profile	Product Description	Contact Information
American Boom & Barrier Corp. – Cape Canaveral, FL	Custom manufacturer of oil spill containment booms including melt-blown polypropylene sorbent and bilge booms. Available in 5 in. dia. x 10 ft. long and 8 in. dia. x 10 ft. long sizes. Features include low ash content and hydrophobic and oleophilic properties.	Melt-blown Polypropylene Sorbent Boom comes with sock and net. It serves as a continuous barrier for land-based spill response. The booms are designed to link together, creating barriers of any length. They are quickly deployed, easily recovered, and capabale of covering large areas quickly and efficiently. The product is ideal for any oil spill, small or large, water or land. Product codes: T-270: 8″ dia. x 10′ long (4 boom per bale) 20 cm x 3 meters; SABB510: 5″ dia. x 10′ long (4 boom per bale) 13 cm x 3 meters	Phone: 1-321-784-2110 Toll free: 1-800-843-2110
New Pig Corp. – Tipton, PA	ISO 9001:2000 certified manufacturer and distributor of containment booms including oil spill containment booms. Length ranges from 50 ft. to 100 ft. Booms with UV resistant shell are available.	Makes a variety of booms; however, most are suitable for small spills.	1-800-468-4647
Applied Fabric Technologies,	Manufacturer of oil spill containment booms. Features include symmetrical	AFTI offers four basic lines of high quality, state of the art oil spill control	Local: (716) 508-4583

Inc. – Orchard Park, NY	design, synthetic plastic components for corrosion free performance, unisex end connectors, four interconnectable draft/freeboard heights, urethane foam-filled cast polyethylene floats for buoyancy characteristics, and high stability in chop, wind, and tidal currents. Oil spill containment booms are suitable for use in oil handling and producing installations, including offshore production platforms, exploratory rigs, marine terminals and other coastal facilities, oil jetties, and settling ponds. Brand Names: AFtrap, FloTemp, GlobeBoom, OILFENCE, PyroBoom, TufTemp	barriers to cover the broadest range of operating conditions. These include the following: GLOBEBOOM® – Five models using various fabrics are available. GlobeBoom® offers "Oilfence tough" emergency response and semi-permanent barriers. GlobeBoom® is based on a conventional design. The paired hemispherical floats allow compact storage, flexibility, and convenient handling. Exclusive Globe fabrics provide hydrodynamic, strength and abrasion resistance compared to comparably priced barriers. OILFENCE™ – Four sizes of durable, functional, compact barriers for permanent or continuous use are available. Many users have also found that OILFENCE™ booms are also effective in emergency response situations under a broad range of sea, wind, and current conditions. CONTRACTOR'S SPECIAL "CS" BOOM – AFTI's "CS" boom is available in various sizes and	Toll Free: (866) 979-9614

Table 4.11 (cont.) Partial list of suppliers.

Distributor	Company Profile	Product Description	Contact Information
		configurations. FIREPROOF BARRIERS – There are two distinct types of fireproof boom. Both of these products are based on proven technology and are designed to assist operators in "In-situ" burning. These two barriers are PyroBoom® and PocketBoom. PyroBoom® is a standalone, solid flotation boom designed to quickly respond to spills, burn them, and then be refurbished for reuse. Developed in 1984 for use on the Alaska North Slope and other remote locations, it is simple, durable, and easy to use. PyroBoom® has been subjected to numerous test protocols. PocketBoom is a re-design of the original Dome Petroleum high seas fireproof barrier. This design is all stainless steel and was developed to be used in conjunction with other booms that, in retrospect, have been shown to be less than durable in	

		actual burn conditions. Eleven companies and agencies participated in the re-design and testing of PocketBoom. Its performance far surpassed all of the test protocol parameters for durability, re-usability and flex and fatigue resistance under fire. PyroBoom® is available in a 30″ overall height. PocketBoom is available in a 36″ overall height. AFTI will tailor packages including containers, handling equipment, and storage facilities for both PyroBoom® and PocketBoom.	
Becker & Associates – Lombard, IL	Stocking distributor of oil spill containment absorbent booms. Features include 5 in. x 10 in. or 8 in. x 10 in. size, 10 gal. or 14 gal. maximum absorption capacity, and 25 lbs. or 40 lbs. weight. Oil booms have sock-and-net design containing shredded poly and allow for full saturation of the boom. Brand Names: Beach, Determan Brownie, Emcee, Enraf, Enterprise Brass Works, FRAM, FUELMASTER, Facet, Others.	Supplies a range of oil select booms and socks with varying absorption capacities. The largest size boom is 8″ X 10″ with 14-gallon maximum absorption capacity.	800-544-6643; 800-325-8205

Table 4.11 (cont.) Partial list of suppliers.

Distributor	Company Profile	Product Description	Contact Information
Brockton Equipment/ Spilldam, Inc. – Brockton, MA	Manufacturer and distributor of oil spill containment booms with 100-feet boom length, 12-inch draft, 6-inch freeboard and a total height of 18 inches and a temperature range from –65 degree F to + 140 degree F. Booms are used in inland waters and harbors with low current velocities.	SPILLDAM – LOG I is a sturdy, all-purpose oil containment boom specifically designed for use in inland waters and harbors with low current velocities, where versatility and effective spill containment are popular design features with oil spill clean-up contractors who are constantly moving and redeploying boom. "Universal" extruded aluminum interlocking connectors allow sections of the boom to be joined without tools. The connector is compatible with many other manufacturers' booms. Boom lengths are typically 100 feet, but other lengths are available. Temperature ratings of booms range from –65°F to +140°F. The Web site provides boom specifications on freeboard, flotation size, stowage volumes, fabric strength, draft, and others. Curtain material is vinyl-coated nylon or polyester. Information on abrasion resistance is also available.	800-699-2374

AbsorbentsOnline.com – Chino Hills, CA	Distributor of booms for oil spill containment. Oil spill containment booms for marine and industrial spill containment. Oil absorbent booms include a strong mesh encasing polypropylene filler. Other features include nylon rope and steel hardware. Booms are available in 5 in. and 8 in. diameters and 10 ft. and others.	Oil Absorbent Booms – These booms can also be used to contain and absorb large industrial hydrocarbon spills on land. These oil only booms repel water, which means they will float even if saturated with oil. Oil absorbent booms can be used to handle all types of hydrocarbon spills such as fuel oils, hydraulic oil, gasoline, diesel, motor oil, jet fuel, and kerosene. The booms are used for skimming oil from water. They are used on the ocean, rivers, wastewater troughs, ponds, or streams. Good for use on land or water for fuel spills or low viscosity oil. Each boom is constructed with a tough outer mesh skin that encases an absorbent polypropylene filler. The filler will not shed. A nylon rope runs through each boom and has carbon steel connectors so you only use the number of booms required to contain the spill. Retrieval is easy with the long-lasting, durable linking system. The oil absorbent booms come in a bright white color to make it easy to see	800-869-9633

Table 4.11 (cont.) Partial list of suppliers.

Distributor	Company Profile	Product Description	Contact Information
		when they are completely fully saturated. The booms are available in 5″ and 8″ diameters and 10-foot or 20-foot lengths.	
Best Value Supply, Inc. – Great Neck, NY	Supplies spill containment and spill response products including spill kits, spill kit refills, absorbent pads, absorbent socks, booms, pallets, berms, pillows, spill containment products, and absorbents for almost every type of chemical spill or oil spill clean-up.	Provides a wide range of sizes and configurations of oil absorbent booms.	877-249-1991
American Pacific Plastic Fabricators, Inc. (APPF, Inc.) – Garden Grove, CA	Custom processors and fabricators of hazardous waste clean-up equipment and supplies including oil spill containment booms. Other containment products include bags, equipment covers, pouches, inflatable bladders, cushions, and medical and sport items serving the military and aviation industries.	The company custom builds and fabricates according to customer.	(714) 891-3191
Impact Absorbents, Inc. – Atascadero, CA	Manufacturer of booms in 5 in. x 10 ft. to 8 in. x 10 ft. size for oil spill containment applications. Booms are constructed of	Supplies several configurations: SKU: XNBBS XSORB Oil Select Bilge Boom 6 in. x 15 in. (Case of 12); SKU:	Local: (805) 468-4789

	two absorbent socks, ropes, netting, and hooks for deployment and disposal.	XNB210S XSORB Oil Select Boom 5 in. x 10 ft. (Case of 02); SKU: XNB310S XSORB Oil Select Boom 8 in. x 10 ft. (Case of 01); SKU: PBS510-4 FiberDuck Oil Booms 5″ x 10′ (Bale of 4); SKU: PBS810-4 FiberDuck Oil Booms 8″ x 10′ (Bale of 4); SKU: PBS512-12 FiberDuck Oil-Only Bilge Boom 5″ x 12″; SKU: PBS818-12 FiberDuck Oil-Only Sump Skimmer Boom 8″ x 18″	Toll Free: (888) 379-6149
GulfRim Navigation – Abbeville, LA	Distributor of oil spill containment anti-pollution booms and barriers.	Supplies a hard containment boom; however, no product details are provided on the Web site.	877-893-0789
Grainger Co., Inc. – Lake Forest, IL	Distributor of material handling products such as forklift booms, oil spill containment booms, telescoping booms, and truck mounted booms.	Oil Sorbent Boom – 10 ft length, 34 gallons sorbed per package, depth of 5 inches, with clip links	See http://www.grainger.com/Grainger/items/5TR21?Pid=search
Lab Safety Supply, Inc. – Janesville, WI	Distributor of industrial safety products including oil spill containment booms. Products include standard booms and mini booms. Booms are made from 100% polypropylene materials in sizes of 10 ft. x 4 in. and 10 ft. x 8 in.	Product provides a wide variety of hazmat response equipment; however, the Web site does not provide details of oil booms.	800-356-0783

Table 4.11 (cont.) Partial list of suppliers.

Distributor	Company Profile	Product Description	Contact Information
Interstate Products, Inc. – Sarasota, FL	Manufacturer and distributor of spill containment systems including spill berms and spill control products.	The company carries a broad line of durable absorbents. These absorbents are made of pure meltblown polypropylene fibers, which absorb up to 25 times their weight. They supply a variety of booms, pillows, and socks that are designed to absorb strictly oil-based liquids. Biodegradable pads are also available for environmentally safe cleaning operations.	800-474-7294
BTL Liners – Multiple Locations	Manufacturer and distributor of standard and custom oil spill containment booms. Materials used include ethylene inter-polymer alloys (Elvaloy®). Range in thickness from 30 mils to 45 mils. Brand Names: CoolGuard, XR-5	Primarily a provider of liners, the company also provides custom made booms.	800-280-0712 • 541-447-0712 • Fax: 541-447-0759
Maxi Container, Inc. – Detroit, MI	Distributor of new and reconditioned, secondary containment oil spill containment booms. Secondary containment products also include mobile drum dispensing spill scooters, drum spill sumps, intermediate bulk container spill sumps,	Spill kits range from 5 to 95 gallons. The company advertises booms, but there are no descriptions on the Web site.	313.891.3880 • Fax: 313.891.6476

	poly tank containment units, 2, 4, 6, and 8 drum workstations, and spill kits.		
The Cary Co. – Addison, IL	Offers a full line of spill control products for proper clean-up and containment, including absorbent mats, pads, rolls, spill socks, hydrophobic pillows, spill kits, salvage drums, and spill containment pallets. Offers Teal-Sorb super absorbent powders.	Absorbent socks, sorbent pillows, and oil spill booms – All meltblown booms are non-linting and are sock-and-net construction to ensure that all particulate stays safely inside the boom. Each boom comes complete with end rings, snaps, and sturdy rope running their length. They contain high quality, all-synthetic, high absorbency fill blended with a flotation agent. Booms are available in a wide range of sizes.	630-376-2400
Carolina Equipment & Supply Co., Inc. – North Charleston, SC	Manufacturer of vacuum booms for oil spill containment applications. Vacuum booms are available with 5 in. dia. hollow flexible tube, 25 ft. vacuum hose, and portable wet vacuum recovery system.	Vacuum Boom – A hollow flexible tube 5″ in diameter placed directly on a hard surface to form a downslope side dam or to completely encircle a cleaning area. During use, the boom is connected by a 25 ft. vacuum hose to a portable wet vac recovery system.	800 - 394 - 4987
Mycelx Technologies Corp. – Gainesville, GA	Developer, manufacturer, and marketer of patented MYCELX® technology for full range of inter-connecting booms and pads designed for the removal of oils	Company offers engineered turn-key solutions to spill response and containment.	Local: (678) 897-4773 Toll Free: (866) 756-9039

Table 4.11 (cont.) Partial list of suppliers.

Distributor	Company Profile	Product Description	Contact Information
	and tarballs, including sheen, from water and meeting needs of remediation, water treatment, and emergency response.		
Hydraulics, Inc. – Vancouver, WA	Distributor of oil spill containment booms. Capabilities include hydraulic, jack and lifting equipment repairing, pumps, valves and motor rebuilding and testing, crane inspections, and field service repair work. Spill containment booms are suitable for industrial and commercial applications.	No product information is provided on the company Web site.	800-500-4399
John G. Shelley Co., Inc. – Wellesley Hills, MA	Distributor of specialty products including oil spill containment booms and polyethylene cartridges. Rubber fabrication services including mixing, molding, slitting, and die cutting are also available.	John G. Shelley Company is a specialty rubber manufacturer and fabricator offering a wide range of custom designed products. They advertise booms, but no product descriptions are provided on their Web site.	Tel: 781-237-0900 or 1-800-525-0202
Mid Atlantic Industrial Supply & Brokerage – Sandston, VA	Distributor of containment and oil spill containment booms designed for absorption of oil-based spills in marine and industrial applications.	Materials include meltblown polypropylene. Booms can be used for skimming oil off water and are available in sizes including 4 in. x 10 ft. and 8 in. x 10 ft.	804-222-9226

Rockwater Manufacturing Corp. – Freeport, NY	Manufacturer of oil spill containment booms for permanent installation and continuous in-water use.	Booms can handle strong currents and high tension and can be used in shallow streams and creeks, quiet waters, holding ponds, and settling basins, and inner-harbor and calm waters. The industries served include marine, structural rehabilitation.	Phone: 516-868-0500, 800-338-3145 (toll free) Fax: 516-868-0505
General Tool & Supply Co. – Multiple Locations	Distributor of industrial oil spill containment booms including sorbents.	Economy sorbents are available in sizes of 36 x 150 x 1/4. Economy sorbent socks are available in sizes of 3 in x 48 in.	Local: (971) 270-4071 Toll Free: (888) 598-6575
Sanderson Safety Supply Co. - Multiple Locations	Distributor of oil spill containment booms including oil only booms.	Booms are available in sizes of 5 in. x 10 ft. and 8 in. x 10 ft. Each boom contains different attachments for multiple interlocking. Booms are used for removing oil from water or land and soaking up petroleum-based products.	Phone: 503-238-5700, 800-547-0927 (toll free) Fax: 503-238-6443
McCutcheon Enterprises, Inc. – Apollo, PA	Supplier of environmental and waste management products and services to industry and municipalities. Stocking distributors of Acme® oil spill containment booms and floats and SPC® sorbent products, including absorbent	Company offers high-performance oil sorbents for clean-up on land or in water, absorbing 25 times their weight in petroleum products. SPC Universal Plus all-purpose sorbents absorb up to 16 times their weight in	724-568-3623

Table 4.11 (cont.) Partial list of suppliers.

Distributor	Company Profile	Product Description	Contact Information
	blankets, pads, pillows, rugs, sweeps, particulates, and socks for recovery and clean-up.	oil, water, or chemicals and are ideal for hazardous material clean-up. Acme containment booms range from the small Mini-Boom (2″ diameter flotation, 12″, 24″, or 36″ long skirt). Company also supplies other combinations of flotation diameters, skirt lengths, chain ballast sizes, and top tension cable sizes.	
W.P. & R.S. Mars Co. – Bloomington, MN	Distributor of oil spill containment booms and T-270 petroleum sorbent booms in 8 in. x 10 ft. size, and 80 gal/bale sorbent capacity.	Available are scrim-encased tubes of particulate sorbent material secured by a UV resistant netting. Built-in attachment clips are provided at each end of the boom.	See http://www.marssupply.com/VIA/index.jsp
World Pipe, Inc. – Holbrook, NY	Distributor of oil spill containment booms.	Materials include carbon steel, stainless steel, fiberglass, Kynar®, ductile iron, PVC, CPVC, aluminum, polyethylene, and polypropylene.	Phone: 631-738-9872, 877-395-8811 (toll free) Fax: 631-738-9871

MOP Enviromental – Bath, NH	Manufacturer of oil spill containment booms.	Absorption rate ranges from 3.4 gal. or 27 lbs. per boom to 20 gals. or 950 lbs. per boom. Booms are available in sizes from 3 in. x 2 ft. to 8 in. x 50 ft. and can also be made in custom sizes.	Phone: 603-747-2202, 888-747-2000 (toll free)
Flexible Containment Products, Inc. – Spokane, WA	Manufacturer of fabric related products for the hazmat, environmental field and remote site industries. Berms, oil, land and silt curtain booms. Tanks: dip, collapsible, secondary-pump and portable water storage. Remote site decontamination and cold weather tents.	Secondary spill containment berms, oil booms, silt curtain booms, and land booms.	Phone: 509-624-8921, 800-728-2905 (toll free) Fax: 509-624-9780
Slickbar – Seymour, CT	High-end oil spill response equipment ranging from inland to off-shore booms, permanent installations, oil skimming systems, and rapid response recovery vessels.	Distributor and custom manufacturer. Need to discuss applications with technical sales personnel.	http://www.slickbar.com Phone: 203-888-7700 Fax: 203-888-7720

Table 4.12 Table of important terminology.

Type Terminology	Term	Description
Boom Classification Term	Oceanic Boom	Term refers to a class of booms that are large, robust, and function well in waves up to 2 m in height. Above this height, the barrier begins to fail. Generally these booms have a freeboard of 40–60 cm and a skirt of 0.80–1.60. Some are up to 3 meters in total and are stored on huge reels. They were developed for the rough northern seas. Many booms have 300 m of continuous length. Other booms look similar but have smaller sections, which are connected by an inflation tube along the top of the float chambers. A blower is connected to the tow end during the deployment. When the required air pressure is achieve, a valve closes, and the air moves to the next section. The deployment process takes about 20 minutes. Other models have several sections. Each chamber must be inflated separately. These booms are equipped with independent chambers with check valves. If one of the compartments is ruptured, the others will continue to operate. This type of boom requires considerable skill on the part of the worker. These booms usually have smaller flotation sections and are in lengths of 50 m on reels of 300 m. If a section is damaged, there is 250 m left when the other has been removed. Reduction in deployment times and damage can only be achieved by continuous training and reinforcement of rigorous procedures.
Boom Classification Term	Bay Boom	Term refers to a class of booms that are much smaller than the oceanic type. They are designed to function well in locations with waves up to 0.80 m high. They are often used in bays or sheltered areas. They are easily handled and require less training. The permanent booms are larger (about 1 m), heavier, and made from material that is resistant to solar radiation

		(PU-polyurethane). They are designed to withstand incrustations and have floats sized to support this added weight. They may work for up to 3 years permanently deployed but require maintenance mainly to remove marine growth. In many cases where a boom will be deployed on a permanent basis in tidal areas, compensators are used to allow the boom to rise and fall with the tide. Tidal ranges can be as much as 10 meters during the year and the location. The system is usually a stainless steel T section with rollers on the inside of the T giving a seal during the operation. These too have to be cleaned of marine growth regularly. Other systems have roller buoys around pier legs, which keeps marine growth down as they roll up and down with the tide.
Design Feature Term	Skirt	This is a feature of the boom that is the continuous portion below the float. It has the purpose of containing the oil that tends to pass under the boom. The distance measured from the water line to the lower limit of the skirt is called the draft. In shallow water the skirt should not take up more than 20% of the water depth as it causes an increase in current velocity, which will drag the boom under the water with the oil flowing over the top or, in other cases, especially near the shoreline, causing oil to flow under the boom.
Design Feature Term	Ballast	This is a feature of the boom that is the weight added to the bottom of the skirt to maintain the barrier in a position perpendicular to the surface of the water and, in some cases, to provide the tension member for the boom. The stresses of current velocity are transferred away from the boom material to the ballast chain or wire rope.
Functional Term	Collection	Term refers to the role of a boom system. An essential part of a boom is to provide a means to remove the oil from the area inside the boom. This may be done by the use of a skimmer. In some systems, the skimmers are

Table 4.12 (cont.) Table of important terminology.

Type Terminology	Term	Description
		incorporated into the boom itself. Aerial observation is essential in order to define the configuration of the boom arrangement and positioning in order to collect the thickest part of the oil slick at sea. It is essential that thorough consideration be given to how the oil is to be recovered. Adequate access by land or by sea for the deployment of men, boats, ships, and equipment is essential. It is important to have anchors on the booms in order to fix them in position, avoid areas with abrasive surfaces such as barnacles, and be able to deploy skimmers.
Functional Term	Deflection	Term refers to one possible function of booms. When booms are deployed for shoreline protection where sensitive environmental areas are concerned, deflection of the slick is the role and objective. It is necessary to have an adequate point for the collection of oil on the beach. Booms are used to deflect the oil towards that point. After the oil has been deflected, it may be retrieved by skimmers, vacuum pumps, or other methods of recovery. The current speeds on river bends are slower on the inside than the outside. Using this type of boom, oil can be moved to quieter water for collection and removal, provided that there is good access.
Functional Term	Protection	Booms are used to protect sensitive areas. It is important to note that deployment of booms for the purpose of providing protection of every possible area is not feasible due to both cost and environmental factors. As a part of the strategy in deploying booms, decisions will need to be made regarding which areas of protection are prioritized, thereby leaving some areas of shoreline unprotected or left to the mercy of natural attenuation and shore wave

		cleansing action over time. Some of the general sensitive areas that should be given priority include cooling water inlets for electrical energy generating plants and industries, marinas and ports, desalination plants, abstraction points for Mari culture/ aquaculture, bird sanctuaries and sites for the spawning of fish, and accesses to areas of mangrove swamps and estuaries.
Functional Term	Deployment	This term, used earlier in the discussions, refers to the laying out of booms on the sea and along shorelines that are intended to be protected. It is essential that the deployment be executed in a logical and well-organized manner, first to facilitate the work and second, and more importantly, to avoid damage to the boom. The need to deploy booms rapidly sometimes causes a boom to have twists in it, which will allow the oil to pass through these areas. A slower methodical deployment usually saves time in the long run. As a general good work practice, before the boom is deployed the next length required should be estimated and connected, whether on land or on board a vessel. Once the boom is ready, it may be deployed and anchored into position by a vessel. Its final position may be determined by adjusting the appropriate anchors using tripping buoys. Where no anchoring point is available, trial and error will determine the correct positions of the nchors. In order to avoid fouling the anchor cables due to wave action, they should be 3 to 5 times the local water depth. Where a boom is being used to collect oil or to protect a sensitive area, precaution should be taken to guarantee that the boom seals with the beach or bank so that oil is unable to pass by. This can be a difficult problem in tidal waters, especially where athe bottom is rocky. The use of shore sealing booms for beach protection is recommended. Upfront planning is critical in order for booms to provide

Table 4.12 (cont.) Table of important terminology.

Type Terminology	Term	Description
		optimum protection of shorelines. This should include a consideration of the best locations for the collection of the oil. Permanent points of anchorage for booms will accelerate installation.
Functional Term	U Containment Technique	Refers to a specific containment method in which three vessels participate in the operation, two of which secure the extremities of the boom. When the configuration is ready, a third vessel recovers the oil. Good communications are needed so as not to lead to loss of oil in the apex. In most cases, the recovery vessel becomes the command vessel as it is closest to the apex and can monitor what is happening with the oil. Recovery vessels should not enter the U because of possible restrictions to maneuverability and unnecessary contamination of the vessel.
Functional Term	V Containment Technique	Term refers to a specific containment method in which two vessels secure the extremities of the barrier, while at the apex a skimming vessel is attached. All the oil is directed towards this skimmer. Many photographs have been taken of the oil passing under the skimmer vessel due to the speed of the other vessels. In this case, there is a huge mismatch of power in the tow vessels, which can cause a seesaw situation leading to oil loss. Good communications are essential to the success of this technique.
Functional Term	J Containment Technique	Term refers to a specific containment method in which two vessels take part in the operation. The command vessel secures one of the extremities of the boom and also recovers and stores the oil. This configuration usually follows the U Containment Technique (see above), which is used to collect the oil. The J Method concentrates the oil to assist with its recovery. Many

		booms have been destroyed with this maneuver, usually during the early stages of the formation. The damage occurs when the less maneuverable, small tow end vessel is expected to pull the whole length of boom up in front of the main vessel. This usually takes a long time against the current, and the pressure against the boom breaks either the tow rope or the boom itself. This can be avoided simply with the much more maneuverable main vessel deploying the boom down the side of the oil with a buoy at the end of the tow rope. When the boom is completely deployed, the main vessel turns to form the J, and the tow end vessel picks up the buoy.
Functional Term	W Containment Technique	Term refers to a specific containment method in which three vessels are used. A central vessel, equipped with skimmers and storage systems, is positioned in the center. The booms deployed from both sides of this vessel with two smaller vessels secure the extremities. This operational configuration restricts maneuverability and is only used with highly trained crews. The U configuration (see above) on each side is towed alternatively to form Js to recover the oil. Communications between vessels are extremely important.
Functional Term	Single Ship System	Term refers to a self-contained single vessel employed in the operation. A steel or flexible boom system is deployed alongside the vessel, usually incorporating a weir skimmer system that is built in. The ship is much more maneuverable, but to operate it successfully requires aerial supervision. Large storage tanks are required as these systems tend to recover large quantities of water that will need to be separated and decanted. Decanting is not possible in some countries, so this leads to significant waste disposal problems.

Table 4.12 (cont.) Table of important terminology.

Type Terminology	Term	Description
Emergency Term	Boom Failures	This general term refers to the collapse of a boom section or its reduction in efficiency. Booms respond in accordance with the oceanographic and meteor graphic characteristics of the location. Booms will not function well if the speed of the current is greater than 0.7 knots. Above this velocity, failure in their capacity to contain oil in the boom starts to occur. With lighter oils these failures can occur much more readily. Boom failures can be reduced by reducing the angle of the boom in order to reduce the pressure on it. Various innovative types of equipment have been developed for use with excessive current speed.
Condition Term	Splash Over	Splash over refers to oil passing over the boom. This commonly occurs with short frequency waves or harbor chop. It may also occur when an inadequate or sufficient number of anchors are employed, causing the boom belly with the current and the apex extending 90° to the current and thereby causing the oil to flow over the boom.
Condition Term	Drainage	Drainage is a condition that occurs when the quantity of oil contained in the boom surpasses the maximum capacity for which it was designed. When a boom becomes full or saturated and the oil is not removed by skimmers, then the oil will find a way of emptying itself out of the containment area by going under the apex.

Condition Term	Undercut	This term refers to a condition that occurs when the relative velocity between the water and the boom is higher than 0.7 knot. It causes dispersion of the oil into droplets that pass under the boom. When operating this type of formation, if white or agitated water can be observed in the apex, the applicator should slow down the deployment of the boom until the frothing dissipates.
Condition Term	Submergence	Term refers to a condition when the length of the cable tied to the anchor is shorter than five times the depth. This also occurs when the angle of the boom is too obtuse and the weight of water on the boom skirt pulls it under, allowing all the collected oil to float over the top of the boom.
Condition Term	Surfing	Term refers to a condition that occurs when there are strong winds and strong currents running in opposite directions, or when there is insufficient weight of ballast chain weight. In some cases, wires are fitted to the top of booms, and the weight of the ballast chain is not increased to compensate due to manufacturing lack of experience on the part of the applicator.

5

Emerging Technologies

5.1 Clean World Innovations and EncapSol

Two technologies we have identified that are promising for combating oil spills are an enzyme-based technology developed by Clean World Innovations (CWI) and an encapsulation technology marketed under the name EncapSol. Both technologies are in a state of commercial readiness and have been demonstrated at limited commercial scales.

Both technologies were originally developed to target environmental clean-up applications. However, they have evolved to address different target markets that are broader and more sustainable. The CWI technology focuses on water quality and wastewater purification. The EncapSol technology is also capable of tackling this market but is poised to also address oil recovery from tar sands and other media. Both technologies seem applicable to responding to marine oil spills.

Both technologies compete with surfactant-based technologies, which have received significant attention within the past few years. Surfactant-based technologies have the potential to recover oil from tar sands and marginal producing oil wells. However, they have most aggressively been applied to address ground water contamination, which is a most intractable form of pollution. Spilled oil and the biodegradable daughter products are trapped below the water table by colloidal forces. Surfactants have proven successful in dramatically improving contaminated aquifer remediation rates. This same technology has been proposed and is being tested for oil recovery from tar sands and similar applications to those addressed by the two technologies reviewed herein. Principal remediation and oil recovery mechanisms include micellar solubilization and mobilization of the trapped liquids by lowering the oil/water interfacial tension. Surfactant selection is a key to the successful design of a remediation or oil recovery effort and involves consideration of factors including Krafft point, surfactant adsorption onto the aquifer solids, and the phase behavior of the oil/water/surfactant system. Successful field demonstrations have occurred but are limited with many competitive surfactant technologies that claim to be commercialized but are not yet fully demonstrated. Critical research issues remain including acceptable clean-up levels, surfactant/contaminant in situ biodegradation rates, surfactant decontamination and reuse, and efficiency in hydrocarbon recovery. These uncertainties offer potential competitive opportunities for the two alternative technologies considered. Further, surfactant-based technology introduces additional environmental concerns because residual chemicals remain in the subsurface stratus as mild detergents, and few fate and transport studies have addressed potential long-term negative impacts.

While surfactant technologies are promoted by various companies, there is little evidence to support that these technologies are unique or command any monopoly on oil recovery applications or environmental remediation projects. Given the uncertainty in the efficiency of oil separation recovery using this approach, and given the concerns for potential negative environmental impacts, the two technologies reviewed potentially offer significant alternative approaches.

5.1.1 Clean World Innovations Technology

Clean World Innovations (CWI) was formed in 2009 to offer environmentally friendly solutions for the treatment of petroleum and industrial waste streams. Its technologies have been used in hundreds of applications in different parts of the world, including the United States, Canada, Venezuela, Trinidad, and Ecuador. The company focuses on designing new formulations of environmentally friendly products that aim at improving oil production and handling as well as cleaning equipment and structures involved in activities related to the energy sector. These products include pure chemical products and enzyme-based products, all claimed to be biodegradable and non-toxic.

The company's Web site is http://cwisolutions.com/. CWI markets enzymatic compounds that are promoted as being effective in low concentrations under conditions that cause high viscosity. Their surfactant action helps improve viscosity by breaking the emulsions, while their solvent properties aid in breaking down the complex and heavy molecules. Without applying heat the enzyme will break the viscosity of heavy crude oil by creating a short temporary emulsion that degrades the paraffin and asphaltene. During this process, the paraffin and asphaltene will join the oil phase permanently, not allowing water and/or solids to nucleate again. The resulting crude is dry and homogeneous.

An enzyme is a catalyst based on proteins that facilitates the completion of bio-chemical reactions by decreasing a reaction's activation energy requirements. Enzymes are large protein molecules that are comprised of long chains of amino acids.

Enzymes act by binding to the substrate, catalyzing the reaction, and then releasing themselves from the products so that they can react again. Enzymes are more specific and work under milder conditions than chemicals used to control chemical reactions. CWI enzymes are combined by a specific process with non-aromatic, non-aliphatic solvents, and other proprietary compounds to form water-based, non-ionic, non-abrasive detergent. Through further processing, this detergent can be modified to yield compounds that have a wide range of applications in several areas of the oil producing activities. The end result is an extended variety of products that can be used to control different problems associated with oil production, processing, transportation, and refining. These compounds are non-toxic, non-pathogenic, non-flammable, and biodegradable, and they do not contain any petroleum hydrocarbons.

The CWI products are extracted from natural sources, and they degrade once they have performed their function. A concern is that CWI has not published any data on biodegradation rates, which therefore places it in the same grey area of claimed benefits associated with surfactant-based technologies. The CWI products can be used to clean, control, and reduce paraffin and asphaltene deposition, emulsions in production lines, flow lines, and in process, and storage tanks. Combinations of enzymes can be used for downhole stimulation and production control and have been successfully demonstrated by the developer.

The following applications are listed on the company's Web site:

- Pit oil treatment – In drilling and oil production activities, pits in Venezuela are used to receive wastes from non-usable products. Most of these products come from producing wells or used oil, whereas others come from drilling and cleaning activities. Oil companies in Venezuela are faced with the task of dealing with waste that has been deposited in pits for decades. CWI has designed enzymatic products that are used to aid in the separation of oil from the pit water and that work simultaneously on the solids that may be suspended in the liquid mixture. After mixing CWI's enzyme with the pit liquids, different layers of substances are observed in the pit or treating tanks. Oil can be recovered with a minimum degree of contamination, water can be separated while the enzyme will stay in the water phase, and the solids will precipitate. The solids may be further treated or disposed of in a designated landfill. CWI's enzymatic compound can also be formulated to work on oil spills on water since it will disperse the oil and can drive it to an absorbing agent. The advantage of using these products is that the enzyme drives off even the sheen on the water surface. When the spill happens on the ground, the soil can be treated with enzymes in a reactor, and the soil can be put back into its original location unharmed.
- Environmental clean-up projects such as oil spill remediation – The company reports effective separation of oil from surface water contamination in soil washing applications.

Relatively low concentrations of the proprietary enzymes can be used (typically, application concentrations are 800 ppm). The separation described results in an emulsion phase that is less than 3%

water, which makes the product highly efficient and very competitive to superior surfactants. The emulsion phase breaks up over a relatively short period of time, leaving a lighter oil phase that can be skimmed – a characteristic that is attractive to in situ groundwater remediation applications.

The provider lists the following companies as clients in various small applications:

- Exxon
- Texaco
- Mobil
- Phillips
- Conoco
- Unocal
- Venoco, Inc.
- BP
- Repsil YPF
- Arco
- Shell
- PDVSA
- Halliburtan

The application that the technology is currently targeting is the clean-up of pits that contain waste oil. According to the developer, oil companies in Venezuela excavate areas where waste generated by oil extraction, transportation, refining, and distribution is poured temporarily. They note that for a long time these fluids have been deposited in pits, lagoons, and holding ponds due to the lack of strict environmental regulations and accurate technologies that allow for proper treatment. As environmental regulations become more rigorous, oil companies are forced to properly dispose of their waste and eliminate the waste accumulated until now. The recovery of this environmental liability is a complex task that must be performed in three stages:

1. the elimination of hydrocarbons,
2. the treatment or decontamination of water, and
3. the decontamination of the solids.

Currently, the corporation has a plan to remove this environmental liability, mainly pits and holding ponds, which initiated several years ago (2001 in eastern Venezuela). It started with the simpler problems, postponing the complex tasks for later. This plan has allowed for the clean-up of "dry pits" (pits with no liquid) and other critical pits with liquid, which have been transported to the so called "macro pits." This has resulted in the deferment of the problem and has added to the already saturated large pits. Among the major environmental liabilities located in the western area of the country are some pits and canals bordering Lake Maracaibo, which contribute to the pollution of this important water reserve. A very large volume of oil is found in these canals whose properties have changed due to the long exposure to the environment. Mainly, the lighter fractions have evaporated leaving a type of extra heavy crude with very low or no commercial value. Therefore, these canals and pits are not economically attractive to the corporation, but they represent a severe environmental hazard that must be eliminated due to the environmental and social threats they pose to the surrounding communities.

The developer also notes the following areas where the technology could be applied:

- The technology could be used in addressing oil spills, especially during the rainy season. The San Andres canals in Bachaquero are rain water collectors that are released into Lake Maracaibo. Oil-water separation is partially achieved at the pumping station, and oil is pumped along with water into the lake, contributing to its contamination.
- Heavy metals such as rhodium, vanadium, and other contaminants such as barium seep into the ground, contaminating underground water reservoirs, especially around the lake where the water table is high. These metals and compounds are not removed by standard water treatment techniques, thus there is a latent risk of supplying contaminated water to the communities.

5.1.2 EncapSol Technology

The EncapSol technology was separately licensed by Freestone Resources, a public company that purchased exclusive rights to build, own, and operate EncapSol oil-extraction systems.

The product is a non-flammable oil recovery solvent that chemically breaks the molecular bonds that bind hydrocarbon molecules to water, sand, clay, rock, and other materials. As the oil molecules are freed, the EncapSol solvent encapsulates the oil molecules, preventing them from binding to anything, and they remain in the low viscosity EncapSol solution for subsequent extraction and recovery using a proprietary EncapSol Oil Recovery (EOR) machine.

The EOR machines can process various hydrocarbon-containing feedstock, including oil sands, oil shale, and waste oil sludge, producing an end-product that includes saleable crude oil and clean, oil-free sand, clay, rocks, gravel, and agricultural brown-water. The newest generation EORs recover >99% of the EncapSol solvent for reuse.

This technology works on a nano-scale molecular level. The EncapSol molecule's affinity for and attraction to hydrocarbon molecules cause multiple EncapSol molecules to surround the hydrocarbon molecule in enough numbers with sufficient attractive force so as to effectively overpower the molecular bond that binds the hydrocarbon water molecule, thus releasing and freeing the hydrocarbon molecule for controlled collection and recovery. EncapSol molecules surround and encapsulate these released hydrocarbon molecules, holding the liquid hydrocarbon molecules in suspension within the low-viscosity EncapSol solution, which is later processed to separate out the oil and recover the EncapSol solution for reuse in the process.

With the EncapSol molecules' combined attraction forces overpowering the bonds holding the hydrocarbons, the hydrocarbon molecules are released and then encapsulated by the gaseous EncapSol molecules that precipitate back into a liquid phase on contact. This action holds the liquid hydrocarbon molecules in suspension within the liquid phase EncapSol droplets, whose lower viscosity enables the EncapSol + Oil solution to flow more easily through the formation to the lower-pressure production exit well.

No information was found supporting that either product has been extensively tested in seawater applications for oil recovery. However, the nature of the technologies suggests that they would be good candidates.

5.2 Centrifuges

The actor Kevin Costner formed a company with his brother called Ocean Therapy Solutions, and they introduced an incredibly simple concept. Their company developed a compact centrifuge capable of separating oil from seawater. Kevin Costner spent approximately $20 million over three years and hired a team of about 15 engineers and scientists to develop centrifuges that would separate water from oil at up to 200 gallons per minute. In theory, twenty of the centrifuge devices could have cleaned up the Exxon Valdez spill in 5 days. BP has a contract for 32 of the centrifuges from Ocean Therapy Solutions. BP tested the devices for a week.

One of the company's V20 machines can clean up to 210,000 gallons of oily water per day. There are 3 V20 centrifuge machines currently operational in the Gulf. Ten more are planned to become operational.

5.3 Skimmers and Response Vessels

Perhaps the greatest innovation in response equipment is not so much innovation at all but mere size. The "A Whale" ship, owned by the Taiwanese company TMT Group, is the length of 3½ football fields and 10 stories high and has been transformed into the world's largest oil skimmer. The vessel is touted as being capable of working close to a spill source where the oil is more concentrated. Smaller skimmers work from nearer the shore to about three to five miles out where oil can be dispersed, and, hence, they are less effective.

The ship is actually a cargo vessel that was retrofitted at TMT's expense to help with the Gulf oil spill. After water is ingested through the ship's vents, it is pumped into a series of containers where oil rises to the top of the water over several days. The oil is then siphoned into a separate container before being transferred to a tanker. It has never been tested under real-world conditions. In Portugal, the ship tested its intake process on water sprayed with fire foam, since using oil was prohibited. The company reported that it proved effective. TNT has two more ships, "B Whale" and "C Whale," which are in the process of being transformed into skimming vessels.

The existing flotilla of smaller skimming vessels being used in the Gulf are mostly converted fishing boats that capture oily water from the surface of calm seas and bring the contaminated mixture back to shore for disposal. The "A Whale" is equipped to separate oil from water and then discharge cleansed water back into the sea. The vessel's design capacity is about 500,000 barrels of oily water a day. But this figure is disheartening when we consider that U.S. EPA guidelines under the Clean Water Act require that the reclaimed water contain no more than 15 parts per million of oil. It remains to be seen whether the A Whale can meet this standard. In addition, we must bear in mind the following facts:

- The oily water mixture captured by skimmers only contains about 10% oil, i.e., the mixture is about 90% seawater. So, in reality the "A Whale" can, at full design capacity, collect and reclaim only about 50,000 barrels of oil per day. That's equivalent to about 2.1 million gallons of oil per day. Some recent estimates place the release of oil from the Macondo well as high as 60,000 barrels per day, which exceeds the vessel's rated capacity.
- Vast amounts (for which there are no reliable estimates at present) released into the Gulf have not surfaced because of BP's decision to apply historically large quantities of dispersants. Dispersants have caused the oil to break up and sink into the water column, making it impossible to recover.

BP would also like to make claims to being innovative, presumably under the auspices of better late than never. The Ella G is a part of BP's response. It is a supply vessel that has been reconfigured with skimmer and centrifuge technology into an oil spill response platform in response to the Gulf crisis. BP's web site[1] touts that this vessel combines "innovation with cutting-edge technology… the first step toward a new generation of oil spill response." It's quite unfortunate that BP did not consider such technology as a part of its Oil Spill Response Plan (OSRP). Its OSRP claimed that it was not possible to spill more than 160,000 gallons of crude from its deep ocean water well.

The Ella G's four centrifuge devices are described by BP as having the potential to clean up to 800,000 gallons of oily water a day, and her deck capacity is 273,000 gallons. The Ella G is 280 feet in length with a beam of 64 feet and can operate in deep water and rough seas. The vessel employs a V-20 CINC centrifuge and a TransRec 150 skimmer. A J-formation of boom deployment funnels oil to the skimmer

[1]http://www.bp.com/genericarticle.do?categoryId=9033656&contentId=7063564

from an area of 125 meters wide on the surface of the water. The next step planned is to implement the U-formation of boom deployment, which funnels oil to the skimmer from an area of 700 meters across (almost half a mile) the surface of the water. Deployment of the Ella G brings the number of centrifuge devices deployed in the Gulf to 9. BP committed to 32 OTS centrifuges in total.

6

Spill Response and Worker Protection

6.1 Countermeasure Options

The specific tools, protocols, technologies, and levels of training required for emergency responders to oil spills depend entirely on the countermeasures that are applied to controlling and combating the spill. A number of the countermeasures have been discussed already. Some of these will first be reviewed to emphasize certain aspects, while others not previously considered are now covered.

There is an arsenal of tools and technologies for containing and cleaning up oil spills in aquatic environments. Mechanical equipment, such as booms and skimmers, is used to block the spread of oil, concentrate it into one area, and remove it from the water surface. Chemical and biological treatment of oil can be used in place of or in combination with mechanical methods, especially in areas where untreated oil may reach shorelines and sensitive habitats where a clean-up becomes difficult and even more environmentally damaging.

6.1.1 Dispersants

Dispersants have already been described at length in Chapter 3. The reader should recall that dispersants are chemicals that contain surfactants, which are compounds that act to break liquid substances such as oil into small droplets. These droplets disperse into the water column where they are subjected to natural processes such as wind, waves, and currents that help to break them down further. This helps to clear oil from the water surface, making it less likely that the oil slick will reach the shoreline. The disadvantage of this is that these underwater regions can become dead zones where larger marine life can become stressed. Figure 6.1 shows dispersants being applied in the Gulf spill.

To a large extent, the effectiveness of a chemical dispersant depends on the composition of the oil it is being used to treat and the method and rate at which the dispersant is applied. As noted in Chapter 3, heavy crude oils do not disperse as well as light- to medium-weight oils. Furthermore, dispersants are most effective when applied immediately following a spill, before the lightest components in the oil have evaporated from the surface,

Figure 6.1 A C-130 airplane is distributing dispersants over the oil in the Gulf (Source: defenseimagery.mil).

and hence there is a window of opportunity for optimum dispersant performance. Additionally, environmental factors, including water salinity and temperature, and conditions at sea influence the effectiveness of dispersants. Many, if not most, dispersants work best at salinity levels close to that of normal seawater. While dispersants can work in cold water, they work best in warm water.

Dispersants are essential to combating oil spills because frequently rough or choppy conditions at sea make mechanical containment and clean-up too difficult. Despite this, there are difficulties with application and concerns about their effectiveness and the toxicity of the dispersed mixtures to aquatic, terrestrial, and human life. Almost no tests on long-term environmental effects have been conducted after a dispersant application. Hence, the fate and transport of these chemicals and their long-term impacts to the environment are simply ill defined. Essentially all the data that is available on the toxicity of dispersing agents are laboratory-scale and have been conducted on a few marine species. Most suppliers of these chemicals do not disclose the ingredients used in dispersing agents and generally provide the results of minimal laboratory testing on trace priority chemical toxins. The risk to spill responders from dispersants is coming into direct contact with the chemicals and inhaling and ingesting aerosols and dusts during the application of the chemicals.

6.1.2 Biological Agents

Biological agents are nutrients, enzymes, or microorganisms that increase the rate at which natural biodegradation occurs. Biodegradation is a process by which microorganisms such as bacteria, fungi, and yeasts break down complex compounds into simpler products to obtain energy and nutrients. The biodegradation of oil is a natural process that slowly breaks down oil. This process (biodegradation) can take months to even years to occur under natural conditions.

Bioremediation technologies accelerate the biodegradation processes. Bioremediation involves adding fertilizers or microorganisms that increase the rate at which natural biodegradation occurs. Bioremediation is most often intended to be used after mechanical oil recovery methods have been exhausted. Two bioremediation approaches have been for oil spill clean-ups: biostimulation and bioaugmentation.

Biostimulation is the method of adding nutrients such as phosphorus and nitrogen to a contaminated environment in order to stimulate the growth of the microorganisms that eat oil. Limited supplies of these necessary nutrients usually control the growth of native microorganism populations. However, when nutrients are added, the native microorganism population can grow rapidly, thereby increasing the rate of biodegradation.

Bioaugmentation is the addition of microorganisms to the existing native oil-degrading population of microorganisms. Sometimes species of bacteria that do not naturally exist in an area will be added to the native population. Seeding is a technique that is used to increase the population of microorganisms that can biodegrade the spilled oil.

These technologies have been successfully applied to various soil and groundwater contamination problems and are generally considered proven technologies in those applications. Even in these applications, the application of biostimulation and bioaugmentation represents final stages of remediation because excessive or gross oil coverage must first be removed.

Concern for worker exposure to the biological units used in bioremediation technologies is primarily the risk of inhalation and ingestion. Biological agents can be applied as mists or dusts or can generate aerosols during application. Hence, there is concern for respiratory related illnesses among workers that handle these agents for extended periods of time.

6.1.3 Shoreline Cleaners

A variety of chemical cleaners are used to herd and solidify spilled oil and to cleanse hard surfaces along shorelines that have become contaminated. These chemicals are generally more toxic than dispersants and biological agents. Their methods of applications are manpower intensive and require workers to come into direct contact with liquids, sludges, mists, vapors, aerosols, and dusts. Workers face exposure through inhalation, ingestion, and skin adsorption. Therefore, appropriate protocols and personal protection are required in application.

6.1.4 Controlled or In-Situ Burning

This alternative countermeasure carries a trade-off in environmental impacts. The trade-off is between continued environmental impacts on marine life and shorelines from the slick and air pollution.

Figure 6.2 Vessels monitoring an in-situ burn of a collected oil slick retained by fireproof boom. (Source: defenseimagery.mil)

In-situ burning of oil requires ignition and combustion of the oil slick. It has been used in past spills when oil is spilled on a water body or on land. The National Oil and Hazardous Substances Contingency Plan authorizes in-situ burning as a clean-up method but requires approval from the regional response team (RRT) before it can be used. The RRT can provide authorization through pre-authorization plans and agreements among various federal and state agencies. Fire-resistant booms are used to collect and concentrate the oil into a slick that is thick enough to burn. See figure 6.2.

In the ongoing Gulf disaster, the oil escaping from British Petroleum's ruptured oil well contains approximately 40% methane according to news media sources[1]. This is extraordinary because typical oil deposits contain much less than 10%, and most often close to 5%, methane or natural gas. Massive amounts of methane gas have now dispersed about the Gulf, potentially suffocating marine life and creating "dead zones" where oxygen is so depleted that nothing can survive. Methane is a colorless, odorless, and flammable gas that

[1]http://www.google.com/hostednews/ap/article/ALeqM5gIXWYBTpLtSayJtg41LKXpxSxVPAD9GDO2VG1

is the principle component in natural gas. Petroleum engineers typically burn off excess gas attached to crude before the oil is shipped off to a refinery. Hence, BP and the Gulf RRT have chosen this countermeasure as one of the approaches to combating the spill.

According to the same news source, a BP spokesman said the company was burning about 30 million cubic feet of natural gas daily from the source of the leak, which adds up to about 450 million cubic feet since the containment effort started. To place this in some perspective, the amount of gas that has been burned through July 2010 is equivalent to the amount that is needed to heat roughly 450,000 homes for four days.

Microbes that live in the sea have been feeding on the oil and natural gas in the water and are consuming larger quantities of oxygen, which they need to digest food. As they draw more oxygen from the water, it creates two problems. When oxygen levels drop too low, the biodegradation of oil stops. Furthermore, as oxygen is depleted in the water column, much of marine life can't be sustained. In early June of 2010, a research team from the Institute of Undersea Research and Technology at the University of Georgia investigated a 15-mile-long plume drifting southwest from the leak site, and reported methane concentrations of up to 10,000 times higher than normal and oxygen levels that had depleted by more than 40%. Scientists have now reported that some parts of the plume had oxygen concentrations just shy of the level that tips ocean waters into the category of a "dead zone," which is a region uninhabitable to fish, crabs, shrimp, and other forms of marine life.

The potential impact to marine life is so devastating and possibly long-term that the commercial fishing industry could very well disappear. Given these concerns, the countermeasure of in-situ oil slick burning is a viable option, but one that carries other consequences.

In industrial applications, the air emissions from fuel oil combustion are well understood and controlled with both specially designed combustion equipment and air pollution control devices. The emissions generated from combustion depend on the grade and composition of the fuel, the type and size of the combustion device (e.g., a boiler), the firing and loading practices used in the combustion equipment, and even the level of equipment maintenance. Because the combustion characteristics of distillate and residual oils are different, their combustion can produce significantly different emissions. In general, the baseline emissions of criteria and non-criteria pollutants are those from uncontrolled combustion sources.

Uncontrolled sources are those without add-on air pollution control (APC) equipment or other combustion modifications designed for emission control. Baseline emissions for sulfur dioxide (SO_2) and particulate matter (PM) can also be obtained from measurements taken upstream of APC equipment.

Particulate emissions include filterable and condensable material. Filterable emissions are generally the particulates that are trapped by the glass fiber filter in the front half of a Reference Method 5 or Method 17 sampling van. Vapors and particles less than 0.3 microns pass through the filter. Condensable particulate matter is material that is emitted in the vapor state but later condenses to form homogeneous and/or heterogeneous aerosol particles.

Filterable particulate matter emissions depend predominantly on the grade of fuel fired to a combustion system. The combustion of lighter distillate oils results in significantly lower PM formation than does the combustion of heavier residual oils. Among residual oils, the firing of No. 4 or No. 5 oil usually produces less PM than does the firing of heavier No. 6 oil.[2]

According to the EPA, filterable PM emissions depend on the completeness of combustion as well as on the oil ash content. The PM emitted by distillate oil-fired boilers primarily comprises carbonaceous particles that result from the incomplete combustion of oil, and it is not correlated to the ash or sulfur content of the oil. However, PM emissions from residual oil burning are related to the oil sulfur content. This is because low-sulfur No. 6 oil, either from naturally low-sulfur crude oil or desulfurized by one of several refining processes, has substantially lower viscosity and reduced asphaltene, ash, and sulfur contents, which results in better atomization and more complete combustion. In modern combustion equipment, the fuel is atomized into tiny droplets when entering the combustion chamber. This creates almost infinite surface area to promote high efficiency burning of the fuel. But when a layer of oil is burned, the oil exists as a slick and there is no atomization to enhance combustion efficiency. The consequence is while in-situ burning is described as a controlled burn it, in fact, is not controlled at all from a combustion efficiency standpoint. See figure 6.3.

---

[2]Chapter 1, *Environmental Protection Agency AP-42 Emissions Factors*, Washington, DC.

Figure 6.3 Photo showing multiple locations of in-situ burning in the Gulf (Source: defenseimagery.mil).

The load to a combustion system like a boiler also affects filterable particulate emissions in units firing No. 6 oil. At low-load (50 percent of maximum rating) conditions, particulate emissions from utility boilers may be lowered by as much as 60 percent using combustion controls and air pollution technologies. However, no significant particulate emission reductions have been noted at low loads from boilers firing any of the lighter grades. At very low load conditions (approximately 30 percent of maximum rating), proper combustion conditions may be difficult to maintain, and particulate emissions may increase significantly.

Sulfur oxides (SOx) emissions are also generated during oil combustion from the oxidation of sulfur contained in the fuel. The emissions of SOx from conventional combustion systems are predominantly in the form of sulfur dioxide (SO_2). Uncontrolled SOx emissions depend on the sulfur content of the fuel and are not affected by boiler size, burner design, or the grade of fuel being fired. Typically, more than 95% of the fuel sulfur is converted to SO_2, about 1–5% is further oxidized to sulfur trioxide (SO_3), and 1–3% is emitted as sulfate particulate. SO_3 reacts with water vapor to form a sulfuric acid mist, which is most often associated with the phenomenon of acid rain.

Oxides of nitrogen (NOx) formed in combustion processes are due either to the thermal fixation of atmospheric nitrogen in the combustion air ("thermal NOx") or to the conversion of chemically bound nitrogen in the fuel ("fuel NOx"). NOx refers to the composite of nitric oxide (NO) and nitrogen dioxide (NO2). More than 95% of the emitted NOx is in the form of nitric oxide (NO). Nitrous oxide (N2O) is not included in NOx, but it has recently received increased interest because of atmospheric effects.

NOx concentration is exponentially dependent on temperature, and it is proportional to the nitrogen concentration in the flame, the square root of O_2 concentration in the flame, and the residence time. Thus, the formation of thermal NOx is affected by four factors: (1) peak temperature, (2) fuel nitrogen concentration, (3) oxygen concentration, and (4) the time of exposure at peak temperature. The emission trends due to changes in these factors are generally consistent for all types of boilers – an increase in flame temperature, oxygen availability, and/or residence time at high temperatures lead to an increase in NOx production. Modern combustion systems have certain degrees of control over each of these variables and, hence, can minimize the formation of NOx emissions. In contrast, there are no degrees of control affordable when burning an oil slick.

Fuel nitrogen conversion is the more important NOx-forming mechanism in residual oil boilers. It can account for 50% of the total NOx emissions from residual oil firing. However, the percent conversion of fuel nitrogen to NOx varies greatly. Typically, 20–90% of nitrogen in oil is converted to NOx. Fuel NOx generally accounts for over 50% of the total NOx generated.

A number of variables influence how much NOx is formed. One important variable is firing configuration. NOx emissions from tangentially (corner) fired boilers are, on the average, less than those of horizontally opposed units. The firing practices employed during boiler operations are also important. Low excess air (LEA) firing, flue gas recirculation (FGR), staged combustion (SC), reduced air preheat (RAP), low NOx burners (LNBs), burning oil/water emulsions (OWE), or some combination thereof can achieve NOx reductions of 5-60% percent. Load reduction (LR) can likewise decrease NOx production. Nitrogen oxide emissions may be reduced 0.5–1% for each percentage reduction in load from full load operation. However, none of these control options exist when oil is burned from a slick.

In addition to SOx, NOx, and PM, carbon monoxide (CO) is generated as an air pollutant from oil burning. The rate of carbon monoxide (CO) emissions from combustion sources depends on the oxidation efficiency of the fuel. By controlling the combustion process carefully, CO emissions can be minimized. Thus, if a combustion devise is operated improperly or is not well maintained, the resulting concentrations of CO (as well as organic compounds) may increase by several orders of magnitude according to the U.S. EPA. Interestingly, smaller boilers, heaters, and furnaces tend to emit more of these pollutants than larger combustors. This is because smaller units usually have a higher ratio of heat transfer surface area to flame volume than larger combustors have. This leads to reduced flame temperature and combustion intensity and, therefore, to lower combustion efficiency. These poor conditions are similar to the conditions created when an open poll of oil is burned.

The presence of CO in the exhaust gases of combustion systems results principally from incomplete fuel combustion. Several conditions can lead to incomplete combustion, including insufficient oxygen (O_2) availability, poor fuel/air mixing, cold-wall flame quenching, reduced combustion temperature, decreased combustion gas residence time, and load reduction (i.e., reduced combustion intensity). The EPA has noted that since various combustion modifications for NOx reduction can produce one or more of the above conditions, the possibility of increased CO emissions is a concern for environmental, energy efficiency, and operational reasons.

Small amounts of organic compounds are also emitted from combustion. As with CO emissions, the rate at which organic compounds are emitted depends on the combustion efficiency of the boiler. Therefore, any combustion modification that reduces the combustion efficiency will most likely increase the concentrations of organic compounds in the flue gases. Total organic compounds (TOCs) include VOCs, semi-volatile organic compounds, and condensable organic compounds. Emissions of VOCs are primarily characterized by the criteria pollutant class of unburned vapor phase hydrocarbons. Unburned hydrocarbon emissions can include essentially all vapor phase organic compounds emitted from a combustion source. These include emissions of aliphatic, oxygenated, and low molecular weight aromatic compounds that exist in the vapor phase. These emissions include all alkanes, alkenes, aldehydes, carboxylic acids, and substituted benzenes (e.g., benzene, toluene, xylene, and ethyl benzene).

The remaining organic emissions are composed of compounds that are emitted from combustion sources in a condensed phase. These compounds are polycyclic organic matter (POM) and a subset of compounds called polynuclear aromatic hydrocarbons (PAH or PNA). There are also PAH-nitrogen analogs. Information available on POM compounds generally pertains to these PAH groups. Table 6.1 tabulates PAH concentrations from smoke emission data for 40 and 60 cm diameter Prudhoe crude pool fires reported by NIST.[3] These concentrations are high, and all of the toxins carry significant health risks from inhalation exposure. Many of the PAHs measured are carcinogens or suspected carcinogens. They are the same PAHs found in coal tars. The EPA, NIOSH, and OSHA all agree that collectively the chemical constituents found in coal tars are human carcinogens.

Formaldehyde is also formed and emitted during the combustion of hydrocarbon oil. Formaldehyde is present in the vapor phase of the flue gas. Formaldehyde is subject to oxidation and decomposition at the high temperatures encountered during combustion.

Trace elements are also emitted from the combustion of oil. Trace metals included in the list of 189 hazardous air pollutants under Title III of the 1990 Clean Air Act Amendments are released when oil is burned. The quantity of trace elements entering a combustion device depends solely on the fuel composition. The quantity of trace metals emitted from the source depends on combustion temperature, the fuel feed mechanism, and the composition of the fuel, all of which can be controlled to a certain degree in engineered combustion systems like boilers. The temperature determines the degree of volatilization of specific compounds contained in the fuel. The fuel feed mechanism affects the separation of emissions into bottom ash and fly ash. In general, the quantity of any given metal emitted depends on the physical and chemical properties of the element itself, the concentration of the metal in the fuel, the combustion conditions, and the type of particulate control device used. Its collection efficiency is a function of particle size. But again, when oil slicks are burned at sea, they are done so in an uncontrolled environment. In-situ burning is mass burning; hence, the levels of pollution generated are high due to low combustion efficiency.

---

[3]Web site-www.fire.nist.gov/aloft/

Table 6.1 PAH Concentrations measured in smoke emissions from burning oil pools (ug/g).

PAH	H2#1(c)	H2#2	H4#1	H4#2	C5#1(c)	C5#2	C7#1	C7#2	Mean
acenaphthylene			3	2	2		35	59	20
acenaphthene			ND	ND			ND	1	1
fluorene			4	3			130	140	69
phenanthrene	31	21	95	58	910	1040		1220	482
anthracene	ND	ND			260	290	1220		590
fluoranthene	47	37	210	130	700	730	740	720	414
acephenanthrylene	22	22							22
pyrene	65	49	260	150	900	840	800	780	481
benzo[ghi]fluoranthene	69	65			240	220			149
cyclopenta[cd]pyrene	24	110			540	540			304
benz[a]anthracene	63	87	88	180	220	220	240	220	165
chrysene/triphenylene	71	87	95	190	220	230	220	220	167
benzofluoranthenes	190	350	490	520	380	390	420	410	394
benzo[e]pyrene	66	110	92	220	120	120	180	170	135
benzo[a]pyrene	110	200	73	190	240	180	210	200	175
perylene	19	38	18	52	42	41	45	47	38
indeno[1,2,3-cd]pyrene	110	220	250	750	190	190	540	530	348
benzo[ghi]perylene	130	240	87	240	210	210	180	180	185
Total PAH (g)	1020	1636	1760	2680	5172	5240	4960	4897	3,421

In addition to the toxins generated as products of incomplete combustion, we believe the possibility exists that dioxins are created among the mix of air emissions. Dioxins are a family of 75 different compounds commonly referred to as polychlorinated dioxins. These compounds have varying harmful effects but often are referred to as super toxins in the general literature. The family is divided into eight groups of chemicals based on the number of chlorine atoms in the compound. The group with one chlorine atom is called the mono-chlorinated dioxin(s). The groups with two to eight chlorine atoms are called di-chlorinated dioxin (DCDD), tri-chlorinated dioxin (TrCDD), tetra-chlorinated dioxin (TCDD), penta-chlorinated dioxin (PeCDD), hexa-chlorinated dioxin (HxCDD), hepta-chlorinated dioxin (HpCDD), and octa-chlorinated dioxin (OCDD). The chlorine atoms can be attached to the dioxin molecule at any one of eight positions. The name of each CDD indicates both the number and the positions of the chlorine atoms. For example, the CDD with four chlorine atoms at positions 2, 3, 7, and 8 on the dioxin molecule is called 2,3,7,8-tetrachlorodibenzo-p-dioxin or 2,3,7,8-TCDD. 2,3,7,8-TCDD is one of the most toxic of the CDDs to mammals and has received the most attention. Thus, 2,3,7,8-TCDD serves as a prototype for the CDDs. CDDs with toxic properties similar to 2,3,7,8-TCDD are called "dioxin-like" compounds.

In the pure form, CDDs are colorless solids or crystals. CDDs enter the environment as mixtures that contain a variety of individual components and impurities. In the environment they tend to be associated with ash, soil, or any surface with a high organic content such as plant leaves. In air and water, a portion of the CDDs may be found in the vapor or dissolved state, depending on the amount of particulate matter, temperature, and other environmental factors. 2,3,7,8-TCDD is odorless. The odors of the other CDDs are not known.

While we have not identified any studies from literature that reports on the formation of dioxins from burning oil at sea, we note that all the elements are present for dioxin formation. Chlorine is an essential component for the formation of CDDs/CDFs in combustion systems, and there is no shortage of chlorine in seawater. By looking at the live feed of BP's gushing oil well, one can recognize that the release of oil from the top of the blow out preventer violently mixes with the surrounding water. By the time the oil and gas rise to the surface it has mixed and entrained seawater. While oil separates out onto the surface and floats on the water, this separation does not take place under quiescent conditions. Actions of waves continue to cause mixing between oil and water, thus creating more of an emulsion rather than a clean separation of water and oil.

CDD/CDF emissions are understood to be controlled by three mechanisms that are not mutually exclusive. In the first mechanism (referred to as "pass through"), CDDs/CDFs are present as contaminants in the combusted organic material. They pass through a furnace or a boiler and are emitted unaltered. In the second mechanism (referred to as "precursor"), CDDs/CDFs ultimately form from the thermal breakdown and molecular rearrangement of precursor ring compounds. Precursor ring compounds are chlorinated aromatic hydrocarbons that have a structural resemblance to the CDD/CDF molecules. Ringed precursors that emanate from the combustion zone are a result of the incomplete oxidation of the constituents of the feed (i.e., products of incomplete combustion). The third mechanism (referred to as "*de novo* synthesis") is similar to the precursor mechanism. *De novo* synthesis describes a pathway of CDD/CDF formation from heterogeneous reactions on fly ash (particulate matter [PM]) involving carbon, oxygen, hydrogen, chlorine, and usually a transition metal catalyst. With these reactions, intermediate compounds that have an aromatic ring structure are formed. From studies reported a little further on, we know that the burning of oil slicks generates massive amounts of soot. It is entirely logical and expected that the soot contains dioxins.

Aliphatic compounds, which arise as products of incomplete combustion, are thought to play an important role in initially forming ring molecules, which later evolve into complex aromatic precursors. CDDs/CDFs are then formed from the intermediate compounds. In both the second and the third mechanism, formation occurs outside of a furnace (i.e., in the so-called post-combustion zone). Particulate-bound carbon is thought to be the primary reagent in the *de novo* synthesis pathway.

Figure 6.4 is a photo that shows the thick black plumes generated from in-situ oil burning practices by BP in the Gulf. The black smoke plumes are indicative of products of incomplete combustion.

While these plumes are highly toxic, we note that the temperature at which the burning takes place is typically around 1800°F. This high temperature makes the plumes more buoyant. Therefore, plume rise heights would be expected to be high. This

Figure 6.4 Smoke plumes from in-situ oil burning in the Gulf.[4]

factor greatly assists in the dispersion and, hence, the dilution of the pollutants.

A study by Day, et. al[5] reported on an exploratory study of the effects of in situ crude oil spill burning on air quality in the Beaufort Sea region of the Arctic. In this investigation they made estimates of the emissions of soot, CO, SO_2 and metals based on

[4]http://articles.latimes.com/2010/jun/17/nation/la-na-oil-spill-burnbox-20100617

[5]Thomas Day, Donald Mackay, Stuart Nadeau, and Robert Thurier, *Emissions From In Situ Burning Of Crude Oil In The Arctic, Water, Air, and Soil Pollution 11* (1979) 139–152. 0049-6979/79/0112–0139.

literature and experimental work and then applied an air dispersion model. The results of this study showed that the concentrations of soot and metals were undesirably high, within 10 km of the fires. These investigators also reported high levels of PAHs and PANs.

A more extensive study was conducted by NIST (National Institute of Standards and Technology). References from their studies along with others are organized at the end of this section for further reference. NIST conducted research in order to understand, measure, and predict the important features of burning oil on water. Results of that research have been included in nationally recognized guidelines for approval of intentional burning. NIST measurements and predictions have played a major role in establishing in-situ burning as an oil spill response method. Data is given for pool fire burning rates, smoke yield, smoke particulate size distribution, smoke aging, and polycyclic aromatic hydrocarbon content of the smoke for crude and fuel oil fires with effective diameters up to 17.2 m. They also developed software, ALOFT, to quantify the large-scale features and trajectory of wind blown smoke plumes in the atmosphere and to estimate the ground level smoke particulate concentrations. Predictions using the model were tested successfully against data from large-scale tests. ALOFT software is being used by oil spill response teams to help assess the potential impact of intentional burning.

One of the most important studies to-date is that of the Newfoundland Offshore Burn Experiment (NOBE)[6], which provided extensive data on in-situ burning of oil at sea. The study reported on an experiment that consisted of two burns of crude oil conducted off the coast of St. John's, Newfoundland, on August 12, 1993. Most of the sampling of the chemical species produced by the burning was done relatively close to the fire. The University of Washington's Cloud and Aerosol Research Group also performed airborne measurements of the smoke plume from the two burns at distances up to 20 km downwind of the fire. Lidar measurements of the plume cross-section and real-time monitoring of the CO_2 levels in the

[6]Fingas, M. F., *et al.*, 1995. The Newfoundland Offshore Burn Experiment: NOBE. *Proceedings of the 1995 International OilSpill Conference.* American Petroleum Institute, Publication No.4620, pp. 123–132.

plume were made. Taking the smoke yield to be 15% and the elemental carbon mass fraction of the fuel to be 0.8664, NIST estimated that 1 ppm excess CO_2 corresponds to a particulate concentration of 103 μg/m^3. Direct measurements of excess CO_2 in the smoke plume showed volume fractions decreasing to about 1.5 ppm (the equivalent of 150 μg/m^3 particulate) at about 16 km downwind of the burn. Model simulations predicted the plume particulate concentrations in excess of 150 μg/m^3 to extend slightly farther than 20 km downwind.

Allen and Ferek[7] have estimated that a 500-foot fire boom towed in a U-shape configuration provides sufficient oil area to sustain a burn eliminating about 715 bbl/hr. NIST has noted, however, that tests conducted by Mobile are more representative of actual burns and that a burning rate of 400 bbl/hr is probably a reasonable rate to expect from an actual burn. They note this rate to be representative of a coastal environment (i.e., a near-shore in-situ burn). The results of both the modeling efforts and the lidar measurements show that even though an inversion layer was in test burns, the plume penetrated it, and less smoke mixed back to the surface as a result. There is no guarantee, of course, that the plume will always penetrate an inversion layer, and in those instances ground-level concentrations are expected to be higher.

The above information shows that in-situ burning of oil at sea creates significant levels of air pollution and that the countermeasure is potentially environmentally damaging. It may lead to health risk exposures for workers and sensitive populations along shorelines near burn sites. It also represents a significant threat to wildlife, which countermeasure plans most often ignore. It still, however, remains a viable countermeasure option that may offer the lesser of two evils. Its advantages are that it offers logistically simple, rapid, and inexpensive means for reducing the net environmental impact of an oil spill. Because a major portion of the oil is converted to gaseous combustion products, the need for the collection, storage, transport, and disposal of recovered material can be substantially reduced.

---

[7] Allen, A. A. and R. J. Ferek, 1993. Advantages and disadvantages of burning spilled oil. *Proceedings of the 1993 International Oil Spill Conference*. American Petroleum Institute, Publication No.4580, pp. 765–772.

The NOAA[8] (National Oceanic and Atmospheric Administration) states that the following conditions are needed in order to implement this countermeasure:

1. The oil layer has to be thick enough to support combustion. Oil layers thinner than 1 to 2 mm lose too much heat to the water and cannot support combustion.
2. The ignition devices used must be hot enough and last long enough to ignite the oil.
3. The water-in-oil emulsion may not contain more than 30 to 50 percent water to ignite and support combustion.
4. To use currently available fire resistant booms, environmental conditions must be favorable. Specifically, the wind speed should be below 20 knots, and wave height should be below 3 feet.

An in-situ burning technique generally employs towing boats and fire-resistant booms to contain the spilled oil and keep it from spreading. The boom, attached to the boats by towing lines, is pulled by the boats to form a U shape. The open end of the U shape is maneuvered through the oil slick, and a "boomful" of oil is collected. The boom is towed away from the main slick and the oil is ignited. During the burning, the boom is pulled so that it slowly advances ahead of the current to ensure that the oil is concentrated at the back end of the boom and maintains maximum thickness. After the oil is burned, the process may be repeated for as long as in-situ burning is feasible.

As we have noted, burn efficiency is a concern from an air pollution standpoint and also from the viewpoint of the effectiveness of this countermeasure. In-situ burning has been studied under controlled conditions in field tests. The overall conclusions from industry-reported testing is that in-situ burning can be an effective countermeasure, removing 50–99% of the oil collected in the boom.

But our discussions are most concerned with the possible health hazards of in-situ burning to nearby response personnel conducting the burn. The potential exposure of such personnel will be very different from those for the general public at a substantial distance away. Response personnel working close to the burn may be exposed to levels of gases and particulates

---

[8]Barnea, N., Health and Safety Aspects of In-situ Burning of Oil, National Oceanic and Atmospheric Administration
http://response.restoration.noaa.gov/book_shelf/655_health.pdf

Table 6.2 In-situ burning pollutants and exposure limits.

POLLUTANT	OSHA PEL*	ACGIH TLV*	NAAQS+
SO_2	5 ppm	2 ppm [5 ppm]	0.14 ppm averaged over 24 hours, 0.03 ppm annual arithmetic mean
NO_2	5 ppm	3 ppm [5 ppm]	0.05 ppm annual mean
PAH	0.2 mg/m^3 (volatile)	0.2 mg/m^3 (volatile)	–
CO	50 ppm	25 ppm	35 ppm over one hour, 9 ppm over 8 hours
Particulates	5 mg/m^3 for particulates < 3.5 mm	10 mg/m^3 for total dust (new standard in progress)	PM-10: 0.15 mg/m^3 over 24 hr., 0.05 mg/m^3 annual mean

* Time-weighted average concentration over 8 hours. Short-term exposure limits are shown in square brackets.

+ National Ambient Air Quality Standards

that would require them to use personal protective equipment. They should receive the training required to conduct the operation safely. Furthermore, monitoring the responders' work environment should be conducted. Occupational standards such as OSHA's Permissible Exposure Limits (PEL) are applicable to this group. Table 6.2 provides a list of the major in-situ burning pollutants and exposure limits.

6.1.5 Suggested References Concerning In-Situ Burning at Sea

The following references are offered for additional resources for the reader.

1. A. Allen, *Contained controlled burning of spilled oil during the Exxon Valdez oil spill, in Thirteenth Arctic and Marine Oilspill Program Technical Seminar,*

Proceedings June 6–8, 1990, Edmonton, Alberta, Environment Canada, Ottawa, Canada K1A 1C8 (1990) pp. 305–313.

2. C. H. Thompson, G. W. Dawson, and J. L. Goodier, *Combustion: An Oil Spill Mitigation Tool,* U.S. Department of Energy, DOE/EV-1830-1 (1979). Available from National Technical Information Service, Springfield, VA 22161.
3. H. M. Brown, and R. H. Goodman, *In situ burning of oil in ice leads, in Ninth Annual Arctic Marine Oilspill Program Technical Seminar, Proceedings, June 10–12, 1986,* Edmonton, Alberta, Environment Canada, Ottawa, Canada K1A 1C8 (1986).
4. N. K. Smith and A. Diaz, *In-place burning of crude oil in broken ice, 1985 testing at OHMSETT, in Eight Annual Arctic Marine Oilspill Program Technical Seminar, Proceedings, June 18–20, 1985,* Edmonton, Alberta, Environment Canada, Ottawa, Canada K1A 1C8 (1985).
5. N. H. Jason, ed., *Alaska Arctic Offshore Oil Spill Response Technology–Workshop Proceedings, November 29-December 1, 1989,* Anchorage, Alaska, NIST SP 762, U.S. Government Printing Office, Washington, DC (1989).
6. N. H. Jason, ed., *In Situ Burning of Oil Spill Workshop, Proceedings, National Institute of Standards and Technology and Minerals Management Service. January 26–28, 1994,* Orlando, FL, NIST SP 867; U.S. Government Printing Office, Washington, DC 20401-003 (1994).
7. W. D. Walton and N. H. Jason, ed., *In Situ Burning of Oil Spills Workshop.* Proceedings. National Institute of Standards and Technology and Minerals Management Service, November 2–4, 1998, New Orleans, LA, NIST SP 935, Government Printing Office, Washington, DC 20401-0003 (1999).
8. USCG R&D Center, Groton Connecticut, Report of the Interagency Planning Workshop on Oil Spill Research and Development, September 26–27, 1989, Groton, Connecticut, Conference Report 90-30-1509.2, November 1989.

9. U.S. Congress, Office of Technology Assessment, *Coping with an Oiled Sea: An Analysis of Oil Spill Response Technology, OTA-BP-O-63*, Washington, DC, U.S. Printing Office, March 1990.
10. V. Babrauskas, *The cone calorimeter–a new tool for fire safety engineering,* ASTM Standardization News 18, 32-35 (1990).
11. W. D. Walton, D. D. Evans, K. B. McGrattan, H. R. Baum, W. H. Twilley, D. Madrzykowski, A. D. Putorti, Jr., R. G. Rehm, H. Koseki, and E. J. Tennyson, *In situ burning of oil spills: mesoscale experiments and analysis,* in Proceedings of the Sixteenth Arctic and Marine Oil Spill Program Technical Seminar, Ministry of Supply and Services Canada (1993) pp. 679–734.
12. D. D. Evans, W. D. Walton, H. R. Baum, J. R. Lawson, R. G. Rehm, R. Harris, A. Ghoniem, I. Holland, *Measurement of large scale oil spill burns,* in Proceedings of the Thirteenth Arctic and Marine Oil Spill Program Technical Seminar, Ministry of Supply and Services Canada (1990) pp. 1–38.
13. D. D. Evans, H. R. Baum, B. J. McCaffrey, G. W. Mulholland, M. Harkelroad, and W. Manders, *Combustion of Oil on Water, NBSIR 86-3420,* U.S. Department of Commerce, National Bureau of Standards, Gaithersburg, MD, November 1987.
14. G. W. Mulholland, V. Henzel, and V. Babrauskas, *The effect of scale on smoke emission,* in Proceedings of the Second International Symposium on Fire Safety Science, edited by T. Wakamatsu, Y. Hasemi, A. Sekizawa, P.G. Pagni and CE. Grant, Hemisphere, New York (1989) pp. 347–357.
15. D. E. Ward, R. M. Nelson, and D. F. Adams, *Forest fire smoke plume documentation, paper 079-6.3* in Proceedings of the Seventy-Seventh Annual Meeting of the Air Pollution Control Association, Air And Waste Management Assoc., Pittsburgh, PA (1979).
16. C. Huggett, *Estimation of rate of heat release by means of oxygen consumption measurements, Fire Mater. 4, 61–65 (1980).*

17. V. Babrauskas, *Smoke measurement results from the cone calorimeter,* in Eighth Joint United States Japan Natural Resources Panel Meeting on Fire Research and Safety, Tsukuba, Japan, May 13-22, 1985, pp. 420–435.
18. D. E. Ward, D. V. Sandberg, R. D. Ottmar, J. A. Anderson, G. C. Hofer, and C. K. Fitzsimmons, *Measurements of smoke from two prescribed fires in the pacific northwest, in Seventy-Fifth Annual Meeting of the Air Pollution Control Association (1982).* (Available from D.E. Ward, U.S.D.A. Forest Service, Intermountain Research station, Missoula, Mont.)
19. J. R. Lawson, G. W. Mulholland, and H. Koseki, *An airborne smoke sampling package for field measurements of smoke, Fire Technol. 30, 155–173 (1994).*
20. G. W. Mulholland, H. Koseki, and W. A Liggett, *The effect of pool diameter on the properties of smoke produced by crude oil fires, in Twenty-Sixth Symposium (International) on Combustion, The Combustion Institute, Pittsburgh (1996)* pp. 1445–1452.
21. H. Koseki, and G. W. Mulholland, *The effect of diameter on the burning of crude oil pool fires, Fire Technol. 27, 54-65 (1991).*
22. H. Koseki, M. Kokkala, and G. W. Mulholland, *Experimental study of boilover in crude oil fires,* in Proceedings of the Second Fire Safety Science International Symposium, G. Cox and B. Langford, eds, Elsevier, London (1991) pp. 865–874.
23. B. A. Benner, N. P. Bryner, S. A. Wise, and G. W. Mulholland, *Polycyclic aromatic hydrocarbon emission from the combustion of crude oil on water, Environ. Sci. Technol. 24, 1418–1426, (1990).*
24. W. D. Walton, W. H. Twilley, J. McElroy, D. D. Evans, and E. J. Tennyson, *Smoke measurements using a tethered miniblimp at the Newfoundland offshore oil burn experiment,* in Proceedings of the Seventeenth Arctic and Marine Oil Spill Program Technical Seminar, Ministry of Supply and Services Canada, Vol. 2 (1994) pp. 1083–1098.

25. D. Evans, W. Walton, H. Baum, K. Notarianni, J. Lawson, H. Tang, K. Keydel, R. Rehm, D. Madrzykowski, R. Zile, H. Koseki, and E. Tennyson, *In-situ burning of oil spills: mesoscale experiments, in Proceedings of the Fifteenth Arctic and Marine Oil Spill Program Technical Seminar, Ministry of Supply and Services Canada, Cat. # En 40-11/5-1992 (1992)* pp. 593–657.
26. K. K. Laursen, R. J. Ferek, P. V. Hobbs, and R. A. Rasmussen, *Emission factors for particles, elemental carbon, and trace gases from the Kuwait oil fires, J. Geophys. Res. 97, 14,491-14,497 (1992).*
27. W. R. Cofer III, R. K. Steven, E. L. Winstead, J. P. Pinto, D. I. Sebacher, M. Y. Abdulraheem, M. Al-Sahafi, M. A. Mazurek, R. A. Rasmussen, D. R. Cahoon, and J. S. Levine, *Kuwait oil fires: composition of source smoke, J. Geophys. Res. 97, 14,521-14,525 (1992).*
28. K. B. MeGrattan, A. D. Putorti, W. H. Twilley, and D. D. Evans, *Smoke Plume Trajectory of In Situ Burning of Crude Oil in Alaska, Technical Report NISTIR 5273,* National Institute of Standards and Technology, Gaithersburg, Maryland, October 1993.
29. S. V. Hering, ed., *Air Sampling Instruments,* (7th ed.) American Conference of Governmental Industrial Hygienists, Cincinnati, Ohio (1989).
30. L. F. Radke, J. H. Lyons, P. V. Hobbs, and R. E. Weiss, *Smoke from the burning of aviation fuel and their self-lofting by solar heating, J. Geophys. Res. 95, 14,071-14076 (1990).*
31. D. W. Johnson, C. G. Kilsby, D. S. McKenna, R. W. Saunders, G. J. Jenkins, F. B. Smith, and J. S. Foot, *Airborne observations of the physical and chemical characteristics of the Kuwait oil smoke plume, Nature 353, 617–621 (1991).*
32. R. A. Dobbins, R. A. Fletcher, and W. Lu, *Laser microprobe analysis of soot precursor particles and carbonaceous soot, Combust. Flame 100, 301–309 (1995).*
33. E. Boyland, *The toxicology of soot,* in *Soot in Combustion Systems,* J. Lahaye and G. Prado, eds., Plenum Press, New York (1983) pp. 13–24.

34. A. F. Ghoniem, X. Zhang, O. M. Knio, H. H. Baum, and R. G. Rehm, *Dispersion and deposition of smoke plumes generated from massive fires, J, Hazard. Mater. 33, 275–293 (1993).*
35. H. R. Baum, K. B. McGrattan, and R. G. Rehm, *Simulation of Smoke Plumes from Large Pool Fires,* 25th International Symposium (International) on Combustion, The Combustion Institute, Pittsburgh, PA (1994) pp. 1463–1469.
36. K. B. McGrattan, H. R. Baum, and R. G. Rehm, *Numerical simulation of smoke plumes from large oil fires,* Atmospheric Environ. 30, 4125–4136, (1996).
37. M. F. Fingas, G. Halley, F. Ackerman, R. Nelson, M. Bissonnette, N. Laroche, Z. Wang, P. Lambert, K. Li, P. Jokuty, G. Sergy, E. J. Tennyson, J. Mullin, L. Hannon, R. Turpin, P. Campagna, W Halley, J. Latour, R. Galarneau, B. Ryan, D. V. Aurand, and R. R. Hiltabrand, *The Newfoundland offshore burn experiment-NOBE,* in Proceedings of the 1995 International Oil Spill Conference, Publication 4620, American Petroleum Institute (1995) pp. 123–132.
38. W. D. Walton, W. H. Twilley, A. D. Putorti, and R. R. Hiltabrand, *Smoke measurements using an advanced helicopter transported sampling package with radio telemetry,* in Proceedings of the 18th Arctic and Marine Oil Spill Program (AMOP) Technical Seminar, Environment Canada, Emergencies Science Division, Ottawa, Ontario, Canada (1995) pp. 1053–1074.
39. A. A. Allen and R. J. Ferek, *Advantages and disadvantages of burning spilled oil,* in Proceedings of the 1993 International Oil Spill Conference, American Petroleum Institute, Pub. No. 4580 (1993) pp. 765–772.
40. *In Situ Burning Guidelines for Alaska, The Alaska Federal/State Preparedness Plan for Response to Oil and Hazardous Substance Discharges/Releases, Unified Plan Appendix II Annex F,* Revised May 1994.
41. K. B. McGrattan, H. R. Baum, W. D. Walton, and J. J. Trelles *Smoke Plume Trajectory From In Situ Burning of Crude Oil in Alaska: Field Experiments and Modeling of Complex Terrain, NISTIR 5958,* National Institute of Standards and Technology, January 1997.

6.2 Spill Response Protocols and Strategies

6.2.1 Defining Worker Training Requirements

Personnel involved in the post emergency response of oil spills must be trained in accordance with OSHA1910.120, which is a 40-hour course on health and safety for hazardous waste clean-up workers. This is most commonly referred to as HazMat training, where HazMat stands for Hazardous Materials. Programs are also commonly referred to as the HAZWOPER standard. These courses are offered through training organizations and academic institutions with some training modules offered online. Table 6.3 gives the reader a partial list of providers. In addition to training organizations, OSHA's area offices offer a variety of information services such as publications, audiovisual aids, technical advice, and speakers for special engagements. OSHA's Training Institute in Des Plaines, Illinois provides basic and advanced courses in safety and health for federal and state compliance officers, state consultants, federal agency personnel, and private sector employers, employees, and their representatives. The OSHA Training Institute also has established OSHA Training Institute Education Centers to address the increased demand for its courses from the private sector and other federal agencies. These centers are nonprofit colleges, universities, and other organizations that have been selected after a competition for participation in the program. They are located in various parts of the United States.

It is our opinion that that the most thorough programs are those that are supplemented with field or classroom assignments. Note that the programs and providers listed in Table 6.3 have not been evaluated, and we make no claims as to whether they offer the best or even the most suitable training required. The reader needs to first review the training requirements explained in this section and then discuss the needs with the training provider.

We also want to emphasize that, because the training and skills are specialized, the workforce needs to be in place before a spill occurs. Highly qualified, skilled workers and supervisors in emergency response and planning are a part of the critical infrastructure of oil drilling and exploration. If spills occur, this infrastructure must already be in place and ready to mobilize. The team of response specialists cannot be established or built after the fact. The type of training, the level of training, and the numbers of individuals that need to be trained should be determined before any well drilling

Table 6.3 Organizations and companies that provide HazMat training.

Provider	Location	Program Description	Contact Information
American Military University	111 West Congress Street, Charles Town, WV 25414	This institution offers Online Degrees and Certificates in Emergency and Disaster Management. There is a wide range of curriculum offered for specialists responding to hurricanes, terrorist attacks, or hazmat spills. A quick overview of the site shows programs address major areas in depth such as establishing command and control, coordination of communication, evacuation of citizens, and performing clean-up operations. American Military University offers a Bachelor of Arts and a Master of Arts degree in Emergency and Disaster Management as well as a six course graduate level academic certificate in Emergency and Disaster Management. The interdisciplinary curriculum of its emergency and disaster management programs address topics such as incident command, crisis action planning, consequence management, public policy,	Student Payment Issues and Corporate Office: 877-468-6268 Student Services: 877-755-2787 Financial Aid Help Desk: 877-372-3535 Admissions: 877-777-9081

		interagency disaster management, hazard mitigation, health management, and a variety of other topics that will help prepare individuals to start or advance careers.	
Lion Technology Inc.	Nationwide training	Offers both on-site and on-line training programs. Offers broad range of safety training programs but one most pertinent to current topic discussions of the book is the OSHA training program (29CFR1910.120). Programs can be tailored for on-site training. Online programs cover Managing Hazard Communication, Recordkeeping & Reporting, Bloodborne Pathogens. Lion will tailor their training programs or create a new program to cover employees' responsibilities and facility's materials, wastes, and operations. Training specialists will work with client to determine what topics should be focused on. Lion has trained more than 100,000 professionals since 1977.	Phone: 888-546-6511; Web Site: http://www.lion.com/Catalog/OSHTraining.asp

Table 6.3 (cont.) Organizations and companies that provide HazMat training.

Provider	Location	Program Description	Contact Information
Washington State Patrol	Various locations within state of Washington	The Hazardous Materials Training Unit provides hazardous materials training to first responders throughout the State of Washington. As described in RCW 70.136.030, the Washington State Patrol is designated as the Incident Command Agency in any hazardous materials incident on or along any state route or interstate freeway corridors, as well as within any jurisdiction that has not designated an incident command agency. The Washington State Patrol is also directed by RCW 70.136.035 to provide assistance at the scene of hazardous materials incidents where a separate agency is designated as the incident command agency. While this training is specific to overland hazard materials incidents much of the curriculum is applicable to clean-up operations that are required for shoreline management from oil spills. Programs cover Hazardous Materials Awareness,	E-mail: hazmat@wsp.wa.gov Phone: (360) 596-3923 Fax: (360) 596-3936

		Operations, Technician, Chemistry, and On-Scene Incident Command, Other. The training is funded by a federal grant from the Hazardous Materials Emergency Preparedness (HMEP) program.	
Georgia Tech – Professional Education Program	Courses are offered at multiple locations. Following is a list of typical locations: ALABAMA (Birmingham, Huntsville); COLORADO (Boulder, Denver), FLORIDA (Destin, Fort Lauderdale, Jacksonville, Orlando, Pompano Beach), GEORGIA	The Hazmat Training program is the instructional component of Georgia Tech's Center for Emergency Response Technology (CERTIP), which has been recognized as an outstanding program three times by governmental leaders and organizations. The courses are comprehensive and aimed at preparing specialists to safely and effectively respond to hazardous materials incidents. Courses include: **24-Hour Hazmat Technician-Level Emergency Response Course**: Industrial Chemical Spill and Disaster Response (HAZ 1000P)–Learn the basic knowledge and techniques required for personnel responding proactively to hazardous material spills in an industrial setting and by OSHA for demonstrated competency at the Hazmat Technician	Georgia Tech Global Learning Center 84 Fifth St. N.W. Atlanta, GA 30308-1031 404-385-3500 (Phone) 404-385-1408 (Fax)

Table 6.3 (cont.) Organizations and companies that provide HazMat training.

Provider	Location	Program Description	Contact Information
	(Atlanta, Savannah), MISSISSIPPI (Biloxi, Gulfport), NEVADA (Las Vegas), NORTH CAROLINA (Wilmington), SOUTH CAROLINA (Charleston, Columbia), TENNESSEE (Chattanooga, Nashville), WASHINGTON D.C. AREA (Arlington)	Level under 29 CFR 19190.120 (q) (6) (iii). **40-Hour Hazardous Waste Site Worker Course (HAZWOPER) (HAZ 1004P)**–Suit up in protective equipment, including a self-contained breathing apparatus and totally encapsulated chemical protective suits, and go through drills emphasizing a variety of skills necessary on hazardous waste sites. Meet your HAZWOPER 40-hour training requirement. **8-Hour Annual HAZWOPER Refresher (HAZ 1002P)**–Meets HAZWOPER requirements for annual refresher training. **Advanced Hazmat School (HAZ 1006P)**–Gain advanced knowledge and techniques in decontamination procedures, as well as training for atmospheric hazards, basic sampling considerations, and details on the types of instruments most commonly used in hazmat work. **Certified Hazardous Materials Management (CHMM) Review Course (EST**	

		7006)–Prepare for the CHMM exam with this intensive review course. **Hazardous Materials (OTI 2015P)**–Get an overview of OSHA's general industry standards and other consensus and proprietary standards relating to hazardous materials.	
OSHA.com Training	Online training program available worldwide	**HAZWOPER 8hr Refresher Course**: The 8-hour annual refresher course meets the requirements for eight hours of annual refresher training for workers at hazardous waste sites. This course is designed for general site workers who remove hazardous waste or who are exposed or potentially exposed to hazardous substances or health hazards. **HAZWOPER 24hr Training:** The 24-hour course covers broad issues pertaining to the hazard recognition at work sites. HAZWOPER 24 Hour is required for employees visiting an Uncontrolled Hazardous Waste Operation mandated by the Government. **HAZWOPER 40hr Course:** The 40-hour course is specifically designed	http://www.osha.com/courses/hazwoper.html Sales: 1-866-959-6230 E-mail: support@osha.com

Table 6.3 (cont.) Organizations and companies that provide HazMat training.

Provider	Location	Program Description	Contact Information
		for workers who are involved in clean-up operations, voluntary clean-up operations, emergency response operations, and storage, disposal, or treatment of hazardous substances or uncontrolled hazardous waste sites. HAZWOPER 40 Hour is required for employees working on a project consisting of Uncontrolled Hazardous Waste Operation mandated by the government.	
Online Institute of America	Online training program available worldwide	If you are a worker involved in clean-up operations, voluntary clean-up operations, emergency response operations, and storage, disposal and management of hazardous materials and uncontrolled hazardous waste sites, take the 40 hour HAZWOPER training course. If you are a manager or supervisor involved in supervising clean up operations, you are required to take at least 8 Hours of HAZWOPER Supervisor Training. For occasional site workers,	Tel: 1-877-251-4387 http://hazwopertraininginstitute.com/

		OSHA regulations require 24 hour HAZWOPER Training. All courses are 100% online and come with a printable certificate of completion. Courses offered: HAZWOPER 40, HAZWOPER 24, HAZWOPER 8, HAZMAT First Responder Awareness and Operation Courses.	
Safety Unlimited	Online training program available worldwide	This program offers all the HAZWOPER and HAZMAT courses. Upon completion of the course the student takes the final exam and is allowed to print the training completion certificate.	Phone: (888) 309-SAFE (7233) Fax: (866) 869-7097 690A East Los Angeles Ave. Suite 180, Simi Valley, CA 93065 www.safetyunlimited.com

and recovery operations are even implemented. They should be defined at the time projects are envisioned and should be part of the basis for environmental impact statements that justify any type of drilling operations and transport. Note also that any omission of programs and organizations that provide this type of specialized training in Table 6.3 is simply an oversight on our part.

The HAZWOPER standard identifies two basic phases of a response action: emergency response and post-emergency response. Depending on the size of the spill, these phases may be managed quite differently. In addition, workers who participate only in post-emergency response require different training than emergency response workers receive. Much of this chapter focuses on post-emergency response.

According to OSHA[9], emergency response is defined as "a response effort...to an occurrence which results, or is likely to result, in an uncontrolled release of a hazardous substance." For marine oil spills, an uncontrolled release is a situation in which the oil and its associated airborne and surface contamination hazards are releasing into the environment or pose a danger of being released into the environment and represent a worker exposure hazard. For example, oil in grounded ships in which cargo is being released constitutes an emergency response situation. On-water containment, skimming operations, and underwater oil recovery operations also are considered to be emergency response activities because the oil is still in danger of being released into the environment. Shoreline clean-up is normally considered to be a post-emergency response unless the oil is below the high-tide mark or storm surge boundary (active or forecasted) and can reasonably be expected to be re-released into the marine environment.

The term "post-emergency response" refers to those activities that are implemented "after the immediate threat of a release has been stabilized or eliminated and clean-up of the site has begun."[10]

During a response to a large release such as the Gulf incident, it makes sense for emergency response and post-emergency response clean-up activities to be executed simultaneously. In these cases, the boundaries between the emergency response and

---

[9]29 CFR 1910.120(a)(3)
[10]29 CFR 1910.120(a)(3)

the post-emergency response arenas must be well-defined and explained to responders and clean-up workers.

To appreciate the type of training that responders are required to have, a better understanding of how marine oil spill response is organized is important. Marine oil spill response is organized and managed according to the regulations found in 40 CFR 300 in the National Oil and Hazardous Substance Pollution Contingency Plan (NCP). These are regulations that describe procedures for responding to hazardous substance releases and oil discharges. Appendix E of the regulation addresses oil spill response. The U.S. Coast Guard (USCG) and the Environmental Protection Agency (EPA) jointly contributed to the development of the NCP.

Marine oil spill response involves a network of government agencies, community organizations, industry groups, and independent contractors. Federal and/or state agencies usually monitor the responsible party (generally, the owner or operator of the vessel, facility, port, oil rig, or pipeline involved in the spill). The federal government can direct clean-up operations if the responsible party does not respond adequately or is not capable of taking appropriate action. Note that the federal government itself does not have the infrastructure, expertise, or command of technologies to directly respond to emergency responses. While the National Guard and U.S. Coast Guard have the expertise and some equipment to respond to post-emergency response, we have seen in the Gulf incident their inability to assist directly in shutting down on the Deepwater Horizon blow out.

An On-Scene Coordinator (OSC) acts as the leader for response activities. In the coastal areas of the United States, USCG serves as the OSC for oil spill responses. In inland areas, including rivers and other inland waters, the EPA takes the lead.

HAZWOPER requires that a senior official who is present at the response site serves as the Incident Commander. His or her role is to lead an emergency response operation. For marine oil spills, the ranking Coast Guard officer or EPA official at the spill scene usually functions as the On-Scene Incident Commander. The emergency response remains in effect until the Incident Commander declares it completed.

OSHA is responsible for assuring safe working conditions for workers. During marine oil spill response, OSHA provides consultation at the request of other government agencies. If necessary, OSHA uses enforcement action to assure that workers are properly protected.

Response actions conducted under the NCP must comply with the provisions of HAZWOPER, the requirements of which can be found in 40 CFR 300.150. Workers engaged in a response action under the NCP are required to follow an occupational safety and health program consistent with HAZWOPER, and they must be trained according to HAZWOPER's training requirements, regardless of whether the responsible party or a government agency is directing the clean-up.

For marine oil spill emergency response, the HAZWOPER provisions that most directly apply include:

1. emergency response operations in HAZWOPER paragraph (q), and
2. post-emergency response clean-up operations in paragraph (q)(11).

See also emergency response training provisions in paragraph (q)(6) and post-emergency response training requirements in paragraph (q)(11).

Marine oil spill responders face a variety of health and safety hazards, including fire, explosions, oxygen deficiency, exposure to carcinogens and other chemical hazards, heat and cold stress, and safety hazards associated with working around heavy equipment in a marine environment. The following is a list of potential hazards workers face in oil spill responses:

- Biological (e.g., plants, animals, insects, and remediation materials)
- Drowning
- Noise
- Electricity
- Slips and trips
- Biohazardous debris (e.g., syringes on shoreline)
- Ergonomic stresses (e.g., repetitive strain and low back pain)
- Sunburn
- Confined spaces
- Underwater diving
- Falls
- Unguarded equipment cranes
- Fatigue vehicles (e.g., aircraft, boats, cars, and trucks)
- Cutting and welding

- Fire and explosion
- Degreasers
- Heat or cold stress
- Dispersants
- In-situ burning particles
- Exposure to benzene (crude oils high in BTEX, benzene, toluene, ethylbenzene, and xylene) – adverse health effects include irritation to eyes, skin, and respiratory system, dizziness, rapid heart rate, headaches, tremors, confusion, unconsciousness, anemia, and cancer
- Exposure to benzo(a)pyrene (a polycyclic aromatic hydrocarbon reproductive, formed when oil or gasoline burns) – adverse health effects include irritation to eyes and skin, cancer, and possible effects
- Exposure to carbon dioxide (inerting atmosphere, and a byproduct of combustion) – adverse health effects include dizziness, headaches, elevated blood pressure, rapid heart rate, loss of consciousness asphyxiation, and coma
- Exposure to carbon monoxide (byproduct of combustion) – adverse health effects include irritation to eyes, skin, and respiratory system, dizziness, confusion, headaches, nausea, weakness, loss of consciousness, asphyxiation, and coma
- Exposure to ethyl benzene (high in gasoline) – adverse health effects include irritation to eyes, skin, and respiratory system, loss of consciousness, asphyxiation, and nervous system effects
- Exposure to hydrogen sulfide (oils high in sulfur, emitted from decaying plants and animals) – adverse health effects include irritation to eyes, skin, and respiratory system, dizziness, drowsiness, cough, headaches, and nervous system effects
- Exposure to methyl tert-butyl ether (MTBE) (octane booster and clean air additive for gasoline, or pure MTBE) – adverse health effects include irritation to eyes, skin, and respiratory system, headaches, nausea, dizziness, confusion, fatigue, weakness, nervous system effects, liver problems, and kidney problems
- Exposure to oolycyclic aromatic hydrocarbons (PAHs) (occur in crude oil and formed during burning of

oil) – adverse health effects include irritation to eyes and skin, cancer, possible reproductive effects, and immune system effects
- Exposure to sulfuric acid (byproduct of the combustion of sour petroleum product) – adverse health effects include irritation to eyes, skin, teeth, and upper respiratory system, severe tissue burns, and cancer
- Exposure to xylenes (high BTEX crude oils) – adverse health effects include irritation to eyes, skin, and respiratory system, dizziness, confusion, change in sense of balance, and nervous system, gastrointestinal system, liver, kidney, and blood effects
- Exposure to toluene (high BTEX crude oils) – adverse health effects include irritation to eyes, skin, and respiratory system, fatigue, confusion, dizziness, headaches, memory loss, nausea, and nervous system, liver, and kidney effects

OSHA's 29 CFR 1910.120(q)(6) defines the HAZWOPER training requirements that apply to marine oil spill emergency responses (see section concerning requirements and training guidance for post-emergency response along with paragraph (q)(11) and in OSHA Instruction CPL 02-02-051, Inspection Guidelines for Post-Emergency Response Operations under 29 CFR 1910.120). The HAZWOPER training requirements are based on the workers' assigned duties during an oil spill. For example, if you have deck hands who are assigned to equipment decontamination during and after an oil spill, those persons must have received the level of HAZWOPER training required for the duties and hazards of the decontamination tasks. In speaking with training providers, it is important that these standards be consulted to make sure that the program meets the needs of the various work force activities.

OSHA provides a detailed series of logic diagrams to assist in the structuring response team training requirements.[11] We have organized the decision trees provided by OSHA and included them in Figures 6.5 and 6.6 to assist the reader in the selection of training

[11]The logic scheme can be accessed at the following web site: http://www.osha.gov/Publications/3172/3172.html.

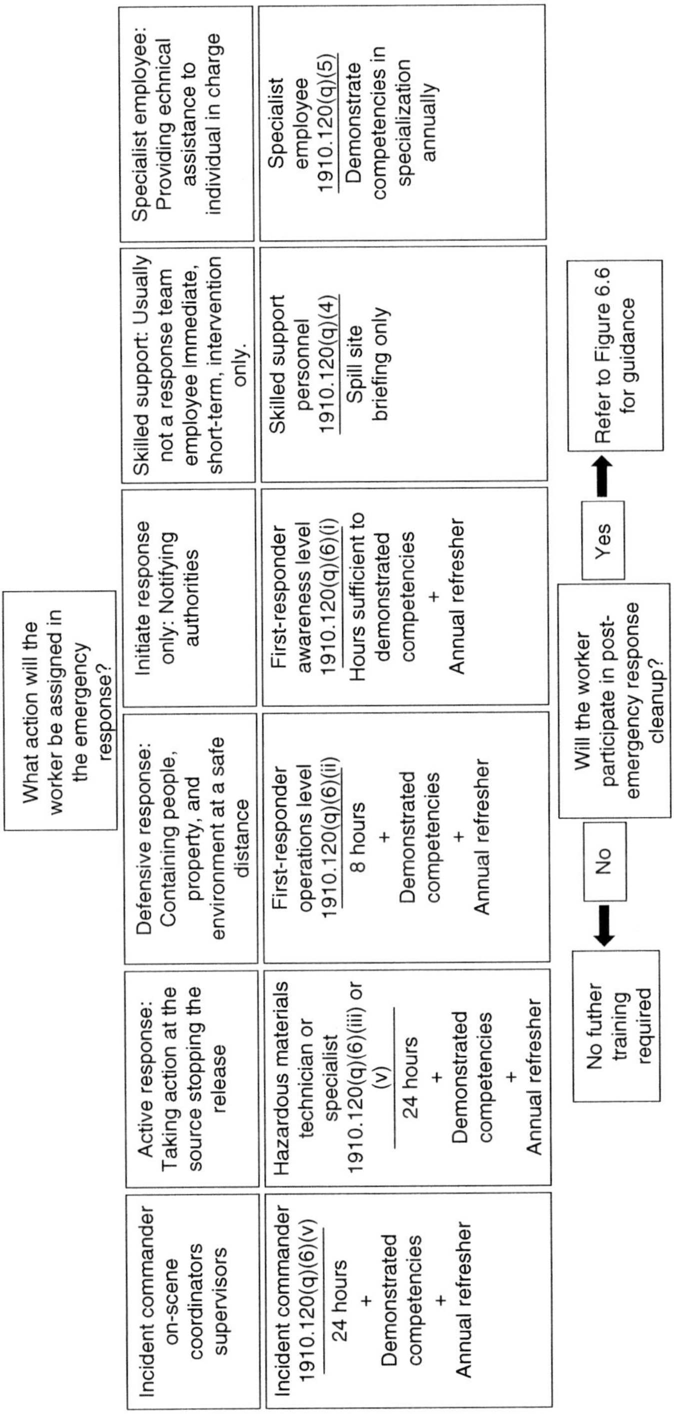

Figure 6.5 Training requirements for workers/supervisors involved in emergency response.

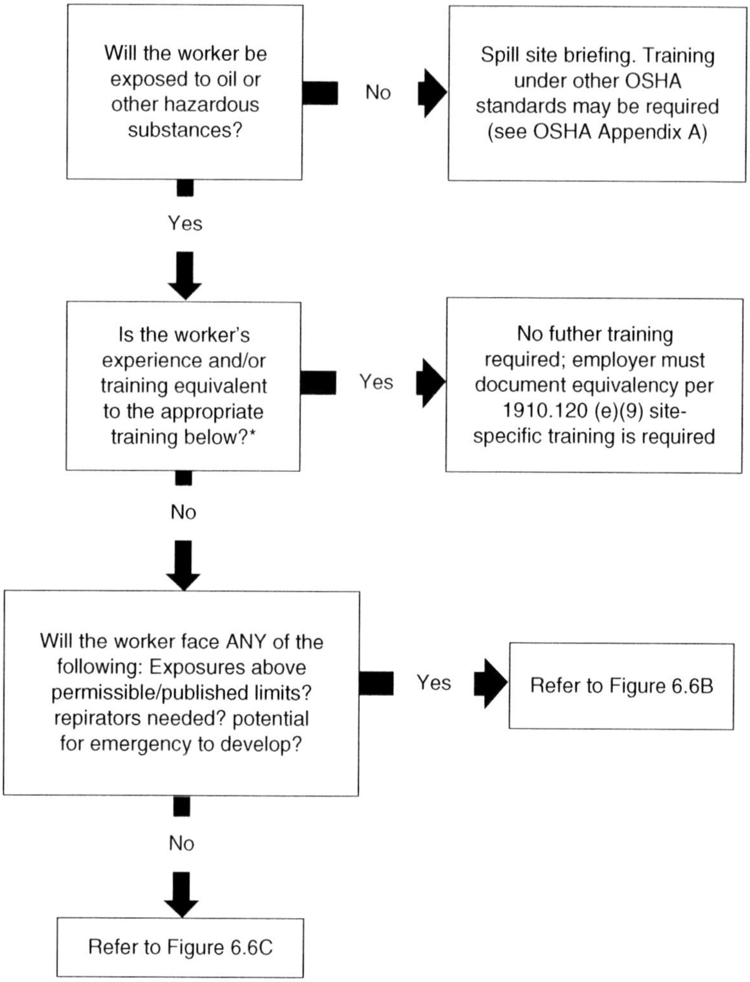

Figure 6.6A Training requirements for workers/supervisors involved in post-emergency response.

needs. In reviewing the logic diagrams in Figures 6.2 and 6.3, make note of OSHA's asterisk comments:

- *Typically, First Responder Awareness Level training (1910.120 (q)(6)(i) will NOT prepare a worker for clean-up operations because it addresses emergency notification procedures only. Workers who could have direct contact with or potential

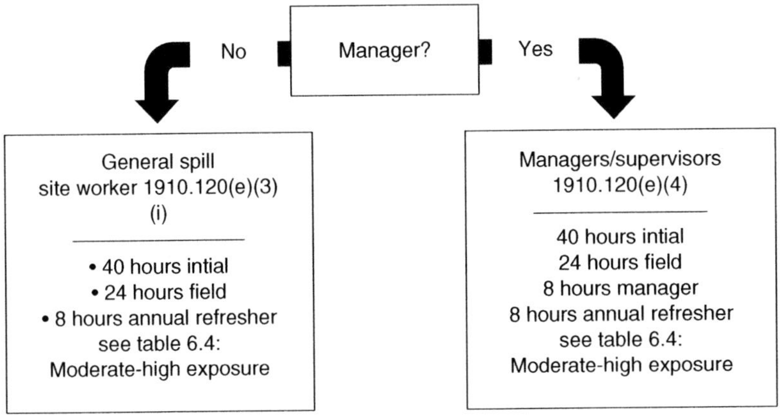

Figure 6.6B Training requirements for workers/supervisors involved in Post-Emergency Response – training requirements if worker faces hazards.

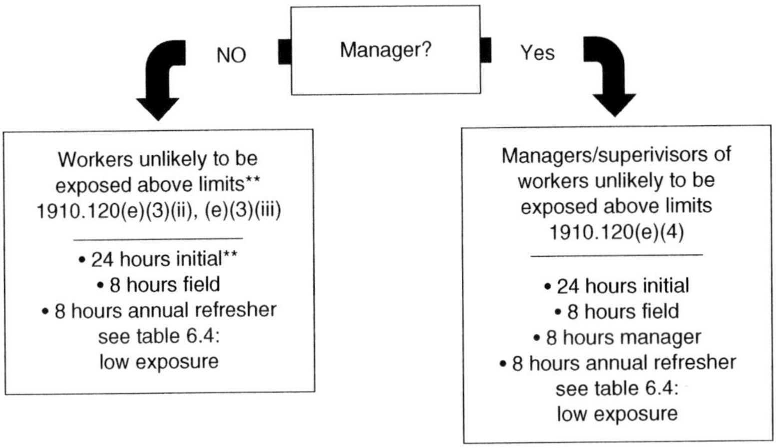

Figure 6.6C Training requirements for workers/supervisors involved in post-emergency response – training requirements if worker does not face hazards.

exposure to hazardous substances often need additional training.

- **If you need to train workers for a specific spill and for tasks that involve minimal exposure (e.g., beach clean-up workers) you may be able to use the reduced training provision OSHA describes in its compliance directive, CPL2-2.51. This directive applies in limited circumstances.

Also, make note of the following list of training topics and competency areas that are recommended by OSHA for marine oil spills. These training topic areas are referred to in Figures 6.2 and 6.3 to help identify the training subjects.

On-Scene Coordinator

1) Know and be able to implement the employer's incident command system.
2) Know how to implement the employer's emergency response plan.
3) Know and understand the hazards and risks associated with employees working in personal protective clothing.
4) Know how to implement the local emergency response plan.
5) Know the state emergency response plan and the Federal Regional Response Team.
6) Know and understand the importance of decontamination procedures.

Active Response

7) Know how to implement the employer's emergency response plan.
8) Know how to use field survey instruments and equipment to classify, identify, and verify known and unknown materials.
9) Be able to function within an assigned role in the Incident Command System.
10) Know how to select and use proper specialized personal protective equipment provided to the hazardous materials technician.
11) Understand and be able to apply hazard and risk assessment techniques.
12) Be able to perform advanced control, containment, and/or confinement operations within the capabilities of the resources and available personal protective equipment.
13) Understand and implement decontamination procedures.
14) Understand termination procedures.

15) Understand terminology and behavior of chemicals and their toxic effects.
16) Know how to use advanced survey instruments and equipment to classify, identify, and verify known and unknown materials.
17) Understand in-depth hazard and risk techniques.
18) Be able to determine and implement decontamination procedures.
19) Know how to implement the local emergency response plan.
20) Know the state emergency response plan.
21) Be able to develop a site safety and control plan.
22) Understand chemical, radiological, and toxicological terminology and behavior.
23) Be able to select and use proper specialized chemical personal protective equipment provided to the hazardous materials specialist.
24) Be able to perform specialized control, containment, and/or confinement operations within the capabilities of the resources and personal protective equipment available.

Defensive Response

25) Know basic hazard and risk assessment techniques.
26) Know how to select and use proper personal protective equipment necessary for the first responder operation level.
27) Understand basic hazardous materials terms.
28) Know how to perform basic control, containment, and/or confinement operations within the capabilities of the resources and available personal protective equipment.
29) Know how to implement basic decontamination procedures.
30) Understand the relevant standard operating and termination procedures.

Initiate Response Only

31) Understand the hazards of oil and the risks in a spill.
32) Understand what happens during an emergency involving spilled oil.

33) Recognize the presence of oil or related hazardous materials in an emergency.
34) Identify hazardous substances, if possible (e.g., appearance, smell, monitoring equipment).
35) Understand individual roles in employer's emergency response plan.
36) Recognize when help is needed and when to request assistance from the response team.

Post-Emergency Clean-up – Exposures Above Exposure Limits

37) Know the name(s) of and how to contact the site safety and health personnel for spill clean-up.
38) Know the safety, health, and other hazards present during oil spill clean-up.
39) Know safe clean-up work practices, including decontamination procedures, to minimize risks.
40) Know how to use available controls and equipment, including contamination control procedures and personal protective equipment, to minimize risks.
41) Know the contents of the safety and health plan for the specific clean-up.
42) Know and be able to recognize signs and symptoms of overexposure to hazards present.
43) Know the medical surveillance requirements.

Post-Emergency Clean-up – Exposures Below Exposure Limits or Non-Recurring Minimal Exposure

44) Know the name(s) of and how to contact site safety and health personnel for spill clean-up.
45) Know the safety, health, and other hazards present during oil spill clean-up.
46) Know safe clean-up work practices, including decontamination procedures, to minimize risks.
47) Know how to use available controls and equipment, including contamination control and personal protective equipment, to minimize risks from hazards present.
48) Know the contents of the safety and health plan prepared for the specific clean-up.

49) Know and be able to recognize signs and symptoms of overexposure to hazards present.
50) Know the medical surveillance requirements.

Supervisor/Manager for Clean-up Operations

51) Know and be able to implement effectively the employer's safety and health program.
52) Know and be able to implement effectively the employer's personal protective equipment plan.
53) Know and be able to implement effectively the employer's spill containment program.
54) Know and be able to implement effectively health hazard monitoring procedures and techniques.

Briefing Topics

55) Purpose of visit or duties to be performed
56) Site personnel, chain-of-command, and communications procedures
57) Chemical/physical hazards involved and signs and symptoms of exposure
58) Emergency alarm system, escape routes, and places of refuge
59) Appropriate contamination control procedures, personal protective equipment, decontamination, and other control measures provided

Table 6.4 may be used in conjunction with Figures 6.2 and 6.3. This table has information that summarizes OSHA's recommended minimum levels of training and experience level according to job classification and the type of work activity. Note that the table is organized according to the two general classes of response, namely Emergency Responders and Post-Emergency Responders. The information in the table will assist in clarifying HAZWOPER (29 CFR 1910.120 or 1926.65) training requirements for marine oil spill response.

OSHA notes that regardless of the job title, all employees should receive pre-entry site briefings covering, at a minimum, training areas 55–59 listed above.

As we have noted earlier, a highly skilled emergency response workforce needs to be trained before they respond. This team of specialists must be trained and certified to the highest level

Table 6.4 Minimum training requirements by job classification (refers to Figures 6.5 and 6.6).

OSHA Job Classifications		Examples	Minimum Requirements for Training/Experience
Training for workers who perform emergency response			
Incident Commander		– On-Scene Coordinators – On-Scene Coordinator Representatives – On-Scene Industry Representatives – ICS Operations Section Chief – ICS Site Safety Enforcement Personnel – Supervisors/Managers – ICS Safety Officer – ICS Group Supervisors	24 hours initial emergency response training covering areas 25–36 AND competency in areas 1–6 Annual refresher training or annual demonstration of competency, ensuring competencies in areas 25–36 and 1–6
Active Response		– Work close to flammable/combustible liquids such as in a pumping operation – Work in confined or enclosed spaces containing volatile oil – Underwater free-floating oil removal – Soil/sand subsurface oil assessment	Technician Level [(q)(6)(iii)]: • 24 hours initial emergency response training covering areas 25–36 AND competency in areas 7–15 • Annual refresher training or annual demonstration of competency, ensuring competencies in areas 25–36 and 7–15 **OR** Specialist Level [(q)(6)(iv)]:

		– Plugging holes/emergency repairs to source – Diving operations close to source – Work adjacent to volatile fuel during initial spill hours (boat operations, etc.) – Site characterization of chemical exposures, air monitoring/sampling – Soil investigation/sampling – Dispersant application and monitoring operations – In-situ burn and monitoring operations – Application of herding agents/emulsion breakers – Marine chemist activities – Salvage response – Booming operations at source – Vessel damage assessment – Accident investigation – Firefighting response	• 24 hours initial emergency response training covering areas 7–15 and 25–36 AND competency in areas 16–24 • Annual refresher training or annual demonstration of competency, ensuring competencies in areas 7–36

Table 6.4 (cont.) Minimum training requirements by job classification (refers to Figures 6.5 and 6.6).

OSHA Job Classifications		Examples	Minimum Requirements for Training/Experience
Defensive Response		– Booming operations at a safe distance* – Surface level shoreline impact assessment – Manual pickup and removal of irritant oil and oily debris for oils that could be re-released into environment – Damming/diking at a safe distance – Loading of oil into receptacles at a safe distance – Emergency medical personnel (if working in contaminated areas) – Staging area managers – Vacuum truck operations at a safe distance – Security operations – Safety zone enforcement – Skimmer/boat operations at a safe distance	• 8 hours initial emergency response training covering areas 31–36 AND competency in areas 25–30 • Annual refresher training or annual demonstration of competency, ensuring competencies in areas 25–36

Initiate Response		– Crew members who discover a release and alert the proper emergency response personnel – Industry and government watch standers	• Sufficient hours to demonstrate competency in areas 31–36 • Annual refresher training including demonstration of competency in areas 31–36 • Additional training required for post-emergency clean-up activities unless there is no exposure to hazardous substances.
Skilled Support		– Immediate, short-term intervention of a specialized nature that cannot be performed by other workers – On-scene resource documentation – Operation of certain equipment such as mechanized digging equipment, cranes, and other hoisting equipment – Aerial assessment – Remote sensing aerial platforms – Photo-documentation – Administrative/finance support	• Spill site briefing in areas 55–59

Table 6.4 (cont.) Minimum training requirements by job classification (refers to Figures 6.5 and 6.6).

OSHA Job Classifications		**Examples**	**Minimum Requirements for Training/Experience**
Specialist Employer		Employer – Dispersion modeling personnel	• Briefing covering areas 55–59 • Training or demonstrated competency in area of specialization annually – Analytical chemical/oil specialist – Professional engineer/spill recovery system or shipbuilding specialist
Training for workers who perform only post-emergency response			
Moderate-High (At/Above Exposure Limits)	• Unknown oil or unknown hazardous substance mixed with oil • Exposures equal or exceed exposure limits or other published limits • Respiratory protection required • Concentrations at or above 10 percent of the Lower	– Manual clean-up of stranded oil with potential skin carcinogens (e.g., benzo (a) pyrene) – Clean-up of stranded oil when toxic chemicals are persistent and above exposure limits – Wildlife capture and treatment depending on explosives – Load and transfer piled oil-saturated decaying plants and animals that provide a hydrogen sulfide risk – Clean-up of stranded oil in confined spaces	• 40 hours of initial training in areas 37–43 and 3 days supervised field experience, or equivalent training certification • At least 8 hours of annual refresher training • Supervisory/management personnel must also receive 8 hours of additional initial training covering at a minimum areas 51–54

	Explosive Limit (LEL) • Oxygen levels [3] 19.5 <22 percent) • Carcinogen: known or suspected • Situations in which oil is known but parameters above cannot be reasonably assessed	– On land marsh burning operation	
Low (Below Exposure Limits)	Routine spill clean-up workers [(e)(3)(iii)]: • Oil and other hazards of spill constituents known • Exposures may cause irritation (skin, eye, respiratory) but are below permissible published limits		

Table 6.4 (cont.) Minimum training requirements by job classification (refers to Figures 6.5 and 6.6).

OSHA Job Classifications		**Examples**	**Minimum Requirements for Training/Experience**
	• Oxygen levels [3] 19.5 <22 percent) • Concentrations less than 10 percent, but more than the LEL • Other significant hazards may be present: Physical, safety, ergonomic, thermal. [(e)(3)(ii)]: • Oil and other hazards of spill constituents known • Exposures below permissible/published limits	– Pressure washing operations of stranded weathered oil – Cutting of contaminated live vegetation – Natural resource damage assessment – Bioremediation operations – Shoreline clean-up assessment – Vessel/equipment decontamination – Underwater stranded oil removal operations – Soil/sand substance oil removal – Containerized/packaged waste handling and disposal for transport operations	• 24 hours of initial training in areas 44–50 and 1 day supervised field experience, or equivalent training • At least 8 hours of annual refresher training • Supervisory management personnel must also receive 8 hours of additional initial training covering at a minimum areas 51–54

Non-Recurring/ Minimal Exposure	• Oil and other spill constituents known • Exposures below permission/published limits • Respirators not required • Oxygen levels [3] 19.5 <22 percent) • No potential for chemical emergency, explosion, or fire • Minimal irritants to respiratory system, eyes, or skin • No significant physical, safety, ergonomic, and thermal hazards	– Workers who do not participate in marine oil spill responses on a frequent recurring basis and who will have minimal exposure – Tarball shoreline clean-up – On scene cost documentation – Operators of large construction vehicles during stranded oil removal	• If conditions described in CPL 02-02-051 are met, reduced initial training or equivalent training certification covering items 44–50 • Up-to-date training consistent with 1910.38(a) and 1910.1200, and health and safety training associated with assigned tasks Supervisory/management personnel must receive: • At least 24 hours of initial training in areas 44–50 and 1 day of supervised field experience; • 8 hours of additional initial training covering at a minimum areas 51–54; and • 8 hours annual refresher training

Table 6.4 (cont.) Minimum training requirements by job classification (refers to Figures 6.5 and 6.6).

OSHA Job Classifications		Examples	Minimum Requirements for Training/Experience
No Exposure	• Area characterized and stable with: • No potential for exposure to hazardous waste or substances by any route (i.e., inhalation, skin absorption, ingestion No safety hazards associated with hazardous waste or hazardous substances	– Aerial photo documentation – Command Post support – Food service personnel – Legal representation – Over-flight assessment (if no potential for exposure) – Resource tracking (if no potential for exposure) – Financial services personnel – Historians – Press	• Briefing covering areas 55–59

of responsibility they might need to assume. Clean-up workers need to be trained for the highest exposure conditions they may encounter. Workers should never be permitted to perform an emergency response or clean-up operation without proper training and certification.

Workers who receive HAZWOPER training must receive a written certificate when they successfully complete the training. This is a requirement of paragraphs (e)(6) and (q)(6) of the OSHA HAZWOPER standard. Certificates that are issued provide proof of competency. Individuals who have completed various training do not need to repeat the initial training if the worker goes to work at a new site. However, the worker must have additional training or site briefings that are needed in order to work safely at the new spill incident or clean-up site. Every year, emergency responders and clean-up workers must receive refresher training to maintain and demonstrate competency.

As a final note, the following is a list of related OSHA standards that some workers and supervisors will need to consult depending on their job classification.[12]

Occupational Safety and Health Standards

1910 Subpart D – Walking-Working Surfaces
1910 Subpart E – Means of Egress
1910 Subpart F – Powered Platforms, Manlifts, and Vehicle-Mounted Work Platforms
1910 Subpart G – Occupational Health and Environmental Control
1910 Subpart H – Hazardous Materials
1910 Subpart I – Personal Protective Equipment
1910 Subpart J – General Environmental Controls
1910 Subpart K – Medical and First Aid
1910 Subpart L – Fire Protection
1910 Subpart M – Compressed Gas and Compressed Air Equipment
1910 Subpart N – Materials Handling
1910 Subpart O – Machinery and Machine Guarding
1910 Subpart P – Hand and Portable Powered Tools and Other Hand-Held Equipment
1910 Subpart Q – Welding, Cutting, and Brazing
1910 Subpart S – Electrical

[12]All of these standards are available on line at http://www.osha.gov.

1910 Subpart T – Commercial Diving Operations
1910 Subpart Z – Toxic and Hazardous Substances

Occupational Safety and Health Standards for Shipyard Employment
1915 Subpart B – Confined and Enclosed Spaces and Other Dangerous Atmospheres in Shipyard Employment
1915 Subpart C – Surface Preparation and Preservation
1915 Subpart D – Welding, Cutting, and Heating
1915 Subpart E – Scaffolds, Ladders, and Other Working Surfaces
1915 Subpart F – General Working Conditions
1915 Subpart G – Gear and Equipment for Rigging and Materials Handling
1915 Subpart H – Tools and Related Equipment
1915 Subpart I – Personal Protective Equipment
1915 Subpart J – Ship's Machinery and Piping Systems
1915 Subpart L – Electrical Machinery
1915 Subpart Z – Toxic and Hazardous Substances

6.2.2 National Contingency Plan

The regulation known as 40 CFR 300, the National Oil and Hazardous Substance Pollution Contingency Plan (NCP), describes procedures for responding to hazardous substance releases and oil discharges. The regulation is extensive and comprehensive. We have captured essential elements of the NCP in rapid look-up tables so that the reader will know which sections to examine and adopt more readily. The title of each subject area notes the section (§) number of the NCP to refer to. The organization is in accordance with major sections that pertain to the incident. We do not dwell on and largely do not address statutory requirements but rather technical subjects.

A number of abbreviations are used in the tables and throughout Annex E of the NCP, which addresses oil spills. A number of these are provided in Table 6.5.

6.2.2.1 Useful Definitions

The following are important definitions applied in Annex E of the NCP and are referred to in discussions throughout the volume.

Table 6.5 Abbreviations used in tables below and in the NCP, Annex E.

Term	Abbreviation/Definition
AC	Area Committee – as provided for by CWA sections 311(a)(18) and (j)(4), means the entity appointed by the President consisting of members from qualified personnel of federal, state, and local agencies with responsibilities that include preparing an area contingency plan for an area designated by the President.
ACP	Area Contingency Plan – as defined by CWA sections 311(a)(19) and (j)(4) means the plan prepared by an Area Committee that is developed to be implemented in conjunction with the NCP and RCP, in part to address removal of a worst case discharge and to mitigate or prevent a substantial threat of such a discharge from a vessel, offshore facility, or onshore facility operating in or near an area designated by the President.
ATSDR	Agency for Toxic Substances and Disease Registry
CDC	Centers for Disease Control
CERCLA	Comprehensive Environmental Response, Compensation, and Liability Act of 1980, as amended (Pub. L. 96–510, as amended).
CWA	Clean Water Act
DOI	Department of Interior
DOL	Department of Labor
DOT	Department of Transportation
DRAT	District Response Advisory Team
DRG	District Response Group – *Coast Guard District Response Group* (DRG) as provided for by CWA sections 311(a)(20) and (j)(3), means the entity established by the Secretary of the department in which the USCG is operating within each USCG district and shall consist of: the combined USCG personnel and equipment, including firefighting equipment, of each port within the district; additional prepositioned response equipment; and a district response advisory team.

EPA	Environmental Protection Agency
ERT	Environmental Response Team
ESF	Emergency Support Functions
FEMA	Federal Emergency Management Agency
FRP	Federal Response Plan – the agreement signed by 25 federal departments and agencies in April 1987 and developed under the authorities of the Earthquake Hazards Reduction Act of 1977 and the Disaster Relief Act of 1974, as amended by the Stafford Disaster Relief Act of 1988.
HHS	Department of Health and Human Services
LEPC	Local Emergency Planning Committee
NCP	National Contingency Plan
NIOSH	National Institute for Occupational Safety and Health
NOAA	National Oceanic and Atmospheric Administration
NPFC	National Pollution Funds Center – entity established by the Secretary of Transportation whose function is the administration of the Oil Spill Liability Trust Fund (OSLTF). Among the NPFC's duties are: providing appropriate access to the OSLTF for federal agencies and states for removal actions and for federal trustees to initiate the assessment of natural resource damages; providing appropriate access to the OSLTF for claims; and coordinating cost recovery efforts.
NRC	National Response Center
NRS	National Response System – mechanism for coordinating response actions by all levels of government in support of the OSC. The NRS is composed of the NRT, RRTs, OSC, Area Committees, and Special Teams and related support entities.
NRT	National Response Team

Table 6.5 (cont.) Abbreviations used in tables below and in the NCP, Annex E.

Term	Abbreviation/Definition
NSF	National Strike Force – special team established by the USCG, including the three USCG Strike Teams, the Public Information Assist Team (PIAT), and the National Strike Force Coordination Center. The NSF is available to assist OSCs in their preparedness and response duties.
NSFCC	National Strike Force Coordination Center – authorized as the National Response Unit by CWA section 311(a)(23) and (j)(2), means the entity established by the Secretary of the department in which the USCG is operating at Elizabeth City, North Carolina, with responsibilities that include administration of the USCG Strike Teams, maintenance of response equipment inventories and logistic networks, and conducting a national exercise program.
OPA	Oil Pollution Act – signed into law in August 1990, largely in response to rising public concern following the Exxon Valdez incident. The OPA was intended to improve the ability to prevent and respond to oil spills by establishing provisions that expand the federal government's ability, and provide the money and resources necessary, to respond to oil spills. The Act also created the national Oil Spill Liability Trust Fund, which is available to provide up to one billion dollars per spill incident. In addition, the OPA provided new requirements for contingency planning both by government and industry. The National Oil and Hazardous Substances Pollution Contingency Plan (NCP) has been expanded in a three-tiered approach: the Federal government is required to direct all public and private response efforts for certain types of spill events; Area Committees – composed of federal, state, and local government officials – must develop detailed, location-specific Area Contingency Plans; and owners or operators of vessels and certain facilities that pose a serious threat to the environment must prepare their own Facility Response Plans. The OPA increased penalties for regulatory noncompliance, broadened the response and enforcement authorities of the Federal government, and preserved State authority to establish law governing oil spill prevention and response.

	Key Provisions of the Oil Pollution Act
	§1002(a) Provides that the responsible party for a vessel or facility from which oil is discharged, or which poses a substantial threat of a discharge, is liable for: (1) certain specified damages resulting from the discharged oil; and (2) removal costs incurred in a manner consistent with the National Contingency Plan (NCP).
	§1002(c) Exceptions to the Clean Water Act (CWA) liability provisions include: (1) discharges of oil authorized by a permit under Federal, State, or local law; (2) discharges of oil from a public vessel; or (3) discharges of oil from onshore facilities covered by the liability provisions of the Trans-Alaska Pipeline Authorization Act.
	§1002(d) Provides that if a responsible party can establish that the removal costs and damages resulting from an incident were caused solely by an act or omission by a third party, the third party will be held liable for such costs and damages.
	§1004 The liability for tank vessels larger than 3,000 gross tons is increased to $1,200 per gross ton or $10 million, whichever is greater. Responsible parties at onshore facilities and deepwater ports are liable for up to $350 millon per spill; holders of leases or permits for offshore facilities, except deepwater ports, are liable for up to $75 million per spill, plus removal costs. The Federal government has the authority to adjust, by regulation, the $350 million liability limit established for onshore facilities.
	§1016 Offshore facilities are required to maintain evidence of financial responsibility of $150 million and vessels and deepwater ports must provide evidence of financial responsibility up to the maximum applicable liability amount. Claims for removal costs and damages may be asserted directly against the guarantor providing evidence of financial responsibility.
	§1018(a) The Clean Water Act does not preempt State Law. States may impose additional liability (including unlimited liability), funding mechanisms, requirements for removal actions, and fines and penalties for responsible parties.
	§1019 States have the authority to enforce, on the navigable waters of the State, OPA requirements for evidence of financial responsibility. States are also given access to Federal funds (up to $250,000 per

Table 6.5 (cont.) Abbreviations used in tables below and in the NCP, Annex E.

Term	Abbreviation/Definition
	incident) for immediate removal, mitigation, or prevention of a discharge, and may be reimbursed by the Trust fund for removal and monitoring costs incurred during oil spill response and clean-up efforts that are consistent with the National Contingency Plan (NCP). **§4202** Strengthens planning and prevention activities by: (1) providing for the establishment of spill contingency plans for all areas of the U.S. (2) mandating the development of response plans for individual tank vessels and certain facilities for responding to a worst case discharge or a substantial threat of such a discharge; and (3) providing requirements for spill removal equipment and periodic inspections. **§4301(a) and (c)** The fine for failing to notify the appropriate Federal agency of a discharge is increased from a maximum of $10,000 to a maximum of $250,000 for an individual or $500,000 for an organization. The maximum prison term is also increased from one year to five years. The penalties for violations have a maximum of $250,000 and 15 years in prison. **§4301(b)** Civil penalties are authorized at $25,000 for each day of violation or $1,000 per barrel of oil discharged. Failure to comply with a Federal removal order can result in civil penalties of up to $25,000 for each day of violation. **§9001(a)** Amends the Internal Revenue Act of 1986 to consolidate funds established under other statutes and to increase permitted levels of expenditures. Penalties and funds established under several laws are consolidated, and the Trust Fund borrowing limit is increased from $500 million to $1 billion.
OSC	On-Scene Coordinator – The OSC works within a framework of the response management structure of the system (e.g., a unified command system) that brings together the functions of the federal government, the state government, and the responsible party to achieve an effective and efficient response, where the OSC maintains authority. Except in a case when the OSC is required to direct the response to a discharge that may pose a substantial threat to the public health or welfare of the United States (including but not limited to fish, shellfish, wildlife, other natural resources, and the public and private beaches and shorelines

	of the United States), the OSC may allow the responsible party to voluntarily and promptly perform removal actions, provided the OSC determines such actions will ensure an effective and immediate removal of the discharge or mitigation or prevention of a substantial threat of a discharge. If the RP does conduct the removal, the OSC shall ensure adequate surveillance over whatever actions are initiated. If effective actions are not being taken to eliminate the threat, or if removal is not being properly done, the OSC should, to the extent practicable under the circumstances, so advise the RP. If the responsible party does not respond properly the OSC shall take appropriate response actions and should notify the responsible party of the potential liability for federal response costs incurred by the OSC pursuant to the OPA and CWA. Where practicable, continuing efforts should be made to encourage response by RPs.
OSHA	Occupational Safety and Health Administration
OSLTF	Oil Spill Liability Trust Fund
RCP	Regional Contingency Plan
RP	Responsible Party – as defined by section 1001 of the OPA means the following: (a) Vessels – In the case of a vessel, any person owning, operating, or demise chartering the vessel. (b) Onshore Facilities – In the case of an onshore facility (other than a pipeline), any person owning or operating the facility, except a federal agency, state, municipality, commission, or political subdivision of a state, or any interstate body, that as the owner transfers possession and right to use the property to another person by lease, assignment, or permit. (c) Offshore Facilities–In the case of an offshore facility (other than a pipeline or a deepwater port licensed under the Deepwater Port Act of 1974 (33 U.S.C. 1501 *et seq.*)), the lessee or permittee of the area in which the facility is located or the holder of a right of use and easement granted under applicable state law or the Outer Continental Shelf Lands Act (43 U.S.C. 1301–1356) for the area in which the facility is located (if the holder is a different person than the lessee or permittee), except a federal agency, state, municipality, commission, or political subdivision of a state, or any interstate body, that as owner transfers possession and right to use the property to another person by lease, assignment,

Table 6.5 (cont.) Abbreviations used in tables below and in the NCP, Annex E.

Term	Abbreviation/Definition
	or permit. (d) Deepwater Ports – In the case of a deepwater port licensed under the Deepwater Port Act of 1974 (33 U.S.C. 1501–1524), the licensee. (e) Pipelines – In the case of a pipeline, any person owning or operating the pipeline. (f) Abandonment – In the case of an abandoned vessel, onshore facility, deepwater port, pipeline, or offshore facility, the person who would have been responsible parties immediately prior to the abandonment of the vessel or facility.
RRT	Regional Response Team
SONS	Spill of National Significance
USFWS	United States Fish and Wildlife Service

Activation means notification by telephone or other expeditious manner or, when required, the assembly of some or all appropriate members of the RRT or NRT.

Bioremediation agents are microbiological cultures, enzyme additives, or nutrient additives that are deliberately introduced into an oil discharge and that will significantly increase the rate of biodegradation to mitigate the effects of the discharge.

Burning agents are additives that, through physical or chemical means, improve the combustibility of the materials to which they are applied.

Chemical agents are elements, compounds, or mixtures that coagulate, disperse, dissolve, emulsify, foam, neutralize, precipitate, reduce, solubilize, oxidize, concentrate, congeal, entrap, fix, make the pollutant mass more rigid or viscous, or otherwise facilitate the mitigation of deleterious effects or the removal of the oil pollutant from the water. Chemical agents include biological additives, dispersants, sinking agents, miscellaneous oil spill control agents, and burning agents, but do not include solvents.

Clean natural seawater means that the source of this seawater must not be heavily contaminated with industrial or other types of effluent.

Coastal waters, for the purpose of classifying the size of discharges, means the waters of the coastal zone except for the Great Lakes and specified ports and harbors on inland rivers.

Coastal zone, as defined for the purpose of the NCP, means all United States waters subject to the tide, United States waters of the Great Lakes, specified ports and harbors on inland rivers, waters of the contiguous zone, other waters of the high seas subject to the NCP, and the land surface or land substrata, ground waters, and ambient air proximal to those waters. The term coastal zone delineates an area of federal responsibility for response action. Precise boundaries are determined by EPA/USCG agreements and are identified in federal regional contingency plans.

Contiguous zone means the zone of the high seas established by the United States under Article 24 of the Convention on the Territorial Sea and Contiguous Zone, which is contiguous to the territorial sea and which extends nine miles seaward from the outer limit of the territorial sea.

Discharge includes but is not limited to any spilling, leaking, pumping, pouring, emitting, emptying, or dumping of oil, but the term excludes discharges in compliance with a permit under section 402 of the CWA, which are discharges resulting from circumstances identified, reviewed, and made a part of the public record with respect to a permit issued or modified under section 402 of the CWA. For purposes of the NCP, discharge also means substantial threat of discharge.

Dispersants are chemical agents that emulsify, disperse, or solubilize oil into the water column or promote the surface spreading of oil slicks to facilitate the dispersal of the oil into the water column.

Exclusive economic zone means the zone established by Presidential Proclamation Numbered 5030, dated March 10, 1983, which includes the ocean waters of the areas referred to as "eastern special areas" in Article 3(1) of the Agreement between the United States of America and the Union of Soviet Socialist Republics on the Maritime Boundary, signed June 1, 1990.

Facility means any structure, group of structures, equipment, or device (other than a vessel) that is used for one or more of the following purposes: exploring for, drilling for, producing, storing, handling, transferring, processing, or transporting oil. This term includes any motor vehicle, rolling stock, or pipeline used for one or more of these purposes.

Inland waters, for the purposes of classifying the size of discharges, means those waters of the United States in the inland zone, waters of the Great Lakes, and specified ports and harbors on inland rivers.

Inland zone means the environment inland of the coastal zone excluding the Great Lakes and specified ports and harbors on inland rivers. The term inland zone delineates an area of federal responsibility for response action. Precise boundaries are determined by EPA/USCG agreements and are identified in federal regional contingency plans.

Lead administrative trustee means a natural resource trustee who is designated on an incident-by-incident basis for the purpose of pre-assessment and damage assessment and is chosen by the other trustees, whose natural resources are affected by the incident. The lead administrative trustee facilitates effective and efficient communication during response operations between the OSC and other natural resource trustees conducting activities associated with damage assessment. The lead administrative trustee is responsible for applying to the OSC for access to response operations resources on behalf of all trustees for initiation of damage assessment.

Miscellaneous oil spill control agent is any product, other than a dispersant, sinking agent, surface washing agent, surface collecting agent, bioremediation agent, burning agent, or sorbent that can be used to enhance oil spill clean-up, removal, treatment, or mitigation.

Natural resources means land, fish, wildlife, biota, air, water, groundwater, drinking water supplies, and other such resources belonging to, managed by, held in trust by, appertaining to, or otherwise controlled by the United States (including the resources of the exclusive economic zone defined by the Magnuson Fishery Conservation and Management Act of 1976), any state or local government, any foreign government, any Indian tribe, or, if such resources are subject to a trust restriction on alienation, any member of an Indian tribe.

Navigable waters, as defined by 40 CFR 110.1, are the waters of the United States, including the territorial seas. The term includes the following:

- All waters that are currently used, were used in the past, or may be susceptible to use in interstate or foreign commerce, including all waters that are subject to the ebb and flow of the tide
- Interstate waters, including interstate wetlands
- All other waters such as intrastate lakes, rivers, streams (including intermittent streams), mudflats, sandflats, and wetlands, the use, degradation, or destruction of which would affect or could affect interstate or foreign commerce, including any such waters that are or could be used by interstate or foreign travelers for recreational or other purposes, from which fish or shellfish are or could be taken and sold in interstate or foreign commerce, and that are used or could be used for industrial purposes by industries in interstate commerce, all impoundments of waters otherwise defined as navigable waters under this section, tributaries of waters including adjacent wetlands, and wetlands adjacent to waters (waters of the United States do not include prior converted cropland)

Offshore facility, as defined by section 311(a)(11) of the CWA, means any facility of any kind located in, on, or under any of the navigable waters of the United States and any facility of any kind that is subject to the jurisdiction of the United States and is located in, on, or under any other waters, other than a vessel or a public vessel.

Oil, as defined by section 311(a)(1) of the CWA, means oil of any kind or in any form, including, but not limited to, petroleum, fuel oil, sludge, oil refuse, and oil mixed with wastes other than dredged spoil. Oil, as defined by section 1001 of the OPA, means oil of any kind or in any form, including, but not limited to, petroleum, fuel oil, sludge, oil refuse, and oil mixed with wastes other than dredged spoil, but does not include petroleum, including crude oil or any fraction thereof, which is specifically listed or designated as a hazardous substance under subparagraphs (A) through (F) of section 101(14) of the Comprehensive Environmental Response, Compensation, and Liability Act (42 U.S.C. 9601) and which is subject to the provisions of that Act.

Oil Spill Liability Trust Fund means the fund established under section 9509 of the Internal Revenue Code of 1986 (26 U.S.C. 9509).

Onshore facility, as defined by section 311(a)(10) of the CWA, means any facility (including, but not limited to, motor vehicles and rolling stock) of any kind located in, on, or under any land within the United States other than submerged land.

On-site means the areal extent of contamination and all suitable areas in very close proximity to the contamination necessary for implementation of a response action.

Public vessel, as defined by section 311(a)(4) of the CWA, means a vessel owned or bareboat-chartered and operated by the United States, or by a state or political subdivision thereof, or by a foreign nation, except when such vessel is engaged in commerce.

Remove or removal, as defined by section 311(a)(8) of the CWA, refers to containment and removal of oil or hazardous substances from the water and shorelines or the taking of such other actions as may be necessary to minimize or mitigate damage to the public health or welfare (including, but not limited to, fish, shellfish, wildlife, public and private property, and shorelines and beaches) or to the environment. For the purpose of the NCP, the term also includes monitoring of action to remove a discharge.

Sinking agents are additives that are applied to oil discharges in order to sink floating pollutants below the water surface.

Size classes of discharges refers to the following size classes of oil discharges that are provided as guidance to the OSC and serve as the criteria for the actions delineated in subpart D. They are not meant to imply associated degrees of hazard to public health or welfare, nor are they a measure of environmental injury. Any oil discharge that poses a substantial threat to public health or welfare or the environment or results in significant public concern shall be classified as a major discharge regardless of the following quantitative measures: (a) minor discharge means a discharge in inland waters of less than 1,000 gallons of oil or a discharge to the coastal waters of less than 10,000 gallons of oil, (b) medium discharge means a discharge of 1,000 to 10,000 gallons of oil to the inland waters or a discharge of 10,000 to 100,000 gallons of oil to the coastal waters, and (c) major discharge means a discharge of more than 10,000 gallons of oil to the inland waters or more than 100,000 gallons of oil to the coastal waters.

Sorbents are essentially inert and insoluble materials that are used to remove oil and hazardous substances from water by adsorption, in which the oil or hazardous substance is attracted to the sorbent surface and then adheres to it, by absorption, in which the oil or hazardous substance penetrates the pores of the sorbent material, or by a combination of the two. Sorbents are generally manufactured in particulate form for spreading over an oil slick or as sheets, rolls, pillows, or booms.

6.2.2.2 *Planning and Coordination Structure (§ 300.205)*

Table 6.6 summarizes the levels and areas of responsibilities for planning and the coordination of responses to a maritime oil spill.

6.2.2.3 *Operational Response Phases for Oil Removal*

Relevant sections include the following:

- § 300.300 Phase I – Discovery or notification
- § 300.305 Phase II – Preliminary assessment and initiation of action
- § 300.310 Phase III – Containment, countermeasures, clean-up, and disposal
- § 300.315 Phase IV – Documentation and cost recovery

Table 6.7 and Figures 6.7 and 6.8 summarize required response actions for Operational Response Phases I through IV.

To better illustrate the procedures to be followed, the following paraphrases an oil spill scenario that OSHA[13] has published.

---

[13]http://www.osha.gov/Publications/3172/3172.html

Table 6.6 Planning and coordination management structure.

Level	Responsibilities
National	NRT is responsible for national planning and coordination.
Regional	RRTs are responsible for regional planning and coordination.
Area	Prepare an ACP for their areas.
	Work with appropriate federal, state, and local officials to enhance the contingency planning of those officials and to assure pre-planning of joint response efforts, including appropriate procedures for mechanical recovery, dispersal, shoreline clean-up, protection of sensitive environmental areas, and protection, rescue, and rehabilitation of fisheries and wildlife.
	Work with appropriate federal, state, and local officials to expedite decisions for the use of dispersants and other mitigating substances and devices.
State	Designate emergency planning districts, appoint Local Emergency Planning Committees (LEPCs), supervise and coordinate their activities, and review local emergency response plans, which are described in §300.215. The SERC also is to establish procedures for receiving and processing requests from the public for information generated by Title III reporting requirements and to designate an official to serve as coordinator for information.
Local	Emergency planning districts are designated by the SERC in order to facilitate the preparation and implementation of emergency plans. Each LEPC is to prepare a local emergency response plan for the emergency planning district and establish procedures for receiving and processing requests from the public for information generated by Title III reporting requirements. The LEPC is to appoint a chair and establish rules for the LEPC. The LEPC is to designate an official to serve as coordinator for information and designate in its plan a community emergency coordinator.

The Incident – A tugboat is pushing a barge loaded with a sour crude that contains sulfur, benzene, toluene, and xylene to an anchorage for lightering (See Figure 6.9). The anchorage is within sight of the local Coast Guard Marine Safety Office. Contractor resources also are located in the port within view of the anchorage. The tug pushing the barge miscalculates the turning radius, causing the

Table 6.7 Elements of operational phases of response and responsibilities.

Section	Responsible Party	Actions Required
§ 300.300 Phase I – Discovery or notification	Any person in charge of a vessel or a facility that has knowledge of a discharge or spill	Immediately notify the NRC, USCG and/or U.S.EPA pre-designated OSC
	USCG and/or U.S.EPA pre-designated OSC	Relay all such reports promptly to the NRC.
	NRC	Notify the OSC
	OSC	Ensure notification of the appropriate state agency expected to be affected by the discharge. The OSC shall then proceed with next phases of response.
§ 300.305 Phase II – Preliminary assessment and initiation of action	OSC	Initiate a preliminary assessment. The preliminary assessment shall be conducted using available information, supplemented where necessary and possible by an on-scene inspection. The OSC shall undertake actions to:
		(1) Evaluate the magnitude and severity of the discharge or threat to public health or Sovereign, or the environment
		(2) Assess the feasibility of removal
		(3) Identify potentially responsible parties (RPs)

		(4) Exercise authority over the RP as appropriate to address immediate removal and clean-up. See definition of OSC at beginning of this section. The OSC has authority to initiate and direct the following actions:
		(i) Remove or arrange for the removal of a discharge, and mitigate or prevent a substantial threat of a discharge.
		(ii) Direct or monitor all federal, state, and private actions to remove a discharge.
		(iii) Remove and, if necessary, destroy a vessel discharging, or threatening to discharge, by whatever means are available.
		If the discharge results in a substantial threat the OSC must direct all response efforts. The OSC should declare to spill response participants that the federal government will direct the response.
		Ensure that the natural resource trustees are notified.
		Consult with the affected trustees on the appropriate removal actions.
	OSC and Trustees	Coordinate assessments, evaluations, investigations, and planning with respect to appropriate removal actions.

Table 6.7 (cont.) Elements of operational phases of response and responsibilities.

Section	Responsible Party	Actions Required
	Trustees	Provide advice concerning recommended actions with regard to trustee resources potentially affected. Communicate with OSC on their activities in natural resource damage assessment s. Assure that all data from the natural resource damage assessment activities that may support more effective operational decisions are provided to the OSC. The lead administrative trustee is also responsible for applying to the NPFC for funding for initiation of damage assessment for injuries to natural resources.
§ 300.310 Phase III – Containment, countermeasures, clean-up, and disposal	OSC	Oversight
§ 300.315 Phase IV – Documentation and cost recovery	OSC	Oversight

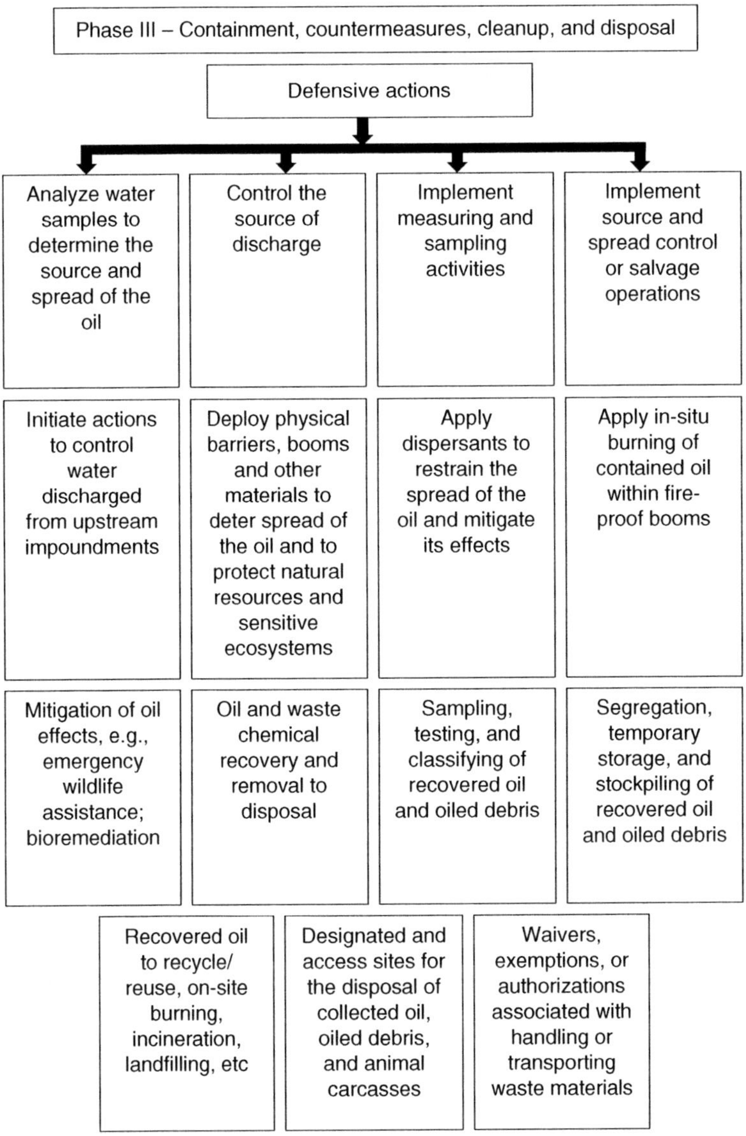

Figure 6.7 Elements of Phase III response – containment, countermeasures, clean-up and waste disposal.

barge to collide with the stern of a tank vessel. The vessels lock together. As a result, two of the barge's port cargo tanks are breached, releasing several thousand gallons of crude oil into the water. Prevailing winds and currents carry the oil away from the

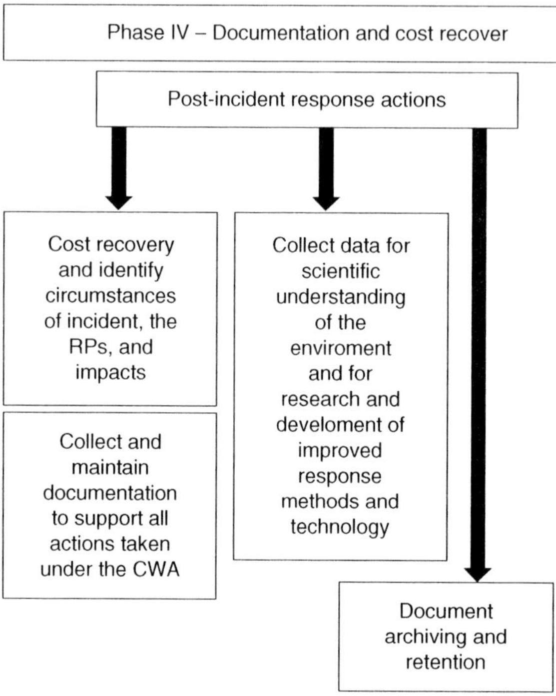

Figure 6.8 Elements of Phase IV post-incident response – documentation and cost recovery.

vessels. Some oil is trapped between the barge and the tank vessel. The incident occurs midday on a sunny summer day. The wind is 5 knots. The water temperature is around 70°F and the ambient temperature is 84°F.

Discovery – Members of the tug crew immediately notify the Coast Guard of the incident. Personnel that are initially engaged in the matter have been trained and certified at the First Responder Awareness Level. The tug and tank vessel crews remain with their vessels to ensure control of their ships. Crew members report that vapors from the fresh oil begin to irritate their eyes. Both captains order crew members to remain upwind of the oil and avoid the area between the vessels where the oil is trapped.

Initial Emergency Response Actions – The owner of the barge is located across the country. According to the company's emergency response plan, the owner calls an Oil Spill Removal Organization (OSRO) to clean up the spill. The tug captain is designated by the

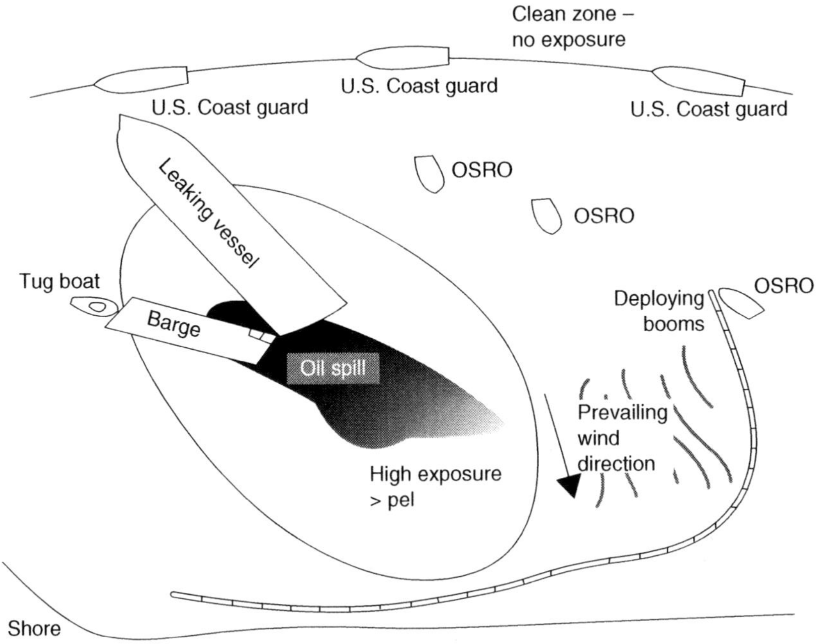

Figure 6.9 Illustration explaining OSHA spill scenario.

owner to be the company's representative. Because the owner is not on-scene and does not participate in spill response activities, he is not required to have HAZWOPER training. The tug captain is on-scene making decisions in the response. Therefore, the tug captain must have Incident Commander training for emergency response.

The Coast Guard radios the tug captain to confirm the oil type, the same type listed on the cargo manifest faxed by the owner. The owner also faxes the cargo manifest to the OSRO. The Coast Guard uses references and advice from federal health professionals to estimate a safe distance from the spill.

After consultation with references, mathematical computer models, and federal health and safety professionals, the Coast Guard creates a safety zone around the spill and uses its boats to enforce it. The safety zone is conservatively large to maximize public safety. Support staff, news media, and other workers who will have no exposure to hazards will access the safety zone to perform their jobs.

The Initial Assessment – An hour after the initial oil release, two OSRO boats approach the spill from downwind to characterize the plume and determine the release zone boundaries. Their crews monitor exposure levels using air-monitoring equipment. The personnel in this case are trained and certified Hazardous Materials Technicians or Specialists. Because exposure to hydrogen sulfide and benzene is expected, these responders must wear self-contained breathing apparatus (SCBA) while confirming estimated exposures. They may be able to switch to air-purifying respirators after the assessment.

The Hazardous Materials Technicians continue monitoring as the OSRO boats approach the spill until reaching permissible exposure limits (PELs). The boats work around the spill area to identify the limits of the High Exposure zone caused by vapors escaping the oil. Colorimetric tubes confirm air concentrations of 0.5 ppm of benzene and 4 ppm of hydrogen sulfide at 200 feet (61 meters) downwind from the slick and 100 feet (30 meters) across the widest part of the plume.

The responders complete site characterization in areas where oil could be trapped between vessels, beneath piers, in dock corners, and other bounded spaces.

Outside the high exposure zone (referred to as the hot zone), workers prepare deflection and exclusion booms to divert the oil to a recovery site. Personnel performing these tasks are trained and certified at the First-Responder Operations Level.

The response team sets up a command center to coordinate response and clean-up activities. This area supports the Incident Commander, supervisors of the clean-up crews with hourly requirements dependent on exposure, and other personnel (training depends on duties and exposures).

Mechanical Recovery of Oil – Four hours have passed. Air measurements of hydrogen sulfide, benzene, toluene, and xylene, taken with colorimetric tubes, register well below permissible exposure limits near the leading edge of the slick. An industrial hygienist and a marine chemist conduct a more detailed site characterization using advanced air-monitoring equipment such as a portable infrared analyzer and portable gas chromatograph. These specialists confirm that the air exposures from the oil slick are below permissible limits. However, air concentrations are above permissible limits near oil trapped between the vessels and oil remaining in the damaged tanks.

Oil recovery skimmers deployed earlier arrive to skim free-floating oil. Containment boom and skimmer operators work from a boat. These operators wear goggles and protective clothing because the oil could irritate their skin and eyes. Because air concentrations are below permissible limits, operators are not required to wear respirators. Personnel performing tasks are trained and certified at the First-Responder Operations Level.

The ships' support crews remain upwind of the spill where they are unlikely to be exposed over limits.

Shoreline Clean-up Activities – Six hours have now passed since the spill notification. Oil has now reached the shoreline. Skimmers near the shore remove oil that was deflected into recovery areas. Crews are deployed to clean the shoreline.

The oil continues to irritate skin and eyes, so workers who are cleaning the shoreline wear chemical protective clothing, gloves, booties, and goggles. Because of the increased clothing requirements, site safety officer assistants should begin monitoring for heat stress, and the workers should be considered or classified as Skilled Support Personnel. They are unlikely to be exposed over limits, depending on the assistants' job duties.

Responders form decontamination lines at each shoreline clean-up area. Each worker decontaminates before a break period and at the end of the shift. Workers dispose of outer booties, wash or remove oiled clothing, dispose of outer gloves, wash goggles, and remove inner gloves. To minimize the spread of contamination, workers conduct their own decontamination while a worker stands by to assist. The decontamination assistants take responsibility for maintaining and dismantling the decontamination line.

Vessel Securement – A skimming team recovers oil trapped between the vessels. Vessel personnel in SCBAs measure chemical and oxygen concentrations and the percentage of the lower explosive limit (LEL) for the space between the vessels. Personnel performing these duties are trained and certified Hazardous Materials Technician/Specialists. Results of the monitoring indicate that workers conducting skimming operations must also wear SCBA. The workers decide to wait until exposure levels register below permissible limits before removing the oil.

Aboard the barge, a pumping team prepares to pump oil from the damaged cargo tanks into the tank vessel. Explosion is a risk during pumping operations, so responders measure the LEL percentage and oxygen concentration throughout the operation. Before initiating the pumping operations, workers in SCBAs put vapor recovery systems in place. Once the vapor recovery systems are in place, measured chemical concentration levels drop below permissible exposure limits, and workers begin pumping.

Final Response Stages – Two days later, no free-floating oil remains. Pumping operations are completed. No hazardous chemicals are detected by air-monitoring equipment. The oil is weathered and is no longer an irritant. However, slip and trip hazards persist. Shoreline clean-up operations continue.

No repair facility is located within a reasonably safe distance for transport, so workers conduct initial, temporary repairs for safe sailing. They clean the damaged tanks inside and out before patching and welding them. There is a risk of explosion, oxygen deficiency, and overexposures to chemicals. Forced ventilation makes the holds safe, and a competent person tests them before the cleaning begins. A marine chemist tests and certifies the areas Safe for Hot Work before cutting and/or welding operations begin.

The company's site safety and health plan does not cover the disposal of waste containers. Because of this oversight, oiled debris and waste decay in the containers are releasing hydrogen sulfide. A hazardous materials response team removes the contaminated waste from the containers. This personnel is a General Spill Site Worker. After a week has passed, efforts are made to speed up the removal of remaining debris on shore, so training is offered to a team of local volunteers (these are workers for job activities that have been assessed as unlikely to expose individuals above PELs). Trained supervisors oversee the volunteers. As shoreline clean-up progresses, shoreline clean-up assessment teams begin to verify that shorelines are clean.

In the above OSHA spill scenario, we are dealing with a small spill. Efforts to respond to the spill of a couple of thousand gallons over a small area requires several weeks of effort, dozens of highly trained hazard materials specialists and emergency responders, and dedicated equipment including skimmers, booms, air monitoring instrumentation, and more. The current Gulf oil spill disaster is on the order of 50,000 times larger and is engaging tens of thousands of personnel, millions of feet of boom, and considerably more complex logistical efforts in an attempt to keep thousands of miles of shorelines from being impacted.

Table 6.8 highlights important elements of a few other sections of the NCP.

Table 6.8 Additional highlights from the NCP.

Section	Title	Key Elements
300.317	National response priorities	Safety of human life is 1st priority during response action.
		Stabilizing the situation to preclude the event from worsening is the 2nd priority.
		Response must use all necessary containment and removal tactics in a coordinated manner.
		All parts of the response strategy should be addressed concurrently.
300.320	General pattern of response	When the OSC receives a report of a discharge, actions are taken in the following sequence:
		(1) Investigate the report to determine pertinent information such as the threat posed to public health or welfare, the type and quantity of polluting material, and the source of the discharge.
		(2) Officially classify the size (i.e., minor, medium, major) and type. Some discharges that are classified as substantial threats to the public health or welfare may be further classified as a spill of national significance.
		When discharge and threat are confirmed, the OSC notifies the RRT and the NRC.
		OSC recommends the activation of the RRT.
		When the investigation shows that a minor discharge exists, the OSC monitors the situation to ensure that proper removal action is taken.
		(3) If the OSC determines that effective and immediate removal, mitigation, or prevention of a discharge can be achieved by private party efforts, and where the discharge does not pose a substantial threat, OSC determines whether the responsible party is properly carrying out removal. Removal is being done properly when:

Table 6.8 (cont.) Additional highlights from the NCP.

Section	Title	Key Elements
		The responsible party applies the resources called for in its response plan to remove, minimize, or mitigate threat(s) to public health and welfare and the environment
		Removal efforts must be accordance with applicable regulations.
		(4) Where appropriate, determine whether a state or political subdivision thereof has the capability to carry out any or all removal actions. If so, the OSC may arrange funding to support actions.
		(5) Ensure prompt notification of the trustees of affected natural resources in accordance with the applicable RCP and ACP.
		Removal to be considered complete when determined by the OSC in consultation with the Governor or Governors of affected states. When the OSC considers removal complete, OSLTF removal funding shall end. This determination shall not preclude additional removal actions under applicable state law.
300.322	Response to substantial threats to public health or welfare of the United States	OSC to determine whether a discharge results in a substantial threat to public health or welfare.
		If a substantial threat identified, the OSC directs all federal, state, or private actions to remove the discharge or to mitigate or prevent the threat of such a discharge. Following actions are directed by OCS:
		Remove or arrange for the removal of the discharge;
		Mitigate or prevent the substantial threat of the discharge; and
		Remove and, if necessary, destroy a vessel discharging, or threatening to discharge, by whatever means are available.

Table 6.8 (cont.) Additional highlights from the NCP.

Section	Title	Key Elements
		Assess opportunities for the use of various special teams and other assistance including the use of the services of the NSFCC, as appropriate.
		Request immediate activation of the RRT.
		Take whatever additional response actions are deemed appropriate, including, but not limited to, implementation of the ACP as required by section 311(j)(4) of the CWA or relevant tank vessel or facility response plan required by section 311(j)(5) of the CWA. When requested by the OSC, the lead agency or RRT shall dispatch appropriate personnel to the scene of the discharge to assist the OSC. This assistance may include technical support in the agency's areas of expertise and disseminating information to the public. The lead agency shall ensure that a contracting officer is available on scene, at the request of the OSC.
300.324	Response to worst case discharges	Notify the NSFCC
		The implementation of the worst case portion of an approved tank vessel or facility response plan required by section 311(j)(5) of the CWA
		Implement the worst case portion of the ACP required by section 311(j)(4) of the CWA
		Take whatever additional response actions are deemed appropriate.
		Under the direction of the OSC, the NSFCC shall coordinate use of private and public personnel and equipment, including strike teams, to remove a worst case discharge and mitigate or prevent a substantial threat of such a discharge

Subpart J of the NCP addresses the arsenal of technologies for combating oil spills. Subpart J is titled "Use of Dispersants and Other Chemicals." This subpart states that the EPA is required to maintain a schedule of dispersants and other chemical or bioremediation products that may be authorized for use on oil discharges. The current list of chemicals is included in Table 4.1 of Chapter 4 of this volume.

In the event of a spill, RRTs and Area Committees will consider the use of these products as part of their planning activities. This consideration stretches over a broad category of product types that include dispersants, surface-washing agents, surface collecting agents, bioremediation agents, or miscellaneous oil spill control agents listed on the NCP Product Schedule. A problem is that, although there are several categories of products and application areas for combating spills, there are not many products from which to choose from as pointed out in Chapter 4.

The decisions as to which types of chemicals to apply to combat a spill are based in part on consideration of factors such as the potential sources and types of oil spilled, the existence and location of environmentally sensitive resources that might be impacted by spilled oil, available product and storage locations, available equipment and adequately trained operators, and the available means to monitor product application and effectiveness. Throughout the decision making process, the EPA applies the Hazard Ranking System (HRS). The HRS is the principal mechanism the U.S. EPA uses to place sites on the National Priorities List (NPL). The HRS serves as a screening device to evaluate the potential for releases of uncontrolled hazardous substances to cause human health or environmental damage. The HRS provides a measure of relative rather than absolute risk. It is designed so that it can be consistently applied to a wide variety of sites and situations.

6.2.3 Environmental and Health and Safety Definitions

The following are important definitions that apply to health risk considerations, the safe handling of chemicals and waste products, marine toxicity, and site management. In the sections following these definitions, we get into specific elements of best management practices and technologies applicable to worker safety.

- *Acute toxicity* is the measure of toxicological responses that result from a single exposure to a substance or from multiple exposures within a short period of time (typically several days or less). Specific measures of acute toxicity used within the HRS include lethal $dose_{50}$(LD_{50}) and lethal $concentration_{50}$(LC_{50}), typically measured within a 24-hour to 96-hour period.
- *Ambient Aquatic Life Advisory Concentrations (AALACs)* are the EPA's advisory concentration limits for acute or chronic toxicity to aquatic organisms as established under section 304(a)(1) of the Clean Water Act, as amended.
- *Ambient Water Quality Criteria (AWQC)* is the EPA's maximum acute or chronic toxicity concentrations for protection of aquatic life and its uses, as established under section 304(a)(1) of the Clean Water Act, as amended.
- *Bioconcentration factor (BCF)* is the measure of the tendency for a substance to accumulate in the tissue of an aquatic organism. BCF is determined by the extent of the partitioning of a substance, at equilibrium, between the tissue of an aquatic organism and water. As the ratio of the concentration of a substance in the organism divided by the concentration in water, higher BCF values reflect a tendency for substances to accumulate in the tissue of aquatic organisms. (It is unitless.)
- *Biodegradation* is a chemical reaction of a substance induced by enzymatic activity of microorganisms.
- *Chronic toxicity* is the measure of toxicological responses that result from repeated exposure to a substance over an extended period of time (typically 3 months or longer). Such responses may persist beyond the exposure or may not appear until much later in time than the exposure. HRS measures of chronic toxicity include Reference Dose (RfD) values.
- *Contract Laboratory Program (CLP)* is an analytical program developed for CERCLA waste site samples to fill the need for legally defensible analytical results supported by a high level of quality assurance and documentation.

- *Contract-Required Detection Limit (CRDL)* is the term equivalent to contract-required quantitation limit, but it is used primarily for inorganic substances.
- *Contract-Required Quantitation Limit (CRQL)* is the substance-specific level that a CLP laboratory must be able to routinely and reliably detect in specific sample matrices. It is not the lowest detectable level achievable but rather the level that a CLP laboratory should reasonably quantify. The CRQL may or may not be equal to the quantitation limit of a given substance in a given sample. For HRS purposes, the term CRQL refers to both the contract-required quantitation limit and the contract-required detection limit.
- *Curie (Ci)* is the measure used to quantify the amount of radioactivity. One curie equals 37 billion nuclear transformations per second, and one picocurie (pCi) equals 10^{-12}Ci.
- *Decay product* is the isotope formed by the radioactive decay of some other isotope. This newly formed isotope possesses physical and chemical properties that are different from those of its parent isotope and may also be radioactive.
- *Detection Limit (DL)* is the lowest amount that can be distinguished from the normal random "noise" of an analytical instrument or method. For HRS purposes, the detection limit used is the method detection limit (MDL) or, for real-time field instruments, the detection limit of the instrument as used in the field.
- *Dilution weight* is the parameter in the HRS surface water migration pathway that reduces the point value assigned to targets as the flow or depth of the relevant surface water body increases. (It is unitless.)
- *Distance weight* is the parameter in the HRS air migration, ground water migration, and soil exposure pathways that reduces the point value assigned to targets as their distance increases from the site. (It is unitless.)
- *Distribution coefficient (K* d*)* is the measure of the extent of the partitioning of a substance between geologic materials (e.g., soil, sediment, and rock) and water (also called partition coefficient). The distribution coefficient is used in the HRS in evaluating the mobility of

a substance for the ground water migration pathway. (Measured in ml/g.)

- *ED 10 (10 percent effective dose)* is the estimated dose associated with a 10% increase in response over control groups. For HRS purposes, the response considered is cancer. (milligrams toxicant per kilogram body weight per day (mg/kg-day)).
- *Food and Drug Administration Action Level (FDAAL)* is, under section 408 of the Federal Food, Drug and Cosmetic Act, as amended, the concentration of a poisonous or deleterious substance in human food or animal feed at or above which FDA will take legal action to remove adulterated products from the market. Only FDAALs established for fish and shellfish apply in the HRS.
- *Half-life* is the length of time required for an initial concentration of a substance to be halved as a result of loss through decay. The HRS considers five decay processes: biodegradation, hydrolysis, photolysis, radioactive decay, and volatilization.
- *Hazardous substances* are the CERCLA hazardous substances, pollutants, and contaminants as defined in CERCLA sections 101(14) and 101(33), except where otherwise specifically noted in the HRS.
- *Hazardous waste stream* is material containing CERCLA hazardous substances (as defined in CERCLA section 101[14]) that was deposited, stored, disposed, placed in, or that otherwise migrated to a source.
- HRS *"factor"* is the primary rating element internal to the HRS.
- HRS *"factor category"* is the set of HRS factors (i.e., likelihood of release or exposure, waste characteristics, and targets).
- HRS *"migration pathways"* are HRS ground water, surface water, and air migration pathways.
- HRS *"pathway"* is the set of HRS factor categories combined to produce a score to measure relative risks posed by a site in one of four environmental pathways (e.g., ground water, surface water, soil, and air).
- HRS *"site score"* is the composite of the four HRS pathway scores.

- *Henry's law constant* is the measure of the volatility of a substance in a dilute solution of water at equilibrium. It is the ratio of the vapor pressure exerted by a substance in the gas phase over a dilute aqueous solution of that substance to its concentration in the solution at a given temperature. For HRS purposes, use the value reported at or near 25°C. (atmosphere-cubic meters per mole (atm-m^3 /mol)).
- *Hydrolysis* is the chemical reaction of a substance with water.
- *LC 50 (lethal concentration, 50 percent)* is the concentration of a substance in air (typically micrograms per cubic meter (µg/m^3)) or water (typically micrograms per liter (µg/l)) that kills 50% of a group of exposed organisms. The LC_{50} is used in the HRS in assessing acute toxicity.
- *LD 50 (lethal dose, 50 percent)* is the dose of a substance that kills 50% of a group of exposed organisms. The LD_{50} is used in the HRS in assessing acute toxicity (milligrams toxicant per kilogram body weight (mg/kg)).
- *Maximum Contaminant Level (MCL)* is, under section 1412 of the Safe Drinking Water Act, as amended, the maximum permissible concentration of a substance in water that is delivered to any user of a public water supply.
- *Maximum Contaminant Level Goal (MCLG)* is, under section 1412 of the Safe Drinking Water Act, as amended, a non-enforceable concentration for a substance in drinking water that is protective of adverse human health effects and allows an adequate margin of safety.
- *Method Detection Limit (MDL)* is the lowest concentration of analyte that a method can detect reliably in either a sample or blank.
- *National Ambient Air Quality Standards (NAAQS)* are the primary standards for air quality established under sections 108 and 109 of the Clean Air Act, as amended.
- *National Emission Standards for Hazardous Air Pollutants (NESHAPs)* are the standards established for substances listed under section 112 of the Clean Air Act, as amended. Only those NESHAPs promulgated in ambient concentration units apply in the HRS.

- *Octanol-water partition coefficient (K* ow *[or P])* is the measure of the extent of partitioning of a substance between water and octanol at equilibrium. The K_{ow} is determined by the ratio between the concentration in octanol divided by the concentration in water at equilibrium. (It is unitless.)
- *Organic carbon partition coefficient (K* oc*)* is the measure of the extent of partitioning of a substance, at equilibrium, between organic carbon in geologic materials and water. The higher the K_{oc}, the more likely a substance is to bind to geologic materials than to remain in water. (Measured in ml/g.)
- *Photolysis* is the chemical reaction of a substance caused by direct absorption of solar energy (direct photolysis) or caused by other substances that absorb solar energy (indirect photolysis).
- *Radiation* is defined by particles (alpha, beta, neutrons) or photons (x- and gamma-rays) emitted by radionuclides.
- *Radioactive decay* is the process of spontaneous nuclear transformation, whereby an isotope of one element is transformed into an isotope of another element, thus releasing excess energy in the form of radiation.
- *Radioactive half-life* is the time required for one-half the atoms in a given quantity of a specific radionuclide to undergo radioactive decay.
- *Radioactive substance* is a solid, liquid, or gas that contains atoms of a single radionuclide or multiple radionuclides.
- *Radioactivity* is the property of those isotopes of elements that exhibit radioactive decay and emit radiation.
- *Radionuclide/radioisotope* is the isotope of an element that exhibits radioactivity. For HRS purposes, "radionuclide" and "radioisotope" are used synonymously.
- *Reference dose (RfD)* is the estimate of a daily exposure level of a substance to a human population below which adverse non-cancerous health effects are not anticipated (milligrams toxicant per kilogram body weight per day (mg/kg-day)).

- *Removal action* is the action that removes hazardous substances from the site for proper disposal or destruction in a facility permitted under the Resource Conservation and Recovery Act, the Toxic Substances Control Act, or by the Nuclear Regulatory Commission.
- *Roentgen (R)* is the measure of external exposures to ionizing radiation. One roentgen equals that amount of x-ray or gamma radiation required to produce ions carrying a charge of 1 electrostatic unit (esu) in 1 cubic centimeter of dry air under standard conditions. One microroentgen (μR) equals 10^{-6}R.
- *Sample quantitation limit (SQL)* is the quantity of a substance that can be reasonably quantified given the limits of detection for the methods of analysis and sample characteristics that may affect quantitation (e.g., dilution and concentration).
- *Screening concentration* is the media-specific benchmark concentration for a hazardous substance that is used in the HRS for comparison with the concentration of that hazardous substance in a sample from that media. The screening concentration for a specific hazardous substance corresponds to its reference dose for inhalation exposures or oral exposures, as appropriate, and to excess cancer risk for inhalation exposures or for oral exposures, as appropriate, if the substance is a human carcinogen with a weight-of-evidence classification of A, B, or C to that concentration that corresponds to its 10^{-6}individual lifetime.
- *Site* is the area(s) where a hazardous substance has been deposited, stored, disposed, or placed, or has otherwise come to be located. Such areas may include multiple sources and may include the area between sources.
- *Slope factor (also referred to as cancer potency factor)* is the estimate of the probability of response (e.g., cancer) per unit intake of a substance over a lifetime. The slope factor is typically used to estimate upper-bound probability of an individual developing cancer as a result of exposure to a particular level of a human carcinogen with a weight-of-evidence classification of A, B, or C (mg/kg-day)$^{-1}$for non-radioactive substances and $(pC_i)^{-1}$for radioactive substances).

- *Source* is any area where a hazardous substance has been deposited, stored, disposed, or placed, plus those soils that have become contaminated from the migration of a hazardous substance. Sources do not include those volumes of air, ground water, surface water, or surface water sediments that have become contaminated by migration. Except in the case of either a ground water plume with no identified source or contaminated surface water sediments with no identified source, the plume or contaminated sediments may be considered a source.
- *Target distance limit* is the maximum distance over which targets for the site are evaluated. The target distance limit varies by HRS pathway.
- *Uranium Mill Tailings Radiation Control Act (UMTRCA) Standards* are the standards for radionuclides established under sections 102, 104, and 108 of the Uranium Mill Tailings Radiation Control Act, as amended.
- *Vapor pressure* is the pressure exerted by the vapor of a substance when it is in equilibrium with its solid or liquid form at a given temperature. For HRS purposes, use the value reported at or near 25°C (atmosphere or torr).
- *Volatilization* is the physical transfer process through which a substance undergoes a change of state from a solid or liquid to a gas.
- *Water solubility* is the maximum concentration of a substance in pure water at a given temperature. For HRS purposes, use the value reported at or near 25°C (milligrams per liter (mg/l)).
- *Weight-of-evidence* is the EPA classification system for characterizing the evidence supporting the designation of a substance as a human carcinogen. EPA weight-of-evidence groupings include:
  - Group A: Human carcinogen; sufficient evidence of carcinogenicity in humans
  - Group B1: Probable human carcinogen; limited evidence of carcinogenicity in humans
  - Group B2: Probable human carcinogen; sufficient evidence of carcinogenicity in animals
  - Group C: Possible human carcinogen; limited evidence of carcinogenicity in animals

- Group D: Not classifiable as to human carcinogenicity; applicable when there is no animal evidence or when human or animal evidence is inadequate
- Group E: Evidence of no carcinogenicity for humans

6.3 Worker Protection

As highlighted in section 6.2, an emergency response is the response effort applied to an uncontrolled release of oil. Uncontrolled releases involving oil create airborne, surface, subsurface, and shoreline contamination hazards that place responders in harm's way. We have also noted that there are two very different responses and skills required to respond to the release: emergency response and post-emergency response. Section 6.2.1 attempts to explain those differences and guide the reader to the resources that can assist in developing the appropriate manpower infrastructure required in preparing an emergency response plan.

This section covers important elements of the post emergency work skills and components that need to be established as a part of an emergency response plan. The subjects covered all fall into the OSHA arena of tools, management practices, technologies, and safe work practices.

6.3.1 Occupational Exposure Standards

Workers will potentially come into contact with a range of different chemicals and not just crude oil. But just considering crude oil, this chemical is a complex mixture of hundreds of chemicals. Safe levels of exposure do not exist for crude oil per se, but they have been written for some of the major ingredients. Table 6.9 lists the allowable occupational exposure limits to some of the major components found in crude oil.

Note in the table the following footnotes:

- (1) – Values reported are the 8-hour TWA (Time Weighted Average), which applies to an 8-hour work shift.
- (2) – Values reported are the 10-hour TWA unless otherwise specified.
- (4) and (5) denote values as coal tar pitch volatiles and cyclohexane extractable fractions, respectively.

Table 6.9 Occupational exposure limits for components found in crude oil

Components	CAS No.	Occupational Exposure Limits			Units
		OSHA[(1)]	ACGIH[(1)]	NIOSH[(2)]	
Benzene	71-43-2	$5^{(1)\text{STEL}}$	0.5 2.5^{STEL}	0.1 1^{STEL}	ppm
Toluene	108-88-3	100 150^{STEL}	20	100 150^{STEL}	ppm
Xylenes	1330-20-7	100 $150^{(3)\text{STEL}}$	100 150^{STEL}	100 150^{STEL}	ppm
Ethylbenzene	100-41-4	100 125^{STEL}	100 125^{STEL}	100 125^{STEL}	ppm
Trimethylbenzene	25551-13-7	25	25	25	ppm
Hydrogen Sulfide	7783-06-4	20^{Ceiling}	10 15^{STEL}	10^{Ceiling}	ppm
Polynuclear Aromatic Hydrocarbons	N/A	$0.2^{)(4)}$	$0.2^{(4)}$	$0.1^{(5)}$	mg/m^3

Because PAHs are complex with more than 75 chemicals within the family, (most of which are linked to caners) occupational exposure standards are expressed in this manner.

- STEL – This abbreviation stands for Short Term Exposure Limit, which is defined by the ACGIH (American Conference of Governmental Hygienists) as the concentration to which workers can be exposed continuously for a short period of time without suffering from irritation, chronic or irreversible tissue damage, or narcosis of a sufficient degree to increase the likelihood of accidental injury, impair self-rescue, or materially reduce work efficiency. STEL's are generally used only when toxic effects have been reported from high acute (short-term) exposures in either humans or animals. An STEL is not a separate independent exposure limit, but it supplements time-weighted average limits where there are recognized acute effects from a substance whose toxic effects are generally chronic (long-term) in nature. For example, one cannot be exposed to an STEL concentration if the TLV-TWA (time-weighted average for an 8-hour shift; see Permissible Exposure Limit (PEL)) discussed is below which would be exceeded. Workers can be exposed to a maximum of four STEL periods per 8-hour shift with at least 60 minutes between exposure periods.

- Ceiling – At no time should this exposure limit be exceeded. Sometimes it is denoted with the letter C.
- Note that N/A means not applicable in Table 6.9.

In addition to the terms use in Table 6.9, the reader needs to convey an understanding of several other safe-level-of-exposure terms now briefly discussed.

A **Permissible Exposure Limit (PEL)** is the maximum amount or concentration of a chemical that a worker may be exposed to under OSHA regulations. A **VPEL** or **Vacated PEL** is an older PEL set by OSHA but later "vacated" (retracted) under court order. However, some states may continue to enforce these lower levels. PEL's can be defined in two different ways as discussed in the OSHA regulation on air contaminants, 29 CFR 1910.1000. The two different ways are with ceiling values (already discussed) and the TWA.

As a further point of clarification, the **8-hour Time Weighted Averages (TWA)** are an average value of exposure over the course of an 8-hour work shift. TWA levels are usually lower than ceiling values. Thus, a worker may be exposed to a level higher than the TWA for part of the day (but still lower than the ceiling value) as long as he or she is exposed to levels below the TWA for the rest of the day. See 1910.1000 for the formulas used in the calculations.

In general, PELs and TLV-STELs appear on MSDSs in reference to substances that may be inhaled, although some can be absorbed through the skin or eyes (STELs will often have "-skin" after them when skin exposure is possible). When working with materials that have listed exposure limits, use proper precautions to minimize the generation of a vapor or dust in the first place. Always insist that workers use appropriate personal protective equipment (PPE) such as gloves, dust masks, and respirators to limit their exposure to chemicals.

When dealing with other chemical products such as dispersants and shoreline cleaners, the TLVs may not always be apparent, and/or condensed/limited information on health hazards are only provided on the supplier's MSDS. The reader may have to compile additional sources of safety related data on a product and not depend entirely on the product's MSDS. The NCP product directory of EPA approved chemical dispersants and other oil spill response chemicals provides barely enough information that is useful for the safe handling of products that have been approved by the EPA for responding to oil spills. The necessary data should be gathered and complied as a part of the emergency response plan so that it is on hand when it is needed. "Right to Know Fact Sheets" are a good source of information prepared by different states.

As an example, the dispersant COREXIT, which is supplied by Nalco Co., provides information that is factual but quite limited. Its MSDS[14] reports that the product contains 30–60 wt. % 2-Butoxyethanol. This is an 11 page document that has considerable information about various properties of the product, but a New Jersey Right to Know Fact Sheet[15] on 2-Butoxy Ethanol is much more extensive and enlightening about how dangerous this one ingredient is.

Nalco and Exxon/Mobil who supply and manufacture COERXIT have a responsibility to list on their product MSDS the most conservative health risk warning. Furthermore, it is not enough to simply provide an MSDS and hazard-warning label to workers and expect them to make a judgment as to how conservative their handling of the product should be. The role of the company using or intending to use a product should be to provide all the relevant information to its employees and to have this information accessible prior to product application. The purpose of an Emergency Response Plan is to be prepared in the event of an emergency. An Emergency Response Plan must have all the levels of worker protection defined beforehand.

End-users should crosscheck the information on MSDSs with other sources. For each individual hazardous ingredient, one should see whether there exists a state Department of Health and Senior Services Hazardous Substance Fact Sheet available through the RTK (Right to Know) Program. For every chemical product that appears on an emergency response plan, the employer should obtain such

---

[14]Nalco Company MSDS for COREXIT® EC9527A, Version Number: 2.0, Date issued : 05/11/2010

[15]New Jersey Right to Know Fact Sheet for 2-Butoxy Ethanol, Date: February 2001 Revision: August 2008

documents along with the MSDSs and create an RTK Central File or a Hazard Communication Standard File. New Jersey's RTK program is fairly extensive (so are a number of other states' programs). We do recommend that the reader go to the New Jersey RTK program Web site[16] because of the extensive safety related information for a large number of chemicals. The ingredients found to be listed in a product's MSDS can be cross-referenced to the RTK fact sheets, and extensive worker protection guidelines far beyond the MSDS will be made available. Workers have a right to this information under the Worker and Community Right to Know Act.

Since the EPA has highlighted COREXIT in the NCP Product Schedule Notebook and on its Web site as a principle dispersant for combating marine oil spills but barely provides adequate information on exposure risks when handling the product, we have decided to incorporate much of the information provided by the New Jersey OSH program in Table 6.10. We would hope that such detailed information eventually filters its way into the NCP literature. Regardless, we think it is important for a company take it upon itself to incorporate as much of the best available information into its emergency response plans and not simply list generic or boiler plate approaches and general classes of products it intends to rely on in the event of a release.

6.3.2 Glossary

The following is a glossary of terms that are useful both from a general knowledge standpoint and in facilitating discussions in subsequent sections.

ACGIH is the American Conference of Governmental Industrial Hygienists. They publish guidelines called Threshold Limit Values (TLVs) for exposure to workplace chemicals.

Acute Exposure Guideline Levels (AEGLs) are established by the EPA. They describe the risk to humans resulting from once-in-a-lifetime or rare exposure to airborne chemicals.

Boiling point is the temperature at which a substance can change its physical state from a liquid to a gas.

A **carcinogen** is a substance that causes cancer.

---

[16]www.nj.gov/health/eoh/rtkweb

Table 6.10 Safety data on 2-Butoxy Ethanol, an ingredient in COREXIT.

IDENTIFIER	INFORMATION
Common Name	2-BUTOXY ETHANOL
Synonyms	Butyl Cellosolve; Ethylene Glycol Monobutyl Ether; EGBE
Chemical Name	Ethanol, 2-Butoxy-
CAS Number	111-76-2
DOT Number	UN 2369
Description	2-Butoxy Ethanol is a colorless liquid with a mild odor. It is used as a solvent for resins, lacquers, varnishes, and enamels, and is found in many hard surface cleaning products.
Odor Threshold	0.1 ppm
	Odor thresholds vary greatly. Do not rely on odor alone to determine potentially hazardous exposures.
Warnings	2-Butoxy Ethanol can affect you by ingestion and may be absorbed through the skin.
	2-Butoxy Ethanol should be handled as a CARCINOGEN – WITH EXTREME CAUTION.
	Contact can irritate the skin and eyes with possible eyedamage.
	Inhaling 2-Butoxy Ethanol can irritate the nose and throat.
	2-Butoxy Ethanol can cause nausea, vomiting, diarrhea, and abdominal pain.
	Exposure can cause headache, dizziness, light-headedness, and passing out.
	2-Butoxy Ethanol may damage the liver and kidneys.
Workplace Exposure Limits	OSHA: The legal airborne permissible exposure limit (PEL) is 50 ppm averaged over an 8-hour work shift.
	NIOSH: The recommended airborne exposure limit (REL) is 5 ppm averaged over a 10-hour work shift.

Table 6.10 (cont.) Safety data on 2-Butoxy Ethanol, an ingredient in COREXIT.

IDENTIFIER	INFORMATION
	ACGIH: The threshold limit value (TLV) is 20 ppm averaged over an 8-hour work shift.
	2-Butoxy Ethanol may be a CARCINOGEN in humans. There may be no safe level of exposure to a carcinogen, so all contact should be reduced to the lowest possible level.
	The above exposure limits are for air levels only. When skin contact also occurs, you may be over-exposed, even though air levels are less than the limits listed above.
Health Hazard Information	**Acute Health Effects** The following acute (short-term) health effects may occur immediately or shortly after exposure to 2-Butoxy Ethanol: Contact can irritate the skin and eyes with possible eye damage. Inhaling 2-Butoxy Ethanol can irritate the nose and throat causing coughing and wheezing. 2-Butoxy Ethanol can cause nausea, vomiting, diarrhea, and abdominal pain. Exposure can cause headache, dizziness, confusion, lightheadedness, and passing out.
	Chronic Health Effects The following chronic (long-term) health effects can occur at some time after exposure to 2-Butoxy Ethanol and can last for months or years: **Reproductive Hazard** 2-Butoxy Ethanol may damage the developing fetus There is limited evidence that 2-Butoxy Ethanol may damage the male reproductive system (including decreasing the sperm count) in animals and may affect female fertility in animals. **Other Effects** 2-Butoxy Ethanol may damage the liver and kidneys.

Table 6.10 (cont.) Safety data on 2-Butoxy Ethanol, an ingredient in COREXIT.

IDENTIFIER	INFORMATION
Medical	For frequent or potentially high exposure (half the TLV or greater), the following are recommended before beginning work and at regular times after that: **Liver and kidney function tests** – Any evaluation should include a careful history of past and present symptoms with an exam. Medical tests that look for damage already done are not a substitute for controlling exposure. Request copies of your medical testing. You have a legal right to this information under the OSHA Access to Employee Exposure and Medical Records Standard (29 CFR 1910.1020). **Mixed Exposures** – More than light alcohol consumption can cause liver damage. Drinking alcohol may increase the liver damage caused by 2-Butoxy Ethanol.
Workplace Controls and Practices	Very toxic chemicals, or those that are reproductive hazards or sensitizers, require expert advice on control measures if a less toxic chemical cannot be substituted. Control measures include: (1) enclosing chemical processes for severely irritating and corrosive chemicals, (2) using local exhaust ventilation for chemicals that may be harmful with a single exposure, and (3) using general ventilation to control exposures to skin and eye irritants. For further information on workplace controls, consult the NIOSH document on Control Banding at www.cdc.gov/niosh/topics/ctrlbanding/.
	The following work practices are also recommended:
	Label process containers.
	Provide employees with hazard information and training.
	Monitor airborne chemical concentrations
	Use engineering controls if concentrations exceed recommended exposure levels.

Table 6.10 (cont.) Safety data on 2-Butoxy Ethanol, an ingredient in COREXIT.

IDENTIFIER	INFORMATION
	Provide eye wash fountains and emergency showers
	Wash or shower if skin comes in contact with a hazardous material.
	Always wash at the end of the work shift.
	Change into clean clothing if clothing becomes contaminated.
	Do not take contaminated clothing home.
	Get special training to wash contaminated clothing.
	Do not eat, smoke, or drink in areas where chemicals are being handled, processed or stored.
	Wash hands carefully before eating, smoking, drinking, applying cosmetics or using the toilet.
Personal Protective Equipment	The OSHA Personal Protective Equipment Standard (29 CFR 1910.132) requires employers to determine the appropriate personal protective equipment for each hazard and to train employees on how and when to use protective equipment. The following recommendations are only guidelines and may not apply to every situation.
	Gloves and Clothing
	Avoid skin contact with 2-Butoxy Ethanol. Wear personal protective equipment made from material that cannot be permeated or degraded by this substance. Safety equipment suppliers and manufacturers can provide recommendations on the most protective glove and clothing material for your operation.
	Safety equipment manufacturers recommend Butyl, Nitrile, Neoprene, Silver Shield®/4H® and Viton for gloves and DuPont Tychem® SL and Responder®; Kappler® Zytron® 500; and Saint-Gobain ONESuit® TEC, or the equivalent, as protective materials for Glycol Ethers.

Table 6.10 (cont.) Safety data on 2-Butoxy Ethanol, an ingredient in COREXIT.

IDENTIFIER	INFORMATION
	All protective clothing (suits, gloves, footwear, headgear) should be clean, available each day, and put on before work.
	Eye Protection
	Wear indirect-vent, impact, and splash-resistant goggles when working with liquids.
	Wear a face shield along with goggles when working with corrosive, highly irritating, or toxic substances.
	Respiratory Protection
	Improper use of respirators is dangerous. Respirators should only be used if the employer has implemented a written program that takes into account workplace conditions, requirements for worker training, respirator fit testing, and medical exams, as described in the OSHA Respiratory Protection Standard (29 CFR 1910.134).
	Where the potential exists for exposure over 5 ppm, use a NIOSH approved full facepiece respirator with an organic vapor cartridge. Increased protection is obtained from full facepiece powered-air purifying respirators. Leave the area immediately if (1) while wearing a filter or cartridge respirator you can smell, taste, or otherwise detect 2-Butoxy Ethanol, (2) while wearing particulate filters abnormal resistance to breathing is experienced, or (3) eye irritation occurs while wearing a full face piece respirator. Check to make sure the respirator-to-face seal is still good. If it is, replace the filter or cartridge. If the seal is no longer good, you may need a new respirator. Consider all potential sources of exposure in your workplace. You may need a combination of filters, pre-filters, or cartridges to protect against different forms of a chemical (such as vapor and mist) or against a mixture of chemicals. Where the potential exists for exposure over 50 ppm, use a NIOSH

Table 6.10 (cont.) Safety data on 2-Butoxy Ethanol, an ingredient in COREXIT.

IDENTIFIER	INFORMATION
	approved supplied-air respirator with a full face piece operated in a pressure-demand or other positive-pressure mode. For increased protection use in combination with an auxiliary self-contained breathing apparatus operated in a pressure-demand or other positive-pressure mode. Exposure to 700 ppm is immediately dangerous to life and health. If the possibility of exposure above 700 ppm exists, use a NIOSH approved self-contained breathing apparatus with a full face piece operated in a pressure-demand or other positive-pressure mode equipped with an emergency escape air cylinder.
Fire Hazards	If employees are expected to fight fires, they must be trained and equipped as stated in the OSHA Fire Brigades Standard (29 CFR 1910.156). 2-Butoxy Ethanol is a COMBUSTIBLE LIQUID. Use dry chemical, CO2, water spray, or alcohol-resistant foam as extinguishing agents. POISONOUS GASES ARE PRODUCED IN FIRE. CONTAINERS MAY EXPLODE IN FIRE. Use water spray to keep fire-exposed containers cool.
Spills and Emergencies	If employees are required to clean-up spills, they must be properly trained and equipped. The OSHA Hazardous Waste Operations and Emergency Response Standard (29 CFR 1910.120) may apply. If 2-Butoxy Ethanol is spilled or leaked, take the following steps: Evacuate personnel and secure and control entrance to the area. Eliminate all ignition sources. Absorb liquids in vermiculite, dry sand, earth, or a similar material and deposit into sealed containers. Ventilate area of spill or leak. DO NOT wash into sewer.

Table 6.10 (cont.) Safety data on 2-Butoxy Ethanol, an ingredient in COREXIT.

IDENTIFIER	INFORMATION
	It may be necessary to contain and dispose of 2-Butoxy Ethanol as a HAZARDOUS WASTE. Contact your state Department of Environmental Protection (DEP) or your regional office of the federal Environmental Protection Agency (EPA) for specific recommendations.
Handling and Storage	Prior to working with 2-Butoxy Ethanol you should be trained on its proper handling and storage. 2-Butoxy Ethanol is not compatible with OXIDIZING AGENTS (such as PERCHLORATES, PEROXIDES, PERMANGANATES, CHLORATES, NITRATES, CHLORINE, BROMINE and FLUORINE) and STRONG BASES (such as SODIUM HYDROXIDE and POTASSIUM HYDROXIDE). 2-Butoxy Ethanol forms Peroxides on exposure to AIR and LIGHT. Sources of ignition, such as smoking and open flames, are prohibited where 2-Butoxy Ethanol is used, handled, or stored in a manner that could create a potential fire or explosion hazard. 2-Butoxy Ethanol attacks some forms of PLASTICS, RUBBER, and COATINGS.

The **CAS number** is a unique, identifying number assigned by the Chemical Abstracts Service to a specific chemical.

CFR is the Code of Federal Regulations, which are the regulations of the United States government.

A **combustible** substance is a solid, liquid, or gas that will burn.

A **corrosive** substance is a gas, liquid, or solid that causes destruction of human skin or severe corrosion of containers.

DOT is the Department of Transportation, the federal agency that regulates the transportation of chemicals.

EPA is the Environmental Protection Agency, the federal agency responsible for regulating environmental hazards.

ERG is the Emergency Response Guidebook. It is a guide for emergency responders for transportation emergencies involving hazardous substances.

Emergency Response Planning Guideline (ERPG) values are intended to provide estimates of concentration ranges where one reasonably might anticipate observing adverse effects.

A **fetus** is an unborn human or animal.

A **flammable** substance is a solid, liquid, vapor, or gas that will ignite easily and burn rapidly.

The **flash point** is the temperature at which a liquid or solid gives off vapor that can form a flammable mixture with air.

IARC is the International Agency for Research on Cancer, a scientific group.

Ionization Potential is the amount of energy needed to remove an electron from an atom or molecule. It is measured in electron volts.

IRIS is the Integrated Risk Information System database maintained by the federal EPA. The database contains information on human health effects that may result from exposure to various chemicals in the environment.

LEL, or **Lower Explosive Limit**, is the lowest concentration of a combustible substance (gas or vapor) in the air capable of continuing an explosion.

mg/m^3 means milligrams of a chemical in a cubic meter of air. It is a measure of concentration (weight/volume).

A **mutagen** is a substance that causes mutations.

A **mutation** is a change in the genetic material in a body cell. Mutations can lead to birth defects, miscarriages, or cancer.

NFPA is the National Fire Protection Association. It classifies substances according to their fire and explosion hazard.

NIOSH is the National Institute for Occupational Safety and Health. It tests equipment, evaluates and approves respirators, conducts studies of workplace hazards, and proposes standards to OSHA.

NTP is the National Toxicology Program, which tests chemicals and reviews evidence for cancer.

OSHA is the federal Occupational Safety and Health Administration, which adopts and enforces health and safety standards.

Permeated is the movement of chemicals through protective materials.

PIH is a DOT designation for chemicals that are Poison Inhalation Hazards.

ppm means parts of a substance per million parts of air. It is a measure of concentration by volume in air.

A **reactive** substance is a solid, liquid, or gas that releases energy under certain conditions.

STEL is a Short Term Exposure Limit, which is usually a 15-minute exposure that should not be exceeded at any time during a workday.

A **teratogen** is a substance that causes birth defects by damaging the fetus.

UEL or **Upper Explosive Limit** is the highest concentration in air, above which there is too much fuel (gas or vapor) to begin a reaction or explosion.

Vapor density is the ratio of the weight of a given volume of one gas to the weight of another (usually hydrogen), at the same temperature and pressure.

Vapor pressure is a measure of how readily a liquid or a solid mixes with air at its surface. A higher vapor pressure indicates a higher concentration of the substance in air and, therefore, increases the likelihood of breathing it in.

6.3.3 Medical Surveillance

A medical surveillance program must be provided for employees and contractors who are a part of the post-emergency clean-up operations. As we have noted earlier, both the emergency and post-emergency response actions can be implemented concurrently in large spill situations.

The program must include the following:

- Period examinations
- Employer's responsibilities
- Physician's written opinion
- Recordkeeping

Periodic physical examinations will be needed. All medical examinations and procedures must be performed by or under the supervision of a licensed physician and one who is preferably knowledgeable in occupational medicine. Medical exams must include a medical and work history of individuals and an examination by a physician. It is recommended that the examination include medical monitoring. In the examination, the physician should be required to look for symptoms that might be caused by hazardous chemicals and should consider the fitness for the work, based on a written job description, and for the ability of the individual to wear a respirator.

Medical monitoring will be important especially for those workers who will be engaged for extended periods of time in the countermeasures aimed at clean-up. Measurements of chemicals in the body and changes in body tissues of a worker who may have been exposed to chemicals should be a part of the monitoring program. Following OSHA recommendations, workers are to be examined by a physician:

1. prior to work assignment;
2. at least once every year, unless the physician believes a shorter period is appropriate based on the job activities;
3. at the end of employment;
4. as soon as possible after the employee has developed signs or symptoms showing a possible overexposure to any hazardous substances;
5. if the employee has been injured or exposed to hazardous substances over the permissible exposure limit; and
6. at more frequent times if the physician thinks it is necessary.

It is the employer's responsibility to:

1. pay for medical examinations,
2. provide exams at a reasonable place and time without loss of pay to the employer, and
3. provide the following to the physician for each employee:
    - A description of the worker's duties
    - The worker's exposure levels
    - A description of the personal protective equipment used by the worker
    - Information from previous medical exams not readily available to the physician

The employer must further obtain a copy of the physician's written opinion and give a copy to the employee. The physician's opinion will include the following:

- Any detected medical conditions restricting work
- Recommended limitations of the employee's assigned work
- Results of the medical exam, if requested by the employee
- A statement that the employee has been informed by the physician of the results of the medical examination

The written opinion does not include any findings that are not related to occupational exposures.

Recordkeeping is an important element of the medical surveillance program. Accurate medical records for each employee must be retained and must include the following items:

- Name and social security number of the employee
- Physician's written opinions
- Any employee medical complaints related to exposure to hazardous substances
- A copy of the information the employer provided to the physician

Records are to be kept a minimum of 30 years.

6.3.4 Fitness and Heat Stress

Before countermeasure work activities can be started, the workers must be able to wear a respirator and work in the proper personal protective equipment required for the job. Not everyone can wear a respirator without risk to his or her health or well-being. Some people are claustrophobic and get upset when they put on a respirator. Others are not fit enough. Wearing a respirator puts a strain on a person's heart and lungs because the wearer must work harder than normal just to breath. Doing strenuous work with a respirator is an even greater strain. Therefore, individuals with breathing or health related problems might not be able to work with a respirator.

Wearing other equipment, such as protective clothing, also may be stressful. The protective clothing and gear can be heavy, hard to move around in, and hot. People who are not physically fit may have a difficult time performing strenuous work in protective gear.

A major concern of workers who wear protective equipment is heat stress. How stressed a person is from heat can be influenced by a number of factors such as the following:

- Temperature, humidity, and air movement
- Type of personal protective equipment worn
- Type of work being performed
- Physical fitness of the individual

Hazardous waste and spill response workers often perform job activities in hot, unshaded areas where fresh water may not be available within the work zone. In these and other situations, heat stress is a potential problem.

Wearing heavy equipment that does not let air flow through increases the risk of heat stress. This happens because body heat gets trapped inside the suit, making the person hotter. This causes the worker to sweat more and lose more water. The protective suit worn by workers does not allow the sweat to evaporate and cool the person, so the individual gets still hotter and continues to lose water.

A person can experience heat stress even without wearing clothing that does not allow air-flow. When the work activities are very strenuous, a person's body produces heat and loses significant amounts of water, causing the individual to become dehydrated.

Heat stress can occur with workers at any age. It can occur in young, physically fit workers who are wearing overalls in 75°F temperatures. A common cause for heat stress is that a worker simply does not drink enough fluids to replace the water lost through sweating.

Too much heat can cause a variety of symptoms depending on the amount of heat compared to the person's physical condition and precautions he or she has taken. The types of heat stress are briefly described below, beginning with the least serious and ending with the most serious, which is heat stroke. Heat stroke may be fatal.

- Heat rash (also known as prickly heat) is caused by unrelieved exposure to humid heat with skin continuously wet from un-evaporated sweat. Showering at the end of the shift helps to prevent heat rash. To treat heat rash, wash and thoroughly dry skin.

- Heat cramps are caused by heavy sweating without replacing body salt. To treat heat cramps, drink water with a little salt (1/4–1/2 teaspoon per glass) or products like Gatorade or Sqwincher. Massage cramping muscles and rest in a cool area. Drinking fluids with salt during work will also help to prevent heat cramps.
- Heat exhaustion is caused by over-exertion in the heat without being acclimated to the heat and not replacing body fluids. To treat heat exhaustion, call for medical help, rest in a cool area, open clothing, and give fluids.
- Heat stroke is caused by not being used to working in these conditions, obesity and lack of physical fitness, alcohol or drug intake, and lack of body fluid replacement. Heat stroke can quickly cause permanent brain damage or death. To treat heat stroke, call for medical assistance immediately, try to cool the body using cold water, place an ice bag on the head, increase air movement, and keep a wet towel or sheet around the body.

Note further that one should consult a physician before administering salted fluids to persons with heart trouble or on a low-sodium diet.

The following chart (Figure 6.10) will help to identify signs and symptoms for each type of heat stress.

It is the responsibility of the on-site health and safety officer (HSO) to monitor the condition of workers. The following are some common sense tips on how to prevent heat stress among workers:

- The job site supervisor and/or HSO and the worker need to be able to qualify that the person is as fit as he or she can be before attempting hot work. People who are overweight, out of shape, and on low-salt diets are most likely to suffer from heat stress. Supervisors and workers should ask the physician who performs the medical exam whether or not the individual can perform the tasks over the periods of time required.
- Workers should insist that their employer allow them to become acclimatized or adjusted to the heat. Workers that are required to wear respirators and protective garments must work fewer hours than they normally would when they begin to work in a hot environment.

Heat rash	Heat cramps	Heat exhaustion	Heat stroke
Skin rash	Muscle cramps	Clammy warm skin	Confusion
Tingling or prickling feeling on skin	Excessive sweating	Confusion	Convulsions
		Dizziness	Hot, dry skin
		Fainting on standing	Incoherent speech
		Fatigue	Staggering gait
		Heat rash	Sweating stops
		Lightheadedness	Very high blood pressure
		Nausea	Unconsciousness
		Rapid pulse	

Figure 6.10 Chart to assist in identifying heat stress conditions.

The worker can gradually increase the amount of hot work he or she does each day. It normally takes from one to two weeks to become acclimatized.

- If possible, do the hottest work during the coolest parts of the day.
- Eat moderate amounts of salt in food during acclimatization.
- After a vacation or long break in hot work, a person will need to acclimatize again.
- Respect your own body's abilities. Some people are more likely to suffer from heat stress than others. Some people take longer to acclimatize than others. Pay attention to the signals of your body, and don't compete with others.
- Avoid alcohol before or during hot work. Many cases of heat stroke have occurred the day after a "night on the town."
- Insist that your employer provide plenty of cold water and drink often.
- Insist that your employer allow you to take enough rest breaks. It is best if they are in a cool area.
- Know the symptoms of heat stress and watch for them. A buddy system is highly important in remediation

activities. It is easier to see symptoms in others than in your own self. Bear in mind that there is a big difference between heat exhaustion and heat stroke symptoms: those with heat exhaustion have moist warm skin, and in heat stroke they have hot dry skin because they can no longer sweat.

- If the temperature is greater than 80°F at the job site, employers must monitor workers for heat stress. Workers can also monitor themselves and their buddies by taking one's pulse. Shorten the work period if the pulse is too high. Also, take your own temperature with a mouth thermometer, and shorten the work periods if it is too high.
- Workers should periodically weigh themselves at the beginning and end of a shift. If you lose more than 1.5% of your weight during the day, you need to drink more fluids as you continue working.

6.3.5 Awareness and Recognizing the Hazards

All on-site response personnel should assume from the beginning that they are in a hazardous environment and that they are potentially exposed to multiple hazards that are physical, chemical, biological, and thermal in nature. When working with hazardous materials or at a site where there are potential physical and other hazards, you should consider the following factors, which one can remember as "CARE":

- **Control** – This is the action taken to remove or reduce the hazards present. It is very important that workers record all markings and containers present in order to identify the hazards. Proper identification will enable the project leader to determine the levels of protection needed for the job. Any information that the clean-up workers can relay to the HSO and job site supervisor will help keep the workers safer. If any new hazards or problems are found while cleaning up the site, report them immediately to the onsite commander or supervisor.
- **Anticipation** – Plan your actions before you do anything. Anticipate what hazards may be present and what you would do if an emergency arises.

- **Recognition** – Look at what is around you. Recognize markings, labels, numbers, and types of containers on the site. These clues will tell you what hazards, chemical products, and general dangers may be present.
- **Evaluation** – Think about the hazards that were identified. Decide how the substances and situations will affect your health and the environment. The substances must be identified and researched to find out how hazardous they may be.

There are numerous types of hazards that may be encountered during an oil spill clean-up. In terms of general categories, these health hazards include the following:

- Physical hazards
- Biological hazards
- Chemical hazards

Physical hazards are those caused by the surroundings and may include ditches or holes while working along shorelines, objects that may fall from overhead operations, and there are numerous possibilities of slips, trips, and falls on vessels and during shoreline response activities. Other physical hazards include sharp objects, exposure to electrical hazards, exposure to excessive heat, cold, noises, radiation (ionizing and non-ionizing), working in confined spaces, and hazards associated with handling drums. Certain hazards require special protocols and work practices. Physical hazards, such as drum handling, can be extremely dangerous. If personnel are not properly trained in procedures for drum handling, they can injure themselves and possibly cause an explosion, fire, or other chemical release. Proper drum handling procedures are described further on. Another physical hazard that needs special consideration is confined space entry, which is also discussed later in this chapter.

Biological hazards are caused by living things such as bacteria, parasites and viruses found in wastes, including some of the bioremediation additives that may be applied. Workers along shorelines may encounter other forms of wastes such as medical wastes and sewage, or they could come face to face with rabid animals that have been impacted by the spill.

Chemical hazards can cause harm in several different ways and include fire, explosion, lightning strikes, corrosion/irritation, poisoning/asphyxiation, chronic damage (over a period of time), allergies, and even genetic/reproductive damage from some of the clean-up chemicals handled and from possible exposure to smoke from in-situ burning of oil.

Some exposures to hazardous chemicals may be very high at job sites even when using common chemicals. At the job site, the total amount of substances present, or the level of the exposure, is usually much higher than at home. When using paint thinner in the home, small exposures occur because the total amount of chemicals present may be one quart. But the identical chemicals in a 500-gallon tank could quickly kill someone exposed to the vapors.

The exposure may be repeated many times. A single exposure to a solvent may not have any effect on your health. Many exposures over time can cause fatal diseases.

The work process may also change the nature of the hazard. For example, a relatively safe degreasing solution can give off highly toxic gases when heated.

There are many ways to identify what hazards may be present at the site. Before you approach the site, you should have an idea of what you will find. Employers are required to:

1. know what kind of work is required, not in a general way, but exactly;
2. define for workers the chemicals that are to be used at the job site;
3. determine the hazards of the chemicals to be used or that are involved in the clean-up actions;
4. conduct pre-entry briefings; and
5. perform sampling and monitoring. (Note that some toxic chemicals cannot be detected by human senses. Others may be colorless, odorless, and may not have any immediate affects on the body. Anticipating these hazardous chemicals is especially important. Otherwise, workers may not know that they are being exposed. By knowing what you are dealing with before you enter the site, you will know what levels of protection will be required to minimize the hazards.)

At a bare minimum, the HSO should have the following resources available at the job site and readily accessible to work crews:

- Material Safety Data Sheets (MSDS)
- NIOSH Pocket Guide to Chemical Hazards

- NIOSH Guide to Industrial Respiratory Protection
- Department of Transportation (DOT) Emergency Response Guidebook
- The Site Specific Safety and Health Plan

Employers have a duty to make sure that workers are well-versed in the use of these standard reference materials. It is a legal requirement under OSHA's Hazard Communication Act, commonly known as "Right to Know." The very first piece of information that should be relayed to workers under Right to Know orientation is the use of Material Safety Data Sheets and a review of the specific MSDSs for the chemicals workers will encounter at job sites.

6.3.6 Material Safety Data Sheets and Worker Orientation

The OSHA Hazard Communication Standard, 1910.120, requires chemical manufacturers to determine if there are any health or physical hazards associated with products they produce or import. This information will be found on a Material Data Safety Sheet (MSDS). The following information must appear on any MSDS for chemicals, including the oil that is being cleaned up:

- Identity of the product and hazardous ingredients
  - Manufacturer's name, address, and telephone number
  - The product name
  - Chemical formula
  - Chemical and common names of hazardous ingredients
  - Other names the product is called
- Physical properties data
  - Boiling point
  - Melting point
  - Vapor pressure
  - Vapor density
  - Specific gravity
  - Evaporation rate
  - Solubility in water
  - Appearance and color

- Permissible Exposure Limits (PELs), Threshold Limit Values (TLVs), and other exposure limits
- Fire and explosion data
  - Flash point
  - Flammable limits
  - Lower Explosive Limit (LEL)
  - Upper Explosive Limit (UEL)
  - Extinguishing method
  - Special fire-fighting procedures
  - Unusual fire and explosion hazards
- Reactivity data
  - Stability
  - Incompatible materials
  - Hazardous decomposition or by-products
  - Hazardous polymerization
- Health hazard data including toxicology
  - Routes of entry
  - Acute and chronic health hazards
  - Whether any ingredients can cause cancer
  - Signs and symptoms of exposure
  - Medical conditions aggravated by exposure
  - Emergency first aid procedures
- Precautions for safe handling and use
  - Steps to be taken in case material is spilled or released
  - Waste disposal method
  - Precautions to be taken in handling and storing
- Control measures
  - Respiratory protection needed
  - Ventilation strategies
  - Protective clothing and type
  - Work practices
- Exposure limits

In a worker Right to Know orientation program, it is highly advisable to provide basic information on toxicology. Some workers will be handling dispersants, biological agents, and shoreline cleaners, and many will be coming into contact with the spilled oil. The following is basic information that can be relayed to workers.

All chemicals can cause harm if the dose or the amount taken into the body is large enough. All chemicals have different levels at which they will cause harm to one's health. Each dose creates a response in the body. This dose-response relationship is a basic concept to understand when it comes to defining the toxicity of any chemical or biological agent. It means that the amount of the chemical we are exposed to is either toxic or not toxic. This is influenced by two factors: duration and frequency.

Duration is the amount of time that a person is exposed to the chemical. Frequency is how many times a person is exposed to the chemical. The more you are exposed to chemical hazards, both time and amount, the more likely the chemical will harm you.

In addition, workers need to be oriented on the routes of entry into the body. A chemical can enter your body by four major routes:

- Inhalation (breathing in)
- Absorption (through skin and occasionally eyes)
- Ingestion (eating)
- Injection (sharp objects)

The worker can prevent chemicals from entering the body by following safety procedures and wearing the correct personal protective equipment. If you wear the correct respiratory protective equipment, you will avoid inhalation. When you are wearing the correct chemical protective equipment, you will be protected from the other two routes of entry, namely absorption and injection.

Ingestion can be avoided by using the right respirator as well as by good personal hygiene (cleanliness) and work practices.

When you breathe in a chemical, it can cause harm in many different ways. Even chemicals that do not specifically affect the lungs can cause harm to the body. The chemical will pass through the lungs, enter into the bloodstream, and move to other parts of the body.

Direct contact with the skin and eyes by hazardous substances is another route of exposure. Some chemicals, such as solvents, directly injure the skin. Some chemicals pass through the skin and into the bloodstream where they can affect other parts of the body. Skin absorption is aided if you have cuts or are sweating. For protection, the worker needs to wear the proper chemical protective equipment such as gloves and chemical-protective suits. Workers should keep hands away from their faces and should minimize contact with liquid and solid chemicals. Contact lenses should not be worn in chemically contaminated areas, because the chemicals can become trapped under the lens.

Ingestion should be the easiest route of entry to protect against. Eating chemicals is unlikely, but personal habits may lead to ingestion without one knowing. Workers should never drink, eat, smoke, or chew tobacco or gum in the work area. Impress upon workers that if they do, they will create a risk of ingesting chemicals and causing harm to themselves. Workers should always wash their hands after working and before they eat, smoke, or chew. Following safe operating procedures and wearing the correct protective equipment will also protect personnel from ingesting chemicals.

Injection is a type of exposure that can occur by stepping or falling onto contaminated sharp objects or sticking oneself with them. Wearing safety shoes and taking common sense precautions are important safety measures for protection against this route of entry.

Workers should also be informed of what a Permissible Exposure Limit (PEL) is. A PEL is the maximum concentration one can be exposed to without respiratory protection. Under the OSHA standard, workers must be told what the PELs are for the chemicals they may come into contact with.

Workers should also be educated on the term "immediately dangerous to life and health" (IDLH). The IDLH of a chemical is the maximum exposure from which one could survive without irreversible health effects.

Workers should also be educated on what a Threshold Limit Value (TLV) is. A TLV is the concentration that a worker can be exposed to for a normal 8-hour day and 40-hour workweek and to which nearly all workers may be repeatedly exposed without health effects.

Workers should also be educated on what a Recommended Exposure Limit (REL) is. An REL is the maximum concentration a worker should be exposed to according to NIOSH (National Institute of Occupational Safety and Health). If you find more than one standard or guideline, consider the lowest one the best because it is the most conservative from a protection standpoint.

News media reports indicate that workers currently being employed in the Gulf oil spill are unskilled in the handling of hazardous materials. People who have made their professions and livelihoods based on commercial and sport fishing are participating in various countermeasures. The enlistment of volunteers and the level of training required are somewhat controversial. Volunteers frequently participate in marine oil spill response, but Federal

OSHA standards do not cover uncompensated workers. In states approved to manage their own occupational health and safety program (called OSHA state plan states), volunteers are often covered under state plan HAZWOPER requirements. In states administered by Federal OSHA, volunteers are covered by the EPA HAZWOPER standard (40 CFR 311). EPA's HAZWOPER standard has identical requirements, but the coverage is different from Federal OSHA standard coverage. The EPA standard covers local and state government employees, both compensated and volunteers. In this regard, if news media reports are accurate and private citizens have been enlisted as employees, then they are required to have some level of HazMat training. Even if the activities of volunteers are restricted, personnel need to be given a thorough orientation and minimum set of job skills to perform response functions. A thorough orientation is not simply a walkthrough of a typical MSDS or a review of definitions related to routes of exposure and toxicology. It should be specific, and it should point out potential hazards and risks. The following information (Table 6.11) is what Martin Marietta Materials (MMM)[17] reports on its MSDS for sweet crude oil (sweet is a term that refers to a crude oil with low sulfur content):

The footnote to Section II of MMM's MSDS notes that there are no published exposure limits for crude published by either ACGIH (American Conference of Governmental Industrial Hygienists) or by OSHA. But that does not mean that crude oil is not toxic to humans. MMM knows that and responsibly states on its MSDS the following information under Section III (Hazards Identification):

- "Crude oil is volatile and flammable, and may cause flash fires. If ingested, aspiration may occur, causing lung damage or death."
- "Contact with eyes may cause mild to severe irritation including stinging, watering, redness, and swelling."
- "Mild skin irritation including redness and a burning sensation may follow acute contact. Prolonged contact may cause dermatitis, folliculitis, or oil acne."
- Liquid may be absorbed through the skin in toxic amounts if large amounts of skin are exposed

[17]MSDS 3600-002, Effective Date: 8-07

Table 6.11 Ingredients found in crude oil as reported by Martin Marietta Materials.

COMPONENTS' CHEMICAL NAME	CAS REGISTRY NO	Concentration (by wt)	MSHA/ OSHA PEL	ACGIH TLV-TWA
Crude Petroleum	8002-05-9		*Oil Mist (mineral) 5mg/m^3	*Oil Mist (mineral) 5mg/m^3
Sulfur Compounds	Mixture	0–2%	N/A	N/A
Saturates	Mixture	80–90%	N/A	N/A
Aromatics	Mixture	8–15%	N/A	N/A
Polars	Mixture	1–5%	N/A	N/A
Asphaltene Content	Mixture	0–2%	N/A	N/A
May Contain: Benzene	71-43-2	0 – 0.1%	1.0 ppm	0.5 ppm
Limestone	1317-65-3		(T) 15 mg/m^3, (R) 5 mg/m^3	

* There are no exposure limits for crude oil published by ACGIH or OSHA. The limit for mineral oil mist is to be used only as a reference.

repeatedly. There have been rare occurrences of pre-cancerous warts on the forearm, back of hands and scrotum from chronic prolonged contact."

- "The major threat of ingestion occurs from the aspiration (breathing) of liquid drops into the lungs, particularly from vomiting. Aspiration may result in chemical pneumonia (fluid in the lungs), severe lung damage, respiratory failure, and death. Ingestion may cause gastrointestinal disturbances including irritation, nausea, vomiting and diarrhea. In severe cases, tremors, convulsions, loss of consciousness, coma, respiratory arrest, and death may occur."
- "May cause respiratory and nasal irritation. Central nervous system effects may include headache, dizziness, loss of balance and coordination, unconsciousness, coma, respiratory failure, and death."
- "Medical Conditions Aggravated By Exposure – Skin disorders, respiratory conditions, liver or kidney

dysfunction, male reproductive and peripheral nerve disorders."

Under Section IX (Toxicological Information), MMM's MSDS discusses the routes of entry and states the following:

- "Acute Effect: May cause eye, skin, respiratory and nasal irritation. Ingestion may cause vomiting, resulting in aspiration and chemical pneumonia. Central nervous system effects from inhalation may include headache, dizziness, loss of balance and coordination, unconsciousness, coma, respiratory failure, and death."
- "Chronic Effect/Carcinogenicity: There have been rare occurrences of precancerous warts on the forearm, back of hands and scrotum from chronic prolonged skin contact. These warts were not necessarily on the exposed parts of the body. Crude Oil is not listed as a carcinogen by the NTP, IARC, or OSHA. However, repeated skin contact by laboratory mice produced skin tumors. The tumors reduced in occurrence when the animals' skin was washed between applications."
- "Benzene: *This product may contain 0 – 0.1% benzene.*
    - ACHIH TLV TWA: 0.5 ppm / STEL: 2.5 ppm
    - MSHA and OSHA PEL TWA: 10 ppm / STEL: 25 ppm / Peak: 50 ppm, 10 minutes.
    - Exposure Routes: Inhalation, skin absorption, ingestion
    - Target Organs: Hematopoietic (blood forming) system, lymphatic system, nervous system, reproductive system.
    - Acute Effects: Inhalation (5–10 minutes) of very high levels of benzene (10,000–20,000 ppm) can result in death. Lower levels (700–3,000 ppm) can cause drowsiness, dizziness, rapid heart rate, headaches, tremors, confusion, and unconsciousness. Ingestion can cause vomiting, irritation of the stomach, dizziness, sleepiness, convulsions, rapid heart rate, coma, and death. Skin contact may cause redness and sores. Eye contact may cause irritation and cornea damage.

- Chronic Effects/Carcinogenicity: Benzene is on the NTP, OSHA and IARC carcinogen lists. The IARC and the EPA have determined that benzene is carcinogenic to humans (Group 1 Carcinogen). Chronic inhalation of certain levels of benzene causes disorders in the blood in humans, including leukemia (cancer of blood forming organs). Benzene specifically affects bone marrow (the tissues that produce blood cells). Aplastic anemia, excessive bleeding, and damage to the immune system (by changes in blood levels of antibodies and loss of white blood cells) may develop. Several occupational studies suggest that benzene may impair fertility in women exposed to high levels. However, these studies are limited due to lack of exposure history, simultaneous exposure to other substances, and lack of follow-up."

Let's make no mistake about it – crude oil is toxic and dangerous, there are multiple routes of exposure when in proximity of and in handling the material, and it has been linked to cancers in test animals and humans. Any and all persons that are engaged in a spill response program have the right to know these facts and to be trained and provided with the proper equipment and personal levels of protection to limit their exposure.

Under Section VI (Accidental Release Measures), it provides the following guidance in the event that crude oil is released or spilled:

- "Persons involved in cleaning should first follow the precautions defined in Section VII of the MSDS. In circumstances of emergency response involving an inhalation hazard or potential inhalation hazard, personnel must wear positive self-contained breathing apparatus while engaged in the emergency response operations until it is determined through the use of air monitoring that a decreased level of respiratory protection will not result in hazardous exposures to employees (29 CFR1910.120 (q)(3)(iv))."
- "Isolate and evacuate area. Shut off source if it is safe to do so. Eliminate all sources of ignition in the vicinity of the spill or released vapor. Contain liquid with

vermiculite, sand or clay to prevent further contamination of soil, surface water or ground water. Place contaminated material in disposable containers, and dispose of in a manner consistent with local regulations."

- "Follow prescribed procedures for reporting and responding to large spills. Advise the National Response Center if the substance has entered a waterway (1-800-424-8802)."

Under Section VIII (Exposure Controls/Personal Protection) it provides the following recommended procedures:

- **"EYE/FACE PROTECTION** – Safety glasses with side shields should be worn when splashing is possible.
- **SKIN PROTECTION** – Supported polyvinyl chloride gloves should be worn to prevent skin contact. Protective clothing such as gloves, apron, boots, and facial protection should be worn when engineering controls or work practices are not adequate for prevention of skin contact.
- **RESPIRATORY PROTECTION** – Where it has been determined that there is no hydrogen sulfide exposure hazard (exposure potential below H2S permissible exposure limit), a NIOSH/MSHA-approved air purifying respirator with organic vapor cartridges or canisters may be permissible under certain circumstances where airborne concentrations are or may be expected to exceed limits for odor or irritation. Protection provided by air purifying respirators is limited. Use a positive pressure, supplied-air respirator if there is a potential for uncontrolled release, exposure levels are not known, in oxygen-deficient atmospheres, or any other circumstances where an air-purifying respirator may not provide adequate protection. Respirator use must comply with applicable OSHA standards, which include provisions for a user training program, respirator repair and cleaning, respirator fit testing, medical clearance and other requirements.
- **GENERAL HYGIENE CONSIDERATIONS** – Following the guidelines in this MSDS are recognized as good industrial hygiene practices. Avoid skin and

eye contact. Wash exposed skin with soap and water before eating, drinking, smoking, and using toilet facilities. Wash work clothes after each use. Use care when laundering clothing to prevent formation of vapors which could ignite the washer or dryer."

The employer is obligated to explain each and every term on an MSDS, including the abbreviations. In a proper orientation program, the trainer should include an explanation of the organizations that help to establish safety standards. The program should explain who and what the following organizations are, what information they have, and why workers need to become familiar with some of their resources:

- ACGIH: American Conference of Governmental Industrial Hygienists
- CFR: US Code of Federal Regulations
- DOT: US Department of Transportation
- IARC: International Agency for Research on Cancer
- NIOSH: National Institute for Occupational Safety and Health, US Department of Health and Human Services
- NTP: National Toxicology Program
- OSHA: Occupational Safety and Health Administration, US Department of Labor
- PEL: Permissible Exposure Limit
- SARA Title III: Title III of the Superfund Amendments and Reauthorization Act, 1986
- TLV: Threshold Limit Value
- TWA: Time-weighted Average

6.3.7 Supplementing the Initial Orientation

There are a number of different signs and marking systems that responders will need to be familiar with. These convey information on potential hazards and areas of concern within the various work zone areas. Among these are the following warning labels:

- UN (United Nations) numbers
- NFPA (National Fire Protection Association) guide
- DOT (Department of Transportation) placards

These warning labels identify the hazards that are present in the drum, truck, or other containers used in waste staging areas and in transport to disposal sites.

The UN numbers identify chemicals so that all countries have a uniform identification system. The Department of Transportation requires that a placard must be shown on all cargo that contains hazardous materials. This placard identifies what is present in the truck, tank, or container. It includes the UN number.

UN numbers are four-digit numbers that generally range between 0000 and 3500 and are ideally preceded by the letters "UN" (for example, "UN1005") to avoid confusion with other number codes.

North American (NA) numbers are identical to UN numbers. If a material does not have a UN number, it may be assigned an NA number. These are usually 4-digit numbers starting with 8 or 9 such as 9037 (or ideally NA9037), the NA number for hexachloroethane.

UN/NA numbers are required for the shipment of hazardous materials. Most people have probably seen placards that bear a UN/NA number on railway cars, trucks, shipping containers, etc., like the one below:

Figure 6.11 Example of a placard.

UN numbers are assigned by a committee of the United Nations, the Economic and Social Council (ECOSOC) Committee of Experts (COE) on the Transport of Dangerous Goods, which issues "Recommendations on the Transport of Dangerous Goods" (also called "the Orange Book"). UN Recommendations are not regulations but are recommendations addressing the international transport of dangerous goods by sea, air, road, rail, and inland waterways. These recommendations are reviewed, amended, and updated biennially (every second year) by the UN COE and are distributed to nations throughout the world. These recommendations serve as the basis for national, regional, and international transport regulations such as those prepared by the International Maritime Organization (IMO) and the International Civil Aviation Organization (ICAO). The UN Recommendations are also used as a basis for the development of regional (e.g., NAFTA and European Road and Rail regulations) and national transport regulations, including the U.S. Hazardous Materials Regulations under the U.S. Department of Transportation (DOT).

The UN Recommendations on the Transport of Dangerous Goods addresses the following areas:

- A list of the most commonly carried dangerous goods, their identification, and their classification
- Consignment procedures such as labeling, marking, and transport documents
- Standards for packaging, test procedures, and certification
- Standards for multimodal tank-containers, test procedures, and certification

The "Emergency Response Guidebook 2008: A Guidebook for First Responders During the Initial Phase of a Dangerous Goods/Hazardous Materials Transportation Incident" contains the UN classification numbers for chemicals.

The hazard classes used under the UN system are summarized in Table 6.12.

Responders also need to be familiar with the National Fire Protection Association's (NFPA) color-coded, numerical system for indicating the health, flammability, and reactivity hazards of chemicals, which also includes a special precaution symbol. NFPA labels assist in assessing the degree and type of hazard associated with a particular substance (see Figure 6.12).

The NFPA system uses both color and number designations within a diamond to identify the hazards. Table 6.13 provides the details of the NFPA rating scheme to be explained to workers. In addition to these hazard ratings, the following symbols are used by the NFPA to identify special hazards.

Table 6.12 UN hazard class system.

UN Hazard Class Number	Description
1	Class A, B, and C explosives
2	Non-flammable and flammable compressed gases
3	Flammable liquids
4	Flammable solids, spontaneously combustible substances, and water reactive substances
5	Oxidizing materials
6	Class A and B poisons, irritants, and disease-causing materials
7	Radioactive materials
8	Corrosive materials
9	Miscellaneous hazardous materials not covered under any of the other classes

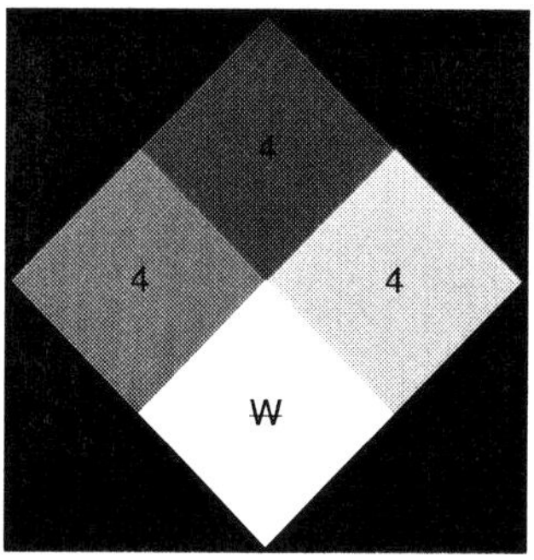

Figure 6.12 NFPA placard.

W   Water Reactive

Ox   Oxidizing Agent

Radioactive

Poison

Table 6.13 NFPA hazard rating and classification system.

Hazard Category	Color	Rating	Rating Designation	Designation Definition	Characteristics
Health	Blue	4	Extreme	Highly toxic material	On very short exposure could cause death or major residual injury even though prompt medical treatment is given.
					A known or suspected human carcinogen, mutagen, or teratogen
		3	Serious	Toxic material	May cause serious temporary or residual injury on short-term exposure even though prompt medical attention is given.
					A known or suspected small animal carcinogen, mutagen, or teratogen
		2	Moderate	Moderately toxic material	Intense or continued exposure could cause temporary incapacitation or possible residual injury unless prompt medical treatment is given.
		1	Slight	Slightly toxic material	May cause irritation but only minor residual injury even without treatment.
					Recognized innocuous material when used with responsible care.

Table 6.13 (cont.) NFPA hazard rating and classification system.

Hazard Category	Color	Rating	Rating Designation	Designation Definition	Characteristics
		0	Minimal	No chemical is without some degree of toxicity	
Flammability	Red	4	Extreme	Extremely flammable	Flash point below 73°F (22.8°C)
		3	Serious	Flammable	Vaporizes readily and can be ignited under almost all ambient conditions
					May form explosive mixtures with or burn rapidly in air
					May burn rapidly due to self-contained oxygen
					May ignite spontaneously in air
					Flash point at or above 73°F (22.8°C) but less than 100°F (37.8°C).
		2	Moderate	Combustible	Must be moderately heated or exposed to relatively high temperatures for ignition to occur.
					Solids that readily give off flammable vapors
					Flash point at or above 100°F (37.8°C) but less than 200°F (93.4°C).

		1	Slight	Slightly combustible	Must be preheated for ignition to occur
					Will burn in air when exposed at 1500°F (815.5°C) for 5 minutes.
					Flash point at or above 200°F (93.4°C).
		0	Minimal		Will not burn
					Will not exhibit a flash point
					Will not burn in air when exposed at 1500°F (815.5°C) for 5 minutes.
Reactivity	Yellow	4	Extreme		Can explode or decompose violently at normal temperature and pressure.
					Can undergo a violent self-accelerating exothermic reaction with common materials or by itself
					May be sensitive to mechanical or local thermal shock at normal temperature and pressure.
		3	Serious		Can detonate or explode but requires a strong initiating force or confined heating before initiation
					Readily promotes oxidation with combustible materials and may cause fires
					Is sensitive to thermal or mechanical shock at elevated temperatures.

Table 6.13 (cont.) NFPA hazard rating and classification system.

Hazard Category	Color	Rating	Rating Designation	Designation Definition	Characteristics
					May react explosively with water without requiring heat or confinement.
		2	Moderate		Normally unstable and readily undergoes violent change but does not detonate
					May undergo chemical change with rapid release of energy at normal temperature and pressure
					May react violently with water
					Forms potentially explosive mixtures with water
		1	Slight		Normally a stable material that can become unstable at high temperature and pressure
		0	Minimal	Normally stable material which is not reactive with water	

Table 6.14 DOT system of placarding.

DOT System	Designation
Color	
Red	Flammable
Green	Non-flammable
Orange	Explosive
Yellow	Reactive
White	Poisonous
White/Red Stripes	Flammable solid
White over black	Corrosive
Blue	Water reactive
Two colors	Two major hazards
Symbol and Hazard Class	
Flame	Flammable
Bursting ball	Explosive
Slash W	Dangerous when wet
Skull and cross bones	Poisonous
Circle with flame	Oxidizing material
Cylinder	Non-flammable gas
Propeller	Radioactive
Test tube with hand	Corrosive

The U.S. Department of Transportation (DOT) relies on placards that make use of a 4-digit ID number (the UN Hazard Class), colors, and symbols to identify waste cargo. The following table (Table 6.14) summarizes the colors and symbols used for the hazard classes.

6.3.8 Safe Handling Of Drums

Shoreline remediation activities in particular have to rely on the use of drums to collect and remove sludges and certain wastes including contaminated protective garments. Metal drums are used to store hazardous wastes. Drums with bung openings on top are used for liquids. Drums with removable tops that have bolted closures are used for solid wastes and powders. Plastic drums are used for chemicals such as solvents, dispersants, and cleaners. Workers must be instructed to label all drums accurately, maintain inventory records, and handle them safely.

While drums seem like simple containers, they present certain hazards to workers ranging from physical to chemical exposure. It is a good idea to familiarize workers with the features of the drums they will be working with and then to explain and train them on drum handling procedures. Figure 6.13 illustrates the key features of a chemical drum, and the following are some common sense safety precautions that should be impressed on workers:

- Do not stand on or work from a drum. Use a ladder if necessary. The OSHA regulation is very strict on this.
- Do not hit or kick a drum. If the chemical inside the drum is a shock-sensitive material or if the drum is weak, serious problems can result.

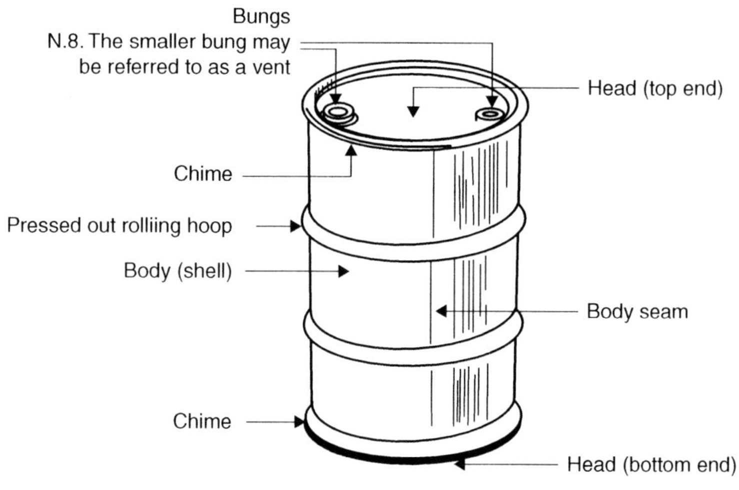

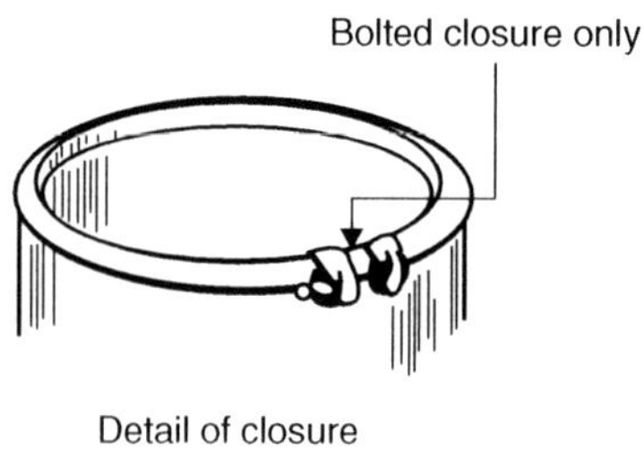

Figure 6.13 Key features of a drum.

- Do not move a drum except when necessary since accidents often occur when the drum is handled. Before any drum is moved, it must be checked to be sure that it can be moved or checked for any signs indicating that moving the drum could cause a leak.

When a drum is to be moved manually, every effort must be made to protect the worker from injury. There are several ways that a drum can be moved. The use of a drum grappler is one way. Manually or the use of a front-end loader, which can be loaded manually or by having a drum sling attached to a forklift or drum cart, are other methods.

The drum grappler is best because it keeps the operator at a distance from the drum. Naturally, the manual movement of a drum puts the worker at the most risk. If the drum is going to be moved manually, workers must make sure that they are taking all of the proper precautions to prevent back injury.

The employer should have a detailed plan for moving drums. The plan should consider the order in which drums should be moved. For example, small containers may be removed first to make room for heavy equipment. Moving a drum too many times only increases the potential for having a spill or an on-site accident.

Although the nature of the chemicals used in oil spills and the type of wastes collected in clean-up make this scenario very low, we mention it for general knowledge to be passed on to workers. Sometimes the chemical found inside of the drum will undergo a change, causing pressure to build-up in the drum. The best hint that this has occurred is if bulging is seen. A bulging drum is extremely dangerous and should be moved only when necessary and in a very safe manner. The site safety and health plan should be thorough enough to provide guidance for this situation. General measures are the same as for shock-sensitive materials. A drum with explosive or shock-sensitive material should only be moved if absolutely necessary. Extreme caution should be used. All unnecessary personnel must leave the area. Use a grappler specially constructed for explosive containment. Stay in constant communication with the Command Post. Signal with a loud signaling device when beginning and ending moving the drums.

Waste drums will have to be opened so that a sample of the contents can be taken for waste classification and manifesting purposes.

The opening of a drum can be dangerous if specific health and safety requirements are not followed. Below are guidelines to open drums safely:

- Only non-sparking tools should be used when opening any drum. Never use a pick, chisel, or firearms to open a drum.
- Only those persons necessary should be present when opening drums. Workers not involved in the actual operation should be kept away at a safe distance. For employees who must work near a drum opening, a shield should be placed between them and the drums.
- Use remote control devices to open drums when contents are unknown or have been staged for long periods in storage conditions poorly defined.
- A bulging drum should only be opened through the use of a remote drum-opening device. If this is not possible, then the use of manual means may be necessary. Excess pressure must be released slowly with the worker doing the opening behind some explosion-resistant barrier.
- After a drum has been opened, reseal it and decontaminate equipment according to site procedures in the Health and Safety Plan.
- If a drum cannot be moved without spilling its contents, then it must be transferred into a sound drum prior to any moving or being placed in an overpack. The transfer of a chemical from one drum to another should be of concern, and procedures should be clearly explained in a site Health and Safety Plan. We review some of the approaches below.

6.3.8.1 Transferring Flammable Liquids

To transfer a flammable liquid from one drum to another, specific safety requirements must be considered in order to protect the workers performing the operation. The transfer can be accomplished relatively quickly and without great effort. However, certain safety precautions need to be strictly followed.

There are many different types of pumps that may be used to transfer a flammable liquid from one drum to another. Pumps can be hand-operated, air-driven, or other. Regardless of the pump used, it must be properly bonded and grounded to avoid static electricity.

Static electricity is generated by the contact and separation of dissimilar materials. During pumping operations, static electricity may be produced by any one or a combination of the following events:

- As a liquid moves through a pipe or hose into a tank or drum
- By splash filling
- Through the agitation of material during transfer

The National Fire Protection Association's (NFPA) *Flammable and Combustible Liquids Code* requires that bonding and grounding be performed before initiating product transfer of liquids from one drum to another. Bonding is the linking of two containers by an electrical connection. Grounding is the elimination of the different electrical charges between the container and the ground.

To bond two containers, one must use a bonding wire, which is a copper wire with alligator clips. To ground the drums, a similar wire is connected from one drum to a grounded object, normally a water pipe, although grounding rods can also be used. In order for the bonding and grounding to be effective, the following steps and considerations should be taken:

- Both containers must be metal.
- At least one of the containers must be grounded.
- Clips on the cables must have direct metal-to-metal contact (scrape off any paint or oxidation that may prevent contact).

Figure 6.14 illustrates for the reader the proper configuration.

Other good practices for workers to adopt when working in or near drum transfer operations include warning personnel about the hazards of drum contents, keeping sorbents on hand in case of spills or leaks, keeping fire extinguishing equipment on hand, and organizing site operations such that movement is minimized.

6.3.9 Chemical Protective Clothing

Chemical protective clothing (CPC) consists of specially designed garments designed to protect workers from being exposed to chemicals from possible contact with skin or eyes. The garments

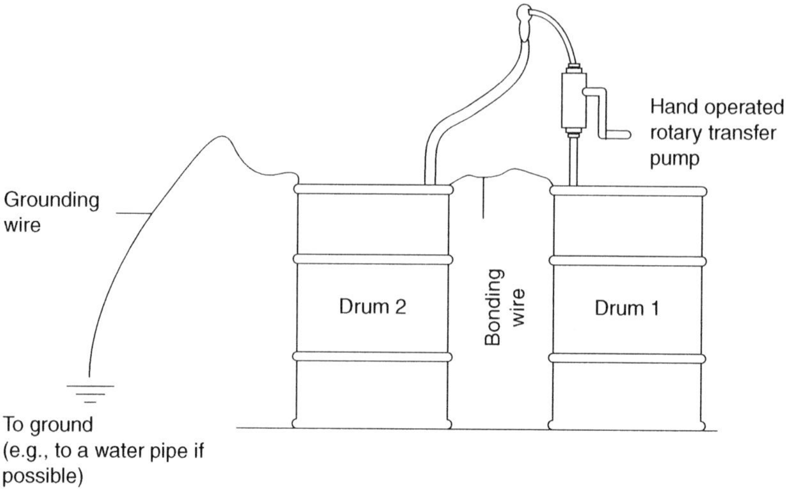

Figure 6.14 Grounding/bonding arrangement for transfer of flammable liquid between drums.

serve as an impervious barrier between the body and the chemicals that may cause harm to the worker. To provide adequate protection from chemicals, the materials from which the garment is made must be selected based on certain criteria to ensure that the wearer is insulated from any contact with the chemical. The clothing that is most resistant to the chemicals present in the work environment should be chosen. Selection is based on three properties, namely permeation, degradation, and penetration. All three of these factors must be considered when selecting CPC to ensure that the garment will afford adequate protection. There are additional factors that are also important to CPC selection, such as ergonomics, maneuverability, and comfort, but these should be considered after the material selection is made.

Permeation refers to the seepage of a chemical through the chemical protective suit material. A very simple analogy that helps to explain the process of permeation is a balloon filled with air. After a few hours an expanded balloon will begin to deflate because the air slowly passes through the rubber skin. In essence the air permeates to the outside. This is a little different than a garment exposed to a chemical because, in the case of the balloon there is a higher pressure inside the balloon than outside, and so the higher pressure serves as a driving force to move the air through the balloon's skin.

But a similar principle applies when we have a garment in which one side of the garment is exposed to a chemical and the other side is not. The high concentration of the chemical on the exposed side of the garment serves as a driving force for the chemical to permeate through the garment. Another simple analogy is to think of waterproof products – some materials have a greater degree of resistance from water passing through their surface to the inner structure and causing damage than others. Chemical permeation through CPC is essentially the same process.

Degradation is a process in which a chemical reaction takes place between the chemical and the CPC material, causing damage to the material. Some degradation is obvious, like acids burning a hole through the material that the suit is made from. However, there are much more subtle forms of degradation that cause not only a mechanical weakness or deterioration of the material but an increase in permeation. Sometimes the more subtle forms become detectable such as a slimy or gummy surface of the garment, or the garment becomes brittle and may show signs of crazing (very tiny cracks), shrinkage, noticeable softness, and even swelling. All of these observations are indicative of some kind of surface reaction between the fabric and chemical to which it is exposed, which will increase the material's permeability.

Penetration refers to the movement of a chemical through an opening in the suit, such as zippers, holes, or stitching in the suit.

When selecting chemical protective clothing, these factors must be considered beforehand. All manufacturers have permeation and degradation charts for their chemical protective clothing. It is important to know what chemicals the worker will be exposed to during work activities. The HSO (Health and Safety Officer) or equivalent should select materials that offer protection from the widest range of chemicals possible.

Other factors to consider when selecting chemical protective garments include the following:

- Resistance to the chemicals present
- Type of exposure (splash, vapor, solids, etc.)
- Risk of being exposed
- Ease of decontamination
- Mobility while wearing the clothing
- Temperature of work area
- Durability of clothing

- Color (bright colors may be important so workers can be seen; also dark colors may get too hot for the wearer)
- Size (it should fit to avoid accidents)
- Style (location of zippers, pockets, etc.)
- Cost (although we argue this should be the least important factor, overall worker protection should never be sacrificed for cost)

Both workers and the HSO should be trained in all aspects of using protective clothing and equipment. This is especially important when wearing fully encapsulating suits and respiratory equipment. Improper use could lead to suffocation. Workers should be instructed to report any problems or malfunctions to his/her supervisor, such as the following:

- Degradation of materials
- Skin irritation
- Unusual residues on material of CPC
- Discomfort in wearing the garment
- Difficulty breathing
- Fatigue due to respirator use
- Interference with vision or communication
- Too much difficulty moving
- Symptoms such as a fast pulse, nausea, or chest pain

A chemical protective suit is intended to protect workers from exposure to chemicals; however, suits are not generally designed to protect the worker from other hazards that may be encountered at the work site. Other hazards can be reduced by selecting other protective equipment, as shown in the Table 6.15.

Tables 6.16 and 6.17 provide some guidance on selection criteria.

It is important that workers recognize that no single combination of protective equipment and clothing is capable of protecting a person against all hazards. Thus, protective clothing should be used in conjunction with other protective methods. For example, engineering or administrative controls to limit chemical contact with personnel should be considered as an alternative measure for preventing chemical exposure. The use of protective clothing can itself create wearer hazards, such as heat stress, physical and psychological stress, in addition to impaired vision, mobility, and

Table 6.15 Guide to personal physical hazards protection.

Protection Against	Exposed Areas of Body	Recommended Clothing
Falling or flying objects	Head	Hard hat
	Feet	Steel-toed and shanked boots
Splashes or flying fragments	Face	Face shields
	Eyes	Face shields or goggles
	Noise	Earmuffs or earplugs
	Fire or blast	Flash suit worn over CPC (chemical protective clothing)
	Explosions	Fragmentation suite

Table 6.16 Physical resistance for chemical protective clothing[18].

Generic Material	Abrade	Cut	Tear	Puncture	Heat	Ozone	Flexibility
Butyl	F	G	G	G	E	E	G
CPC	E	G	G	G	G	E	G
Natural Rubber	E	E	E	E	F	P	E
NBR	E	E	G	E	G	F	E
Neoprene	E	E	G	G	G	E	G
Nitrile	E	E	G	E	G	F	E
Nitrile-PVC	G	G	G	G	F	E	G
Polyethylene	F	F	F	P	F	F	G
Polyurethane	E	G	G	G	G	G	E
PVA	F	F	G	F	G	E	P
PVC	G	P	G	G	P	E	F
SBR	E	G	F	F	G	F	G
Viton	G	G	G	G	G	E	G

E – Excellent; G – Good, F – Fair; P – Poor

[18]A Guide for Evaluating the Performance of Chemical Protective Clothing (CPC), National Institute of Occupational Safety & Health, Morgantown, WV, June 1990.

Table 6.17 Available types of garments for generic materials. (Source: A Guide for Evaluating the Performance of Chemical Protective Clothing (CPC), National Institute of Occupational Safety & Health, Morgantown, WV, June 1990.)

Type Garment	ACR	BUT	CAC	CPE	CPP	CR	CRB	FAB	NBR	NEO	NIT	PE	PP	PST	PU	PVA	PVC	SBR	SLV	SRN	TEF	VIT
Gloves	N	Y	N	-	N	N	N	Y	Y	Y	Y	-	-	-	Y	Y	Y	-	Y	Y	Y	Y
Coat or jacket	N	-	N	Y	N	N	N	Y	Y	Y	Y	Y	-	-	Y	-	Y	Y	-	-	Y	Y
Coveralls	N	-	N	Y	N	N	N	Y	-	Y	-	Y	Y	-	Y	-	Y	-	-	-	Y	-
Apron	N		N	Y	N	N	N	Y	Y	Y	Y	Y	-	-	-	-	Y	Y	-	-	Y	Y
Pants or bib overalls	N	Y	N	Y	N	N	N	Y	-	Y	-	Y	-	-	Y	-	Y	Y	Y	-	Y	Y
Hood	N	Y	-	Y	-	N	N	Y	Y	Y	-	Y	-	-	Y	-	Y	Y	Y	-	-	-
Shoe cover	N	Y	N	-	N	N	N	Y	Y	Y	-	Y	-	-	-	-	Y	-	-	-	-	-
Boot	N	Y	N	-	N	N	N	-	Y	Y	Y	-	-	-	-	-	-	-	-	-	-	-
Face shield	Y	N	Y	-	Y	Y	Y	-	N	N	N	-	N	Y	-	-	-	-	-	-	N	-
Suit material	N	Y	-	Y	-	N	-	Y	Y	Y	-	Y	N	-	Y	-	Y	-	N	-	Y	Y

KEY: Y – Yes; N – No; (-) No data

ACR	acrylic	CR	face shield material	NIT	nitrile	PVC	polyvinyl chloride
BUT	Butyl	CRB	polycarbonate	PE	polyethylene	SBR	styrene butadiene
CAC	cellulose acetate	FAB	nonwoven fabric	PP	polypropylene	SLV	Silver Shield
CPE	chlorinated polyethylene	NBR	nawra1 rubber	PST	polyester	SRN	Saran (PE/polyvinylidene chloride/PE)
CPP	cellulose propionate	NEO	neoprene	PU	polyurethane	TEF	Teflon
				PVA	polyvinyl alcohol	VIT	Viton

communication. In general, the greater the level of chemical protective clothing is, the greater are the associated risks. For any given situation, equipment and clothing should be selected that provide an adequate level of protection. Overprotection as well as underprotection can be hazardous and should be avoided.

The approach in selecting personal protective clothing must encompass an "ensemble" of clothing and equipment items that can be easily integrated in order to provide both an appropriate level of protection and still allow the worker to carry out activities involving chemicals. In many cases, simple protective clothing by itself may be sufficient to prevent chemical exposure, such as wearing gloves in combination with a splash apron and face shield (or safety goggles). The following is a checklist of components that may form a chemical protective ensemble:

- Protective clothing (suit, coveralls, hoods, gloves, boots)
- Respiratory equipment (SCBA, combination SCBA/SAR, air purifying respirators)
- Cooling system (ice vest, air circulation, water circulation)
- Communications device
- Head protection
- Eye protection
- Ear protection
- Inner garment
- Outer protection (overgloves, overboots, flashcover)

6.3.9.1 *Classification of Protective Clothing*

Personal protective clothing includes the following general types:

- Fully encapsulating suits
- Non-encapsulating suits
- Gloves, boots, and hoods
- Firefighter's protective clothing
- Proximity or approach clothing
- Blast or fragmentation suits
- Radiation-protective suits

Firefighter turnout clothing, proximity gear, blast suits, and radiation suits by themselves are not acceptable for providing adequate protection from hazardous chemicals. Table 6.18 provides OSHA's descriptions of different types of protection clothing available and the type of protection they offer, and it lists factors to consider in selection and use.

The following are different descriptors that will assist in the classification of protective garments and selection:

- Design – Categorizing clothing by design is mainly a means for describing what areas of the body the clothing item is intended to protect. In emergency response, hazardous waste site clean-up, and dangerous chemical operations, the only acceptable types of protective clothing include fully or totally encapsulating suits and non-encapsulating or "splash" suits plus accessory clothing items such as chemically resistant gloves or boots. These descriptions apply to how the clothing is designed and not to its performance.
- Performance – The National Fire Protection Association (NFPA) has classified suits by their performance as follows:
    - Vapor-protective suits (NFPA Standard 1991) provide "gas-tight" integrity and are intended for response situations where no chemical contact is permissible. This type of suit would be equivalent to the clothing required in EPA's Level A.
    - Liquid splash-protective suits (NFPA Standard 1992) offer protection against liquid chemicals in the form of splashes, but they do not offer protection against continuous liquid contact or chemical vapors or gases. Essentially, the type of clothing would meet the EPA Level B needs. However, it is important to note that by wearing liquid splash-protective clothing the wearer accepts exposure to chemical vapors or gases because this clothing does not offer gas-tight performance. The use of duct tape to seal clothing interfaces does not provide the type of wearer encapsulation that is necessary for protection against vapors or gases.
    - Support function protective garments (NFPA Standard 1993) must also provide liquid splash protection but offer limited physical protection. These

Table 6.18 OSHA descriptions of types of protective clothing for full body protection.

Description	Type of Protection	Use Considerations
Fully encapsulating suit: One-piece garment. Boots and gloves may be integral, attached and replaceable, or separate.	Protects against splashes, dust gases, and vapors.	Does not allow body heat to escape. May contribute to heat stress in wearer, particularly if worn in conjunction with a closed-circuit SCBA; a cooling garment may be needed. Impairs worker mobility, vision, and communication.
Non-encapsulating suit: Jacket, hood, pants, or bib overalls, and one-piece coveralls.	Protects against splashes, dust, and other materials but not against gases and vapors. Does not protect parts of head or neck	Do not use where gas-tight or pervasive splashing protection is required. May contribute to heat stress in wearer. Tape-seal connections between pant cuffs and boots and between gloves and sleeves.
Aprons, leggings, and sleeve protectors: Fully sleeved and gloved apron. Separate coverings for arms and legs. Commonly worn over non-encapsulating suit.	Provides additional splash protection of chest, forearms, and legs.	Whenever possible, should be used over a non-encapsulating suit to minimize potential heat stress. Useful for sampling, labeling, and analysis operations. Should be used only when there is a low probability of total body contact with contaminants.
Firefighters' protective clothing: Gloves, helmet, running or bunker coat, running or bunker pants (NFPA No. 1971, 1972, 1973, and boots (1974).	Protects against heat, hot water, and some particles. Does not protect against gases and vapors, or chemical permeation or degradation. NFPA	Decontamination is difficult. Should not be worn in areas where protection against gases, vapors, chemical splashes or permeation is required.

Table 6.18 (cont.) OSHA descriptions of types of protective clothing for full body protection.

Description	Type of Protection	Use Considerations
	Standard No. 1971 specifies that a garment consists of an outer shell, an inner liner and a vapor barrier with a minimum water penetration of 25 lb/in^2 (1.8 kg/cm^2) to prevent passage of hot water.	
Proximity garment (approach suit): One- or two-piece overgarment with boot covers, gloves, and hood of aluminized nylon or cotton fabric. Normally worn over other protective clothing, firefighters' bunker gear, or flame-retardant coveralls.	Protects against splashes, dust, gases, and vapors.	Does not allow body heat to escape. May contribute to heat stress in wearer, particularly if worn in conjunction with a closed-circuit SCBA; a cooling garment may be needed. Impairs worker mobility, vision, and communication.
Blast and fragmentation suit: Blast and fragmentation vests and clothing, bomb blankets, and bomb carriers.	Provides some protection against very small detonations. Bomb blankets and baskets can help redirect a blast.	Does not provide for hearing protection.
Radiation-contamination protective suit: Various types of protective clothing designed to prevent contamination of the body by radioactive particles	Protects against alpha and beta particles. Does not protect against gamma radiation	Designed to prevent skin contamination. If radiation is detected on site, consult an experienced radiation expert and evacuate personnel until the radiation hazard has been evaluated.

Table 6.18. (cont.) OSHA descriptions of types of protective clothing for full body protection.

Description	Type of Protection	Use Considerations
Flame/fire retardant coveralls: Normally worn as an undergarment.	Provides protection from flash fires.	Adds bulk and may exacerbate heat stress problems and impair mobility

garments may comprise several separate protective clothing components (i.e., coveralls, hoods, gloves, and boots). They are intended for use in non-emergency, nonflammable situations where the chemical hazards have been completely characterized. Examples of support functions include proximity to chemical processes, decontamination, hazardous waste clean-up, and training. Support function protective garments should not be used in chemical emergency response or in situations where chemical hazards remain uncharacterized.

- The NFPA standards define minimum performance requirements for the manufacture of chemical protective suits. Each standard requires rigorous testing of the suit and the materials that comprise the suit in terms of overall protection, chemical resistance, and physical properties. Suits that an independent certification and testing organization find to be compliant by may be labeled by the manufacturer as meeting the requirements of the respective NFPA standard. Manufacturers also have to supply documentation showing all test results and characteristics of their protective suits.
- Protective clothing should completely cover both the wearer and his or her breathing apparatus. In general, respiratory protective equipment is not designed to resist chemical contamination. Level A protection (vapor-protective suits) requires this configuration. Level B ensembles may be configured either with the SCBA on the outside or inside. However, it is strongly recommended that the wearer's respiratory equipment be worn inside

the ensemble to prevent its failure and to reduce decontamination problems. Level C ensembles use cartridge or canister-type respirators, which are generally worn outside the clothing.

- Service Life
    - A clothing item's service life is an end user decision depending on the costs and risks associated with clothing decontamination and reuse. For example, a Saranex/Tyvek garment may be designed as a coverall (covering the wearer's torso, arms, and legs) intended for liquid splash protection, which is disposable after a single use.
    - Protective clothing may be labeled as reusable for multiple wearings, or disposable and for one-time use.

Disposable clothing is generally lightweight and inexpensive. Reusable clothing is often more rugged and costly. Nevertheless, extensive contamination of any garment may render it disposable. The basis of this classification really depends on the costs involved in purchasing, maintaining, and reusing protective clothing versus the alternative of disposal following exposure. If an end user can anticipate obtaining several uses out of a garment while still maintaining adequate protection from that garment at a lower cost than its disposal, the suit becomes reusable.

6.3.9.2 *Garment Selection Factors*

Design Considerations

Manufacturers sell clothing in a variety of styles and configurations. The following are important factors:

- Clothing configuration
- Components and options
- Sizes
- Ease of donning and doffing
- Clothing construction
- Accommodation of other selected ensemble equipment
- Comfort
- Restriction of mobility

Material Chemical Resistance

The chosen material(s) must resist permeation, degradation, and penetration by the respective chemicals. It is important to note that no material protects against all chemicals and combinations of chemicals, and that no currently available material is an effective barrier to any prolonged chemical exposure. Sources of information include the following:

- *Guidelines for the Selection of Chemical Protective Clothing*, 3rd Edition: This reference provides a matrix of clothing material recommendations for approximately 500 chemicals based on an evaluation of chemical resistance test data, vendor literature, and raw material suppliers. The major limitation for these guidelines is their presentation of recommendations by generic material class. Numerous test results have shown that similar materials from different manufacturers may give widely different performances. That is to say, manufacturer A's butyl rubber glove may protect against chemical X, but a butyl glove made by manufacturer B may not.
- *Quick Selection Guide to Chemical Protective Clothing*: This pocket size guide provides chemical resistance data and recommendations for 11 generic materials against over 400 chemicals. The guide is color-coded by material-chemical recommendation. As with the "Guidelines..." above, the major limitation of this reference is its dependence on generic data.

Vendor data and recommendations are the best source of current information on material compatibility and should be available from the manufacturer of the selected clothing. Many vendors supply charts that show actual test data or their own recommendations for specific chemicals. However, unless vendor data or the recommendations are well documented, end users must approach this information with caution. Material recommendations must be based on data obtained from tests performed to standard ASTM methods. Simple ratings of "poor," "good," or "excellent" give no indication of how the material may perform against various chemicals.

Note that combinations or mixtures of chemicals can be significantly more aggressive towards protective clothing materials than any single chemical alone. One permeating chemical may pull another with it through the material. Very little data is available for chemical mixtures. Other situations may involve unidentified substances. In both the case of mixtures and unknowns, serious consideration must be given to deciding which protective clothing is selected. If clothing must be used without test data, garments with materials having the broadest chemical resistance should be worn, i.e., materials that demonstrate the best chemical resistance against the widest range of chemicals.

Physical Properties

Manufacturer materials offer wide ranges of physical qualities in terms of strength, resistance to physical hazards, and operation in extreme environmental conditions. Comprehensive manufacturing standards, such as the NFPA Standards, set specific limits on these material properties but only for limited applications, i.e., emergency response.

End users in other applications may assess material physical properties by posing the following questions:

- Does the material have sufficient strength to withstand the physical strength of the tasks at hand?
- Will the material resist tears, punctures, cuts, and abrasions?
- Will the material withstand repeated use after contamination and decontamination?
- Is the material flexible or pliable enough to allow end users to perform needed tasks?
- Will the material maintain its protective integrity and flexibility under hot and cold extremes?
- Is the material flame-resistant or self-extinguishing (if these hazards are present)?
- Are garment seams in the clothing constructed so they provide the same physical integrity as the garment material?

Ease of Decontamination

The degree of difficulty in decontaminating protective clothing may dictate whether disposable or reusable clothing is used or a combination of both.

Vapor and Splash Suits

Vapor protective suits also provide liquid splash and particulate protection. Liquid splash-protective garments also provide particulate protection. Many garments may be labeled as totally encapsulating but do not provide gas-tight integrity due to inadequate seams or closures. Gas-tight integrity can only be determined by performing a pressure or inflation test and a leak detection test of the respective protective suit. This test involves (1) closing off suit exhalation valves, (2) inflating the suit to a pre-specified pressure, and (3) observing whether the suit holds the above pressure for a designated period. ASTM Standard Practice F1052 (1987 Edition) offers a procedure for conducting this test.

Splash suits must still cover the entire body when combined with the respirator, gloves, and boots. Applying duct tape to a splash suit does not make it protect against vapors. Particulate protective suits may not need to cover the entire body, depending on the hazards posed by the particulate. In general, gloves, boots, and some form of face protection are required. Clothing items may only be needed to cover a limited area of the body, such as gloves on hands. The nature of the hazards and the expected exposure will determine if clothing should provide partial or full body protection.

The following are guidelines for determining whether clothing provides full body protection:

- Vapor-protective or totally encapsulating suits will meet this requirement by passing a gas-tight integrity test.
- Liquid splash-protective suits are generally sold incomplete (i.e., fewer gloves and boots).
- Missing clothing items must be obtained separately and match or exceed the performance of the garment.
- Buying a PVC glove for a PVC splash suit does not mean that you obtain the same level of protection. This determination must be made by comparing chemical resistance data.

How to Evaluate Manufacturer's Chemical Resistance Data

- Manufacturers of vapor-protective suits should provide permeation resistance data for their products, while liquid and particulate penetration resistance data should accompany liquid splash and particulate protective garments respectively. Ideally data should be provided for every primary material in the suit or clothing item. For suits, this includes the garment, visor, gloves, boots, and seams.

- Permeation data should include the following:
  - Chemical name;
  - Breakthrough time (shows how soon the chemical permeates);
  - Permeation rate (shows the rate that the chemical comes through);
  - System sensitivity (allows comparison of test results from different laboratories); and
  - A citation that the data was obtained in accordance with ASTM Standard Test Method F739-85.
- OSHA recommends that if no data are provided or if the data lack any one of the above items, the manufacturer should be asked to supply the missing data. Manufacturers that provide only numerical or qualitative ratings must support their recommendations with complete test data.
- Liquid penetration data should include a pass or fail determination for each chemical listed, and a citation that testing was conducted in accordance with ASTM Standard Test Method F903-86. Protective suits which are certified to NFPA 1991 or NFPA 1992 will meet all of the above requirements.
- Particulate penetration data should show some measure of material efficiency in preventing particulate penetration in terms of particulate type or size and percentage held out. Unfortunately, no standard tests are available in this area and end users may have little basis for company products.
- Suit materials that show no breakthrough or no penetration to a large number of chemicals are likely to have a broad range of chemical resistance. (Breakthrough times greater than one hour are usually considered to be an indication of acceptable performance.) Manufacturers should provide data on the ASTM Standard Guide F1001-86 chemicals. Manufacturers should also provide test data on other chemicals as

well. If there are specific chemicals within an operating area that have not been tested, ask the manufacturer for test data on these chemicals.

Additional Selection Guidance

OSHA recommends the following. Responsibility in selecting appropriate protective clothing should be vested in a specific individual who is trained in both chemical hazards and protective clothing use such as a safety officer or industrial hygienist. Only chemical protective suits labeled as compliant with the appropriate performance requirements should be used. In cases where the chemical hazards are known in advance or encountered routinely, clothing selection should be predetermined. That is, specific clothing items should be identified in specific chemical operations without the opportunity for individual selection of other clothing items.

Clothing Donning, Doffing, and Use

Table 6.19 provides sample procedures for donning and doffing vapor protective or liquid splash protective suit ensembles. The manufacturer's instruction manual should be consulted for garment specific procedures. Workers must be thoroughly trained on the procedures specific to the CPC they will use.

6.3.9.3 *Decontamination*

Decontamination involves removing or neutralizing contaminants that have accumulated on workers and equipment. Decontamination protects workers from hazardous substances that may contaminate and eventually permeate the protective clothing, respiratory equipment, tools, vehicles, and other equipment used in the vicinity of the chemical hazard. It protects all plant or site personnel by minimizing the transfer of harmful materials into clean areas, helps prevent the mixing of incompatible chemicals, and protects the community by preventing uncontrolled transportation of contaminants from the site. Two types of decontamination are gross decontamination, which allows the worker to safely exit or doff the chemical protective clothing, and decontamination for the reuse of chemical protective clothing.

The first step in decontamination is to establish Standard Operating Procedures (SOPs) that minimize contact with chemicals

Table 6.19 Sample donning and doffing procedures.

STEP	DONNING PROCEDURES		STEP	DOFFING PROCEDURES
1	Inspect clothing and respiratory equipment before donning (see Paragraph on Inspection).		1	Remove any extraneous or disposable clothing, boot covers, outer gloves, and tape.
2	Adjust hardhat or headpiece if worn, to fit user's head.		2	Have an assistant loosen and remove the wearer's safety shoes or boots.
3	Open back closure used to change air tank (if suit has one) before donning suit.		3	Have an assistant open the suit completely, lift the hood over the head of the wearer, and rest it on top of the SCBA tank.
4	Standing or sitting, step into the legs of the suit; ensure proper placement of the feet within the suit; then gather the suit around the waist.		4	Remove arms, one at a time, from suit. Once arms are free, have assistant lift the suit up and away from the SCBA backpack – avoiding any contact between the outside surface of the suit and the wearer's body – and lay the suit out flat behind the wearer. Leave internal gloves on, if any.
5	Put on chemical-resistant safety boots over the feet of the suit. Tape the leg cuff over the tops of the boots. If additional chemical-resistant safety boots are required, put these on now. Some one-piece suits have heavy-soled protective feet. With these suits, wear short, chemical resistant safety boots inside the suit.		5	Sitting, if possible, remove both legs from the suit.

Table 6.19 (cont.) Sample donning and doffing procedures.

STEP	DONNING PROCEDURES		STEP	DOFFING PROCEDURES
6	Put on air tank and harness assembly of the SCBA. Don the facepiece and adjust it to be secure, but comfortable. Do not connect the breathing hose. Open valve on air tank.		6	Follow procedure for doffing SCBA.
7	Perform negative and positive respirator facepiece seal test procedures. To conduct a negative-pressure test, close the inlet part with the palm of the hand or squeeze the breathing tube so it does not pass air, and gently inhale for about 10 seconds. Any inward rushing of air indicates a poor fit. Note that a leaking facepiece may be drawn tightly to the face to form a good seal, giving a false indication of adequate fit. To conduct a positive-pressure test, gently exhale while covering the exhalation valve to ensure that a positive pressure can be built up. Failure to build a positive pressure indicates a poor fit		7	After suit is removed, remove internal gloves by rolling them off the hand, inside out.

8	Depending on type of suit: Put on long-sleeved inner gloves (similar to surgical gloves). Secure gloves to sleeves, for suits with detachable gloves (if not done prior to entering the suit). Additional overgloves, worn over attached suit gloves, may be donned later.		8	Remove internal clothing and thoroughly cleanse the body.
9	Put sleeves of suit over arms as assistant pulls suit up and over the SCBA. Have assistant adjust suit around SCBA and shoulders to ensure unrestricted motion		If the low-pressure warning alarm has sounded signifying that approximately 5 minutes of air remain, then follow the procedures below:	
10	Put on hardhat, if needed.		1	Remove disposable clothing.
11	Raise hood overhead carefully so as not to disrupt face seal of SCBA mask. Adjust hood to give satisfactory comfort		2	Quickly scrub and hose off, especially around the entrance/exit zipper
12	Begin to secure the suit by closing all fasteners on opening until there is only adequate room to connect the breathing hose. Secure all belts and/or adjustable leg, head, and waistbands.		3	Open the zipper enough to allow access to the regulator and breathing hose
13	Connect the breathing hose while opening the main valve.		4	Immediately attach an appropriate canister to the breathing hose (the type and fittings should be predetermined). Although this provides some protection against any contamination still present, it voids the certification of the unit.

Table 6.19 (cont.) Sample donning and doffing procedures.

STEP	DONNING PROCEDURES		STEP	DOFFING PROCEDURES
14	Have assistant first ensure that wearer is breathing properly and then make final closure of the suit.		5	Follow Steps 1 through 8 of the regular doffing procedure above. Take extra care to avoid contaminating the assistant and the wearer.
15	Have assistant check all closures.			
16	Have assistant observe the wearer for a period of time to ensure that the wearer is comfortable, psychologically stable, and that the equipment is functioning properly.			

and, thus, the potential for contamination. Examples of procedures include the following:

- Stress work practices that minimize contact with hazardous substances (e.g., do not walk through areas of obvious contamination, do not directly touch potentially hazardous substances).
- Use remote sampling, handling, and container-opening techniques (e.g., drum grapples, pneumatic impact wrenches).
- Protect monitoring and sampling instruments by bagging. Make openings in the bags for sample ports and sensors that must contact site materials.
- Wear disposable outer garments and use disposable equipment where appropriate.
- Cover equipment and tools with a strippable coating that can be removed during decontamination.
- Encase the source of contaminants, e.g., with plastic sheeting or overpacks.
- Ensure all closures and ensemble component interfaces are completely secured and that no open pockets that could serve to collect contaminant are present.

Types of Contamination

Different types of contamination include surface contaminants and permeated contaminants. Surface contaminants may be easy to detect and remove. Contaminants that have permeated a material are difficult or impossible to detect and remove. If contaminants that have permeated a material are not removed by decontamination, they may continue to permeate the material and cause an unexpected exposure.

Factors that affect the extent of permeation include the following:

- Contact time – The longer a contaminant is in contact with an object, the greater the probability and extent of permeation. For this reason, minimizing contact time is one of the most important objectives of a decontamination program.
- Concentration – Molecules flow from areas of high concentration to areas of low concentration. As

concentrations of chemicals increase, the potential for permeation of personal protective clothing increases.
- Temperature – An increase in temperature generally increases the permeation rate of contaminants.
- Physical state of chemicals – As a rule, gases, vapors, and low-viscosity liquids tend to permeate more readily than high-viscosity liquids or solids.

Decontamination Procedures

Decontamination methods either (1) physically remove contaminants, (2) inactivate contaminants by chemical detoxification or disinfection/sterilization, or (3) remove contaminants by a combination of both physical and chemical means.

Gross decontamination is accomplished using detergents (surfactants) in water combined with a physical scrubbing action. This process will remove most forms of surface contamination including dusts, many inorganic chemicals, and some organic chemicals. Soapy water scrubbing of protective suits may not be effective in removing oily or tacky organic substances. This form of decontamination is unlikely to remove any contamination that has permeated or penetrated the suit materials. Using organic solvents such as petroleum distillates may allow easier removal of heavy organic contamination but may result in other problems, including permeation into clothing components pulling the contaminant with it, spreading localized contaminant into other areas of the clothing, and generating large volumes of contaminated solvents that require disposal.

Protective clothing or equipment reuse depends on demonstrating that adequate decontamination has taken place. Decontamination methods vary in their effectiveness and unfortunately there are no completely accurate methods for nondestructively evaluating clothing or equipment contamination levels. Methods that may assist in a determination include the following:

- Visually examine protective clothing for signs of discoloration, corrosive effects, or any degradation of external materials. However, many contaminants do not leave any visible evidence.
- Wipe a sampling of external surfaces for subsequent analysis. This may or may not be effective for determining levels of surface contamination and depends

heavily on the material-chemical combination. These methods will not detect permeated contamination.

- Evaluate the cleaning solution. This method cannot quantify clean method effectiveness since the original contamination levels are unknown. The method can only show if a chemical has been removed by the cleaning solution. If a number of garments have been contaminated, it may be advisable to sacrifice one garment for destructive testing by a qualified laboratory with analysis of contamination levels on and inside the garment.

SOPs must include a written decontamination plan. A decontamination plan should be developed and set up before any personnel or equipment are allowed to enter areas where the potential for exposure to hazardous substances exists. The decontamination plan should:

- determine the number and layout of decontamination stations;
- determine the decontamination equipment needed;
- determine appropriate decontamination methods;
- establish procedures to prevent the contamination of clean areas;
- establish methods and procedures to minimize wearer contact with contaminants during the removal of personal protective clothing; and
- establish methods for disposing of clothing and equipment that are not completely decontaminated.

The plan should be revised when the type of personal protective clothing or equipment changes, the use conditions change, or the chemical hazards are reassessed based on new information. The decontamination process should consist of a series of procedures performed in a specific sequence. For chemical protective ensembles, outer, more heavily contaminated items (e.g., outer boots and gloves) should be decontaminated and removed first, followed by the decontamination and removal of inner, less contaminated items (e.g., jackets and pants). Each procedure should be performed at a separate station in order to prevent cross contamination. The sequence of stations is called the decontamination line.

Stations should be separated physically in order to prevent cross-contamination and should be arranged in order of decreasing contamination, preferably in a straight line. Separate flow patterns and stations should be provided to isolate workers from different contamination zones containing incompatible wastes. Entry and exit points to exposed areas should be conspicuously marked. Dressing stations for entry into the decontamination area should be separate from redressing areas for exit from the decontamination area. Personnel who wish to enter clean areas of the decontamination facility, such as locker rooms, should be completely decontaminated.

All equipment used for decontamination must be decontaminated and/or disposed of properly. Buckets, brushes, clothing, tools, and other contaminated equipment should be collected, placed in containers, and labeled. Also, all spent solutions and wash water should be collected and disposed of properly. Clothing that is not completely decontaminated should be placed in plastic bags, pending further decontamination and/or disposal.

Decontamination of workers who initially come in contact with personnel and equipment leaving exposure or contamination areas will require more protection from contaminants than decontamination workers who are assigned to the last station in the decontamination line. In some cases, decontamination personnel should wear the same levels of protective clothing as workers in the exposure or contaminated areas. In other cases, decontamination personnel may be sufficiently protected by wearing one level lower protection (e.g., wearing Level B protection while decontaminating workers are wearing Level A).

Due to the difficulty in assessing contamination levels in chemical protective clothing before and after exposure, the HSO must determine if the respective clothing can be reused. This decision involves considerable risk in determining clothing to be contaminant-free. Reuse can be considered if, in the estimate of the HSO, no "significant" exposures have occurred. Decontamination methods have been successful in reducing contamination levels to safe or acceptable concentrations.

Note that contamination by known or suspected carcinogens should warrant automatic disposal. The use of disposable suits is highly recommended when extensive contamination is expected.

A final area worth covering is procedures for emergency decontamination. In an emergency the primary concern is to prevent the loss of life or severe injury to personnel. If immediate medical treatment is required to save a life, decontamination should be delayed until the victim is stabilized. If decontamination can be performed without interfering with essential life-saving techniques or first aid, or if a worker has been contaminated with an extremely toxic or corrosive material that could cause severe injury or loss of life, decontamination should be continued. If an emergency due to a heat-related illness develops, protective clothing should be removed from the victim as soon as possible to reduce the heat stress. During an emergency, provisions must also be made for protecting medical personnel and disposing of contaminated clothing and equipment.

Inspection, Storage, and Maintenance

The end user in donning protective clothing and equipment must take all necessary steps to ensure that the protective ensemble will perform as expected. During emergencies is not the right time to discover discrepancies in the protective clothing. Emphasizing a standard program for inspection, proper storage, and maintenance along with realizing protective clothing/equipment limitations is the best way to avoid chemical exposure during emergency response. Table 6.20 provides summary notes from OSHA guidelines that the reader may refer to.

6.3.10 Levels of Protection

The U.S. EPA defines four levels of protection: Levels A, B, C, and D. These levels are considered starting points for creating ensembles. Each ensemble must be tailored to the specific situation in order to provide the most appropriate level of protection. For example, if an emergency response activity involves a highly contaminated area, or if the potential of contamination is high, it may be advisable to wear a disposable covering, such as Tyvek coveralls or PVC splash suits, over the protective ensemble. Table 6.21 summarizes the levels of protection and recommended ensembles.

6.3.10.1 Respiratory Protection

OSHA's respirator standard is 29 CFR 1910.134. Respirator selection requires matching the respirator with the hazard, the degree of the hazard, and the user. The respirator selected must be adequate to effectively reduce the exposure of the respirator user under all conditions of use, including reasonably foreseeable emergency

Table 6.20 OSHA guidance on inspection, storage, and maintenance.

SOP	Important Elements
Inspection	Inspection and operational testing of equipment received as new from the factory or distributor.
	Inspection of equipment as it is selected for a particular chemical operation.
	Inspection of equipment after use or training and prior to maintenance.
	Periodic inspection of stored equipment
	Periodic inspection when a question arises concerning the appropriateness of selected equipment, or when problems with similar equipment are discovered.
	Each inspection will cover different areas with varying degrees of depth. Those personnel responsible for clothing inspection should follow manufacturer directions; many vendors provide detailed inspection procedures
	Records must be kept of all inspection procedures. Individual identification numbers should be assigned to all reusable pieces of equipment (many clothing and equipment items may already have serial numbers), and records should be maintained by that number. At a minimum, each inspection should record: Clothing/equipment item ID number; Date of the inspection; Person making the inspection; Results of the inspection; and Any unusual conditions noted. Periodic review of these records can provide an indication of protective clothing that requires excessive maintenance and can also serve to identify clothing that is susceptible to failure.
Storage	Clothing must be stored properly to prevent damage or malfunction from exposure to dust, moisture, sunlight, damaging chemicals, extreme temperatures, and impact. Procedures are needed for both initial receipt of equipment and after use or exposure of that equipment. Many manufacturers specify recommended procedures for storing their products. These should be followed to avoid equipment failure resulting from improper storage

Table 6.20 (cont.) OSHA guidance on inspection, storage, and maintenance.

SOP	Important Elements
	1. Some guidelines for general storage of chemical protective clothing include: Potentially contaminated clothing should be stored in an area separate from street clothing or unused protective clothing; Potentially contaminated clothing should be stored in a well-ventilated area, with good air flow around each item, if possible; Different types and materials of clothing and gloves should be stored separately to prevent issuing the wrong material by mistake (e.g. many glove materials are black and cannot be identified by appearance alone). Protective clothing should be folded or hung in accordance with manufacturer instructions
Maintenance	Manufacturers frequently restrict the sale of certain protective suit parts to individuals or groups who are specially trained, equipped, or authorized by the manufacturer to purchase them. Explicit procedures should be adopted to ensure that the appropriate level of maintenance is performed only by those individuals who have this specialized training and equipment. In no case should you attempt to repair equipment without checking with the person in your facility who is responsible for chemical protective clothing maintenance.
	1. The following classification scheme is recommended to divide the types of permissible or non-permissible repairs: Level 1: User or wearer maintenance, requiring a few common tools or no tools at all. Level 2: Maintenance that can be performed by the response team's maintenance shop, if adequately equipped and trained. Level 3: Specialized maintenance that can be performed only by the factory or an authorized repair person.
	Each facility should adopt the above scheme and list which repairs fall into each category for each type of protective clothing and equipment. Many manufacturers will also indicate which repairs, if performed in the field, void the warranty of their products. All repairs made must be recorded on the records for the specific clothing along with appropriate inspection results.

Table 6.21 EPA levels of worker protection.

Level Protection	OSHA/NIOSH Recommended Ensemble	Optional Safety Equipment	Protection Provided	Recommended Conditions to Apply	Limitations
A	Vapor protective suit (meets NFPA 1991) Pressure-demand, full facepiece SCBA Inner chemical-resistant gloves, chemical-resistant safety boots, two-way radio communication	Cooling system, outer gloves, hard hat	Highest available level of respiratory, skin, and eye protection from solid, liquid and gaseous chemicals	The chemical(s) have been identified and have high level of hazards to respiratory system, skin, and eyes. Substances are present with known or suspected skin toxicity or carcinogenicity. Operations must be conducted in confined or poorly ventilated areas.	Protective clothing must resist permeation by the chemical or mixtures present. Ensemble items must allow integration without loss of performance.
B	Liquid splash-protective suit (meets NFPA 1992) Pressure-demand, full facepiece SCBA nner chemical-resistant gloves, chemical-Iresistant safety boots,	Cooling system, outer gloves	Provides same level of respiratory protection as Level A, but less skin protection. Liquid splash protection, but no protection against chemical vapors or gases.	The chemical(s) have been identified but do not require a high level of skin protection. Initial site surveys are required until higher levels of hazards are	Protective clothing items must resist penetration by the chemicals or mixtures present. Ensemble items must

	two-way radio communications hardhat			identified. The primary hazards associated with site entry are from liquid and not vapor contact.	allow integration without loss of performance.
C	Support Function Protective Garment (meets NFPA 1993) Full facepiece, air-purifying, canister-equipped respirator Chemical resistant gloves and safety boots Two-way communications system, hard hat	Face shield, escape SCBA	The same level of skin protection as Level B, but a lower level of respiratory protection. Liquid splash protection but no protection to chemical vapors or gases.	Contact with site chemical(s) will not affect the skin. Air contaminants have been identified and concentrations measured. A canister is available which can remove the contaminant. The site and its hazards have been completely characterized.	Protective clothing items must resist penetration by the chemical or mixtures present. Chemical airborne concentration must be less than IDLH levels. The atmosphere must contain at least 19.5% oxygen. Not Acceptable for Chemical Emergency Response

Table 6.21 (cont.) EPA levels of worker protection.

Level Protection	OSHA/NIOSH Recommended Ensemble	Optional Safety Equipment	Protection Provided	Recommended Conditions to Apply	Limitations
D	Coveralls, safety boots/shoes, safety glasses or chemical splash goggles	Gloves, escape SCBA, face-shield	No respiratory protection, minimal skin protection.	The atmosphere contains no known hazard. Work functions preclude splashes, immersion, potential for inhalation, or direct contact with hazard chemicals.	This level should not be worn in the Hot Zone. The atmosphere must contain at least 19.5% oxygen. Not Acceptable for Chemical Emergency Response

situations. Selection involves choosing a device that fully protects the worker from the respiratory hazards to which he or she may be exposed and permits the worker to perform the job with the least amount of physical burden.

There are many factors that can influence respirator selection. In choosing the appropriate respirator, consider the nature and extent of the hazard, work requirements and conditions, and the characteristics and limitations of the respirators available. The following categories of information must be taken into account:

- Nature of the hazard and the physical and chemical properties of the air contaminant
- Concentrations of contaminants
- Relevant permissible exposure limit or other occupational exposure limit
- Nature of the work operation or process
- Time period the respirator is worn
- Work activities and physical/psychological stress
- Fit testing
- Physical characteristics, functional capabilities, and limitations of respirators

Once the above factors have been taken into consideration, the HSO must select a NIOSH-certified respirator. Where NIOSH has not specifically certified any respirator for use against the particular contaminant present, the employer must select a NIOSH-certified respirator that has no limitation prohibiting its use for that contaminant. The respirator must be appropriate for the contaminant's physical form and chemical properties and the conditions under which it will be used. All respirators must be chosen and used according to the limitations of the NIOSH certification, which appears on the NIOSH certification label. Figures 6.15 and 6.16 illustrate typical arrangements for Self Contained Breathing Apparatus (SCBA).

To obtain adequate respiratory protection, there must be a proper match between respirator and wearer. Respirators that don't seal properly around the face offer only the illusion of protection. All users must be fit tested. The purpose of fit testing is to identify the specific make, model, style, and size of respirator best suited for each worker. Fit testing also provides an opportunity to check on problems with respirator wear, and it reinforces respirator training

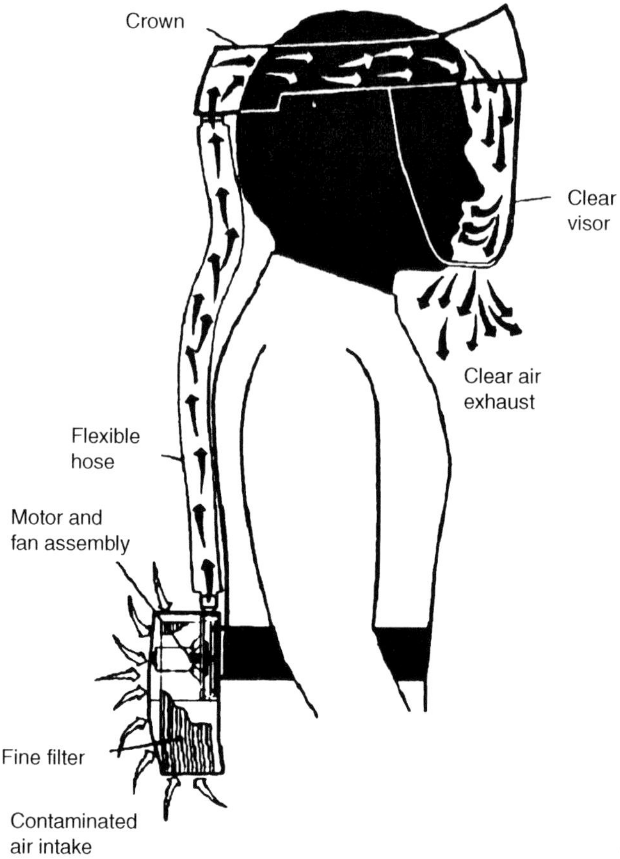

Figure 6.15 Powered air-purifying respirator providing eye, face, and respiratory protection.

by having wearers review the proper methods of donning and wearing the respirator.

Fit testing is required for all negative or positive pressure tight-fitting facepiece respirators. The OSHA respiratory protection standard requires that fit testing be performed before an employee first starts wearing a respirator in the work environment, whenever a different respirator facepiece is used, and at least annually thereafter. Prior to the actual fit test, the employee must be shown how to put on a respirator, position it on the face, set the strap tension, and determine an acceptable fit. Next, the employee must be allowed to choose a respirator from a sufficient number of models and sizes so that the employee can find an acceptable and correctly fitting respirator. Once

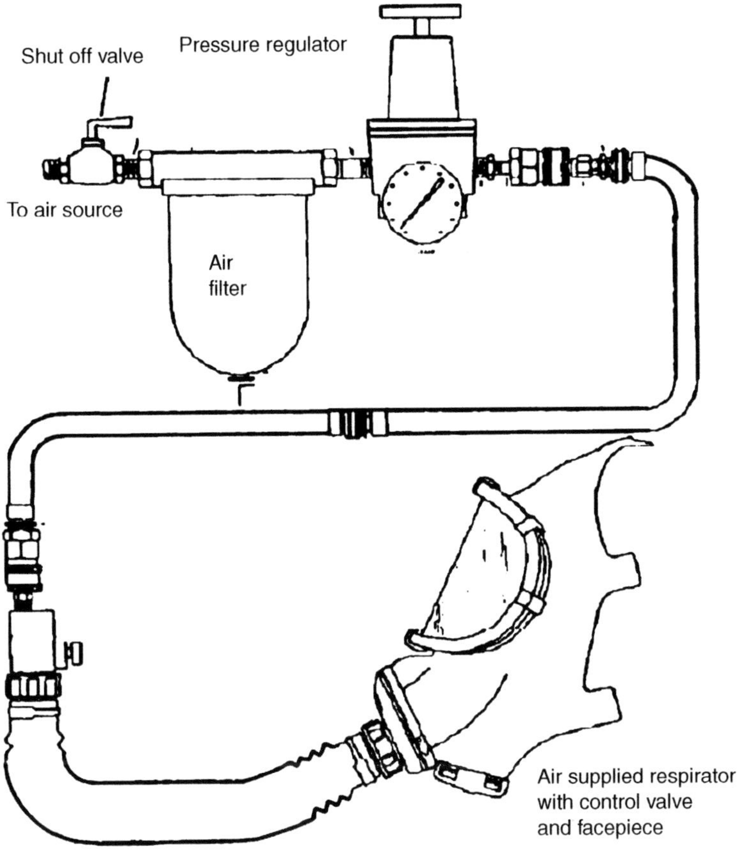

Figure 6.16 Illustrates typical assembly for constant-flow airline respirator.

an acceptable respirator has been found, a user seal check must be conducted (refer to on "Use of Respirators"). Fit testing may be either qualitative (QLFT) or quantitative (QNFT) and must be administered using an OSHA-accepted QLFT or QNFT protocol. These protocols are described in mandatory Appendix A to 1910.134.

Table 6.22 provides OSHA's guidelines on maintenance, inspection, and repairs for respirators.

6.3.10.2 Atmospheres that are Immediately Dangerous to Life or Health (IDLH)

Atmospheres are IDLH when they pose an immediate threat to life, would cause irreversible adverse health effects, or would interfere

Table 6.22 OSHA guidelines for cleaning, maintenance, and repairs to respirators.

PROPER CARE	PURPOSE	OSHA GUIDANCE
CLEANING AND DISINFECTING	Cleaning and sanitizing respirators are necessary to prevent skin irritation, dermatitis, and to encourage worker acceptance. Where the contaminant is a dust, mist, or fume, build-up on the respirator face-to-facepiece seal or within the respirator will reduce the protection provided by the respirator because the contaminant is in the breathing zone or has compromised the seal. In addition, the build-up of contamination on the respirator can contribute to the deterioration of the respirator's materials, which can lead to reduced protection. Full facepieces must be cleaned to ensure that employees can see through the facepiece.	Respirators that are issued for the exclusive use of an employee must be cleaned and disinfected as often as necessary to be maintained in a sanitary condition. Respirators used by more than one employee must be cleaned and disinfected prior to being used by a different individual. Respirators maintained for emergency use as well as respirators used in fit testing and training, must be cleaned and disinfected after each use. The employer must use either the OSHA cleaning and disinfecting procedures recommended in Appendix VIII: 2–3 of this chapter or the procedures recommended by the respirator manufacturer, as long as they are equivalent in effectiveness to the OSHA method.
STORAGE	All respirators must be stored so that they are protected against damage, contamination, dust, sunlight, extreme temperatures, excessive moisture, and damaging chemicals. When respirators are packed or stored, the facepiece and exhalation valve must be stored in a manner that will prevent deformation. Each respirator should be positioned so that it retains its	Respirators intended for emergency use must be kept accessible to the work area, but not in an area that might itself be involved in the emergency because such an area may become contaminated or inaccessible. Emergency-use respirators must be stored in compartments or covers that are clearly marked to indicate that they contain emergency respirators, and stored according to any applicable manufacturer instructions.

	natural configuration. Synthetic materials and even rubber will warp if stored in an unnatural shape, thus affecting the fitting characteristics of the facepiece.	
INSPECTION	To ensure the continued reliability of respiratory equipment, it must be inspected on a regular basis. The frequency of inspection and the procedures to be followed depend on whether the respirator is intended for non-emergency, emergency, or escape use only.	The OSHA standard requires that all respirators used in non-emergency situations be inspected before each use and during cleaning. Respirators designated for use in an emergency situation are to be inspected at least monthly and in accordance with the manufacturer's instructions, and checked for proper function before and after each use. Emergency escape-only respirators must be inspected before being carried into the workplace.
		For all respirators, inspections must include a check of respirator function, tightness of connections, and the condition of the various parts including, but not limited to, the facepiece, head straps, valves, connecting tube, and cartridges, canisters, or filters. In addition, the elastomeric parts must be evaluated for pliability and signs of deterioration
		For SCBA's, which require monthly inspections, the air and oxygen cylinders must be maintained in a fully charged state and recharged when the pressure falls to 90% of the manufacturer's recommended pressure level. In addition, the regulator and warning devices must be inspected to ensure that they function properly.

Table 6.22 (cont.) OSHA guidelines for cleaning, maintenance, and repairs to respirators.

PROPER CARE	PURPOSE	OSHA GUIDANCE
		For respirators that are maintained for use in emergencies, the OSHA standard requires certifying the respirator by documenting the date that the inspection was performed, the name or signature of the inspector, the findings of the inspection, any required remedial action, and a serial number or other means of identifying the inspected respirator. This information must be provided on a tag or label that is attached to the storage compartment for the respirator, is kept with the respirator, or is stored in the form of inspection reports (paper or electronic). The information must be maintained until it is replaced following a subsequent certification.
REPAIR	Respirators that fail to pass inspection or are otherwise found to be defective, must be removed from service, and discarded, repaired, or adjusted	Repairs or adjustments to respirators must be done only by appropriately trained personnel using only the respirator manufacturer's NIOSH-approved parts designed for that respirator. The repairs also must be made in accordance with the manufacturer's recommendations and specifications regarding the type and extent of repairs to be performed. Because components such as reducing and admission valves, regulators, and alarms are complex and essential to the safe functioning of the respirator, they are required to be adjusted and repaired only by the manufacturer or a technician trained by the manufacturer.

with an individual's ability to escape from a dangerous atmosphere. Care must be exercised in these situations since failure of the respirator to provide the appropriate protection may result in serious injury or death. Consequently, the HSO must develop and implement specific procedures for the use of respirators in IDLH atmospheres, and these procedures should include the following provisions recommended by OSHA:

- At least one employee (referred to as the "standby employee") is to be located outside the IDLH atmosphere and should maintain visual, voice, or signal line communication with the employee(s) in the IDLH atmosphere.
- The standby employee(s) located outside the IDLH atmosphere must be trained and equipped to provide effective emergency rescue.
- The employer or authorized designee is to be notified before the standby employee(s) enters the IDLH atmosphere to provide emergency rescue.
- The employer or authorized designee, once notified of such entry, must provide the necessary assistance appropriate to the situation.
- The standby employee(s) must be equipped with pressure demand or other positive pressure SCBA, or other positive pressure supplied-air respirator with auxiliary SCBA.
- The standby employee(s) must be equipped with appropriate retrieval equipment for lifting or removing the employee from the hazardous atmosphere or, when such retrieval equipment cannot be used because it would increase the overall risk resulting from entry, must ensure that equivalent provisions for rescue have been made.

6.3.10.3 Glossary of Respiratory Protection Terms

The following terms are taken from the OSHA standards. They are worth reviewing with workers in the safety program.

Air-purifying respirator is a respirator with an air-purifying filter, cartridge, or canister that removes specific air contaminants by passing ambient air through the air-purifying element.

Atmosphere-supplying respirator is a respirator that supplies the respirator user with breathing air from a source independent of the ambient atmosphere and includes supplied-air respirators (SAR's) and self-contained breathing apparatus (SCBA) units.

Canister or cartridge is a container with a filter, sorbent, catalyst, or a combination of these items that removes specific contaminants from the air passed through the container.

Demand respirator is an atmosphere-supplying respirator that admits breathing air to the facepiece only when a negative pressure is created inside the facepiece by inhalation.

Emergency situation is any occurrence such as, but not limited to, equipment failure, the rupture of containers, or the failure of control equipment that may or does result in an uncontrolled substantial release of an airborne contaminant.

Employee exposure is an exposure to a concentration of an airborne contaminant that would occur if the employee were not using respiratory protection.

End-of-service-life indicator (ESLI) is a system that warns the respirator user of the approach of the end of adequate respiratory protection, e.g., that the sorbent is approaching saturation or is no longer effective.

Escape-only respirator is a respirator intended to be used only for emergency exit.

Filtering facepiece (dust mask) is a negative pressure particulate respirator with a filter as an integral part of the facepiece or with the entire facepiece composed of the filtering medium.

Filter or air purifying element is a component used in respirators to remove solid or liquid aerosols from the inspired air.

Fit factor is a quantitative estimate of the fit of a particular respirator to a specific individual and typically estimates the ratio of the concentration of a substance in ambient air to its concentration inside the respirator when worn.

Fit test is the use of a protocol to qualitatively or quantitatively evaluate the fit of a respirator on an individual. See also "Qualitative fit test (QLFT)" and "Quantitative fit test (QNFT)."

Helmet is a rigid respiratory inlet covering that also provides head protection against impact and penetration.

High efficiency particulate air (HEPA) filter is a filter that is at least 99.97% efficient in removing monodisperse particles of 0.3 micrometers in diameter and larger. The equivalent NIOSH 42 CFR part 84 particulate filters are the N100, R100, and P100 filters.

Hood is a respiratory inlet covering that completely covers the head and neck and may also cover portions of the shoulders and torso.

"**Immediately dangerous to life or health (IDLH)**" is an atmosphere that poses an immediate threat to life, would cause irreversible adverse health effects, or would impair an individual's ability to escape from a dangerous atmosphere.

Interior structural firefighting is the physical activity of fire suppression, rescue, or both inside of buildings or enclosed structures that are involved in a fire situation beyond the incipient stage.

Loose-fitting facepiece is a respiratory inlet covering that is designed to form a partial seal with the face.

Negative pressure respirator (tight fitting) is a respirator in which the air pressure inside the facepiece is negative during inhalation with respect to the ambient air pressure outside the respirator.

Oxygen deficient atmosphere is an atmosphere with an oxygen content below 19.5% by volume.

Physician or other licensed health care professional (PLHCP) is an individual whose legally permitted scope of practice (i.e., license, registration, or certification) allows him or her to independently provide, or be delegated the responsibility to provide, some or all of the health care services required by 29 CFR 1910.134(e), "Medical evaluation."

Positive-pressure is a respirator in which the pressure inside the respiratory inlet covering exceeds the ambient air pressure outside the respirator.

Powered air-purifying respirator (PAPR) is an air-purifying respirator that uses a blower to force the ambient air through air-purifying elements to the inlet covering.

Pressure demand respirator is a positive pressure atmosphere-supplying respirator that admits breathing air to the facepiece when the positive pressure is reduced inside the facepiece by inhalation.

Qualitative fit test (QLFT) is a pass/fail fit test to assess the adequacy of the respirator fit based on the individual's response to the test agent.

Quantitative fit test (QNFT) is an assessment of the adequacy of respirator fit by numerically measuring the amount of leakage into the respirator.

Respiratory inlet covering is the portion of a respirator that forms the protective barrier between the user's respiratory tract and an air-purifying device, breathing air source, or both. It may be a facepiece, helmet, hood, suit, or a mouthpiece respirator with a nose clamp.

Self-contained breathing apparatus (SCBA) is an atmosphere-supplying respirator for which the breathing air source is designed to be carried by the user.

Service life is the period of time that a respirator, filter, sorbent, or other respiratory equipment provides adequate protection to the wearer.

Supplied-air respirator (SAR) or airline respirator is an atmosphere-supplying respirator for which the source of breathing air is not designed to be carried by the user.

Tight-fitting facepiece is a respiratory inlet covering that forms a complete seal with the face.

User seal check is an action conducted by the respirator user to determine if the respirator is properly sealed to the face.

6.4 The Oil Spill Response Plan

Owners/operators of offshore facilities that handle, store, or transport oil are required to prepare an Oil Spill Response Plan (OSRP). OSRPs must demonstrate that the owners/operators of offshore facilities "can respond quickly and effectively whenever oil is discharged."[19]

Under the current statutory system, OSRPs must be submitted to the Mineral Management Services (MMS) agency for approval. Facilities may be operated for up to two years while the MMS reviews and approves an OSRP. If the MMS finds that an OSRP contains inadequate plans, it can require revisions to the plans.

Under Federal regulation, OSRPs may be for a single lease site or facility or for a group of leases or facilities, but all the facilities or sites must have the same owner or operator (including affiliates) and must be located in the same MMS Region. Owners/operators of offshore facilities required to develop OSRPs must review the plans at least every two years and submit all modifications resulting from such reviews to the MMS or inform the MMS in writing that no changes are being made. Revisions to OSRPs must also be submitted to the MMS within 15 days when a number of conditions occur,

[19]30 C.F.R. § 254.1.

including changes that reduce response capabilities, a change in the worst case scenario, a change in the type of oil being handled, or a change in an Area Contingency Plan. Facility owners/operators are required to exercise their OSRPs at least once every three years.

Regulations require an OSRP to contain key information:

- An Emergency Response Action Plan
- Equipment inventory
- Response materials, supplies, services, and equipment locally and regionally available to respond to a spill
- Contractual agreements that provide proof of the responsible party's contracts with OSROs and includes provisions demonstrating that they are available on a 24-hour-per-day basis
- Worst case discharge scenario, including an analysis of the likely volume of the worst case scenario spill and its potential trajectories
- A list of the economic and environmental resources that would be affected
- A discussion of the response that would be undertaken to the worst case scenario in adverse weather conditions
- A dispersant use plan that includes an inventory and the location of the dispersants and related products that would be used in the event of a spill, a discussion of application procedures, and a summary of toxicity data for the dispersants
- An in-situ burning plan that describes the equipment available to conduct in situ oil burns as well as ignition provisions for and environmental effects of an in situ burn
- A training and drills appendix that describes the dates and types of training provided to members of the responsible party's oil spill response management and operational team members

The OSRP must be site-specific and not generic. The plan must explain in detail the clean-up actions as well as the time required for mobilization. Clean-up actions are required to begin as soon as possible to minimize the effect on natural and economic resources. These actions may include locating the source of the discharge and preventing any further spillage, the placement of containment booms to control the spread of oil and to protect sensitive area, measuring and sampling, the physical removal of oil from water and land, the use of chemicals to herd or disperse the oil, and in-situ burning. The coordinating response to the worst-case scenario spill must address many questions, including the following:

- How large an area will the spill cover?
- How thick will the slick be?
- How fast and in what direction will the slick drift?
- When and where will the oil hit the shoreline?
- What will happen to the oil if it is not removed?
- What is the value and sensitivity of the resource at risk?
- What negative impacts may happen from use of dispersants and other chemicals?

The answers to these questions will determine what response actions are to be taken. The following provides a summary of the major sections that are generally required in an OSRP.

- Cover Page
    - Title (e.g., Draft Spill Response Plan)
    - Prepared for – name of owner/operator
    - Prepared by – name of author(s) (some OSRP's are prepared by 3rd party consultants)
    - Project number
    - Date of document
    - Contact information
- Table of Contents – The following is an illustrative Table of Contents listing subject areas that generally should be covered.
    - Section 1. OSRP Rapid Guide
    - Section 2. Preface
        - Executive Summary
        - Table of Contents
        - Record of Revisions
        - Cross Reference Table
    - Section 3. Introduction
        - Companies Covered
        - Purpose and Use

- Types of Leases and Pipelines
- Facility Information Statement
- Coverage Area
- Contract Certification Statement

- Section 4. Organization – This section should detail the chain-of-command responsibilities in the event of an emergency including spill team members, contractors, and outside agencies.
  - Qualified Individual(s)
  - Spill Management Plan
  - Spill Response Operating Team
  - Oil Spill Removal Organizations
- Section 5. Spill Response Operations Center and Communications
  - Spill Response Operations Center
  - Communications
- Section 6. Spill Detection, Source Identification, and Control
  - Spill Detection
  - Pipeline Spill Detection and Location
  - Source Control
- Section 7. QI, SMT, SROT, and OSRO Notifications
  - Reporting Procedures
  - Company Contact Information
  - SROT Contact Information
  - OSRO Contact Information
  - Internal Spill Reporting Forms and Other Documents
  - Note that SMT stands for Spill Management Team. The SMT should be based on the National Interagency Incident Management System (NIIMS) format, which includes the following:
    - Command – Sets objectives and priorities and has overall responsibility at the incident or event
    - Operations – Conducts tactical operations to carry out the plan, develops tactical objectives, and directs all organizational and equipment resources
    - Planning – Develops the action plan to accomplish the objectives, collects and evaluates

  information, and maintains the situation and resource status
  - Logistics – Provides support to meet incident needs, resources, and all other services needed to support the incident
  - Finance/Administration – Monitors cost related to the incident and provides procurement
- Section 8. External Notifications
  - Reporting Procedures
  - External Contact Information
  - External Spill Reporting Forms
- Section 9. Available Technical Expertise and Resources
  - Locating a Spill
  - Determining the Size and Volume of a Spill
  - Prediction Spill Movement
  - Monitoring and Tracking the Spill Movement
- Section 11. Resource Identification
- Section 12. Strategic Response Planning
- Section 13. Resource Protection Methods
- Section 14. Mobilization and Deployment Protocols and Methods
- Section 15. Oil and Debris Removal Procedures
  - Offshore Procedures
  - Shallow Water Procedures
- Section 16. Oil and Debris Disposal Procedures
- Section 17. Wildlife Rehabilitation Procedures
- Section 18. Dispersant Use Plan
  - Dispersant Inventory
  - Toxicity Data
  - Dispersant Effectiveness
  - Application Equipment
  - Application Methods
  - Conditions for Use
  - Approval Procedures and Forms
- Section 19. In Situ Burning Plan
  - In Situ Burning Equipment
  - Procedures
  - Environmental Effects
  - Safety Provisions

- Conditions for Use
- Decision Processes
- Approval Procedures and Forms

- Section 20. Alternative Chemical and Biological Response Strategies
  - Product Inventory
  - Toxicity Data
  - Application Equipment
  - Application Methods
  - Conditions for Use
  - Approval Procedures and Forms
- Section 21. Documentation
- Appendix A. Facility Information
- Appendix B. Training Information
- Appendix C. Drill Information
- Appendix D. Contractual Agreements
- Appendix E. Response Equipment
  - Equipment Inventory
  - Inspection and Maintenance Programs
- Appendix F. Support Services and Supplies
  - Internal Spill Reporting Forms
  - External Spill Reporting Forms
- Appendix H. Worst-Case Discharge Scenarios
  - Worst-Case Discharge Scenario Selection
  - Worst-Case Discharge Scenario Discussion
  - ASA Oil Spill Trajectory Analysis
- Appendix I. Oceanographic and Meteorological Information
  - Oceanographic Information
  - Meteorological Information
- Appendix J. Bibliography

The OSRP must be specific and detailed. It must include the names, roles, and responsibilities and describe the duties, responsibilities, and authorities of each SMT member. Sufficient SMT members will be designated and available, such that the duties of each individual position on the SMT can be fulfilled on a 24-hour per day basis. The facility must have sufficient equipment and personnel to respond to incidental as well as major emergencies at the facility. In the event of any large-scale incident, the owner/operator will enlist the services of outside responders, the response contractor in particular.

The SMT must include a designated trained oil spill response coordinator (OSRC)/ Incident Commander (IC) and alternate, who will be delegated responsibility and authority to direct and coordinate response operations.

The plan should identify and describe all Oil Spill Removal Organizations (OSRO). OSROs provide oil spill response materials and supplies, equipment, and dedicated vessels in the event of an oil spill. The supplied equipment and materials must be of sufficient quantity and recovery capacity to respond effectively to oil spills from the facilities and leases covered by the regional or subregional OSRP.

Communication systems such as telephones, mobile phones, radio communications, and pagers may be used during a spill event. All personnel should be provided with mobile communication devices during work hours. Additional communication devices (telephones, fax machines, computers, etc.) will be located in the Operations Center. Contact information for response team members should be provided in the quick reference guide, such as Section 1 in the outline above.

A number resources are available that can provide additional resources and information:

- National Ocean Service (NOS) Map Finder – Provides images and data including coastal aerial photography, nautical charts, coastal survey maps, environmental sensitivity index atlases, hydrographic survey outlines, historical maps, water level station data, geodetic control points, and estuarine bathymetry data
- NOAA National Weather Service Weather – Provides data, forecasts, warnings, and current conditions
- Weather.gov – Web site providing weather watches, warnings, forecasts, and current conditions
- NOAA Marine Prediction Center – Provides forecasts for oceans including real-time buoy observations
- NOS Center for Operational Oceanographic Products and Services – Provides observations and predictions of water levels and currents
- US Fish and Wildlife Service – Provides assistance in sensitive receptor and resource identification for oil spills that reach shorelines

- National Marine Fisheries Service – Provides assistance in sensitive receptor and resource identification for oil spills that reach shorelines

Spilled materials would be expected to float and will be recoverable via mechanical methods. The OSRP should stress that response times are expected to be quick and will ensure that oils do not reach sensitive areas. A key step in the initial response actions is the determination of whether shorelines require protection. The owner/operator will be responsible for making this assessment, with assistance from the response contractors, environmental and civil engineering consultants, and likely a marine and freshwater environmental modeling firm.

Should shoreline clean-up operations become necessary, the plan should address how the owner/operator intends to work with federal, state, and local authorities to ensure that appropriate techniques are employed in a timely and environmentally sensitive manner. Such techniques may include the following:

- Natural recovery
- Manual removal
- Mechanical removal
- Collection with sorbents
- Vacuum
- Sediment reworking/tilling
- Vegetation cutting/removal
- Water washing
- Sand blasting

The response procedures established in the OSRP should ensure that all releases are quickly identified, addressed, and recovered. In the unusual event that impact to sensitive areas or wildlife occurs, the owner/operator must dedicate the necessary resources (with assistance from the response contractor) and will work with federal, state, and local authorities to fully remedy the impacts.

Rapid mobilization is a key component to response and recovery. In order to minimize the initial response time, the owner/operator will need to maintain workboats and response equipment at its Operations Center. This will allow initial assessment and defensive response actions to occur within approximately two hours or less of spill detection and full spill response to occur within approximately eight to twelve hours of spill detection. The operator must allocate resources and establish an emergency response system that will ensure that adequate containment equipment, recovery equipment, and response personnel are mobilized and deployed at the spill site and projected impact locations. To ensure the shortest possible response time, the owner/operator should plan on stocking spill equipment at its Operation Center and rely on a local response company to provide emergency response and clean-up.

Documentation is also an important part of the OSRP. The following documents will need to be maintained to record spill response actions, impacts, recovery methods, and the final outcome of the emergency:

- Internal Log for incidental spills (may be part of the maintenance log)
- Internal spill report for non-incidental spills
- Telephone conversation logs for oral spill reporting
- Written reports to agencies/authorities (including summary of environmental impacts and discussion of third parties)
- Memos, press releases, and other daily correspondence for non-incidental spills
- Reports of spill recovery and clean-up activities from the response contractor
- Estimates of the total cost for spill clean-up
- Manifest copies for spill clean-up materials requiring disposal as hazardous waste (or a determination that waste is non-hazardous)
- Updated copies of this OSRP

The OSRP should include examples of reporting forms that will be used.

The OSRP needs to be as site specific as possible. Equipment that is to be used for the emergency response needs to be listed by type, make model, capacity, location, and other pertinent information. It should be a detailed roadmap with information, contact numbers, names, and specific instructions on how to respond to the oil release. It must also be a plan that details the type, level, content, and specific personnel to be trained for the emergency response activities.

The OSRP should be based on a worst case scenario, i.e., it should provide the exact countermeasures to be employed and should detail the specific equipment and personnel that are responsible for activities for a scenario that represents the worst possible set of conditions or magnitude of the spill. If, for example, the plan has been prepared for a floating barge and that facility maintains 800,000 barrels of fuel oil at the most at any one time, then the OSRP should be based on a catastrophic release of the entire amount of oil. Even if there are redundant safeguards in place, and the probability that a catastrophic release of the 800,000 barrels is very low, the response plan should still be based on the total release of this product. Unless the probability of release can be eliminated (i.e., zero probability – not a low probability, but rather the elimination of the release of the entire amount), then the worst-case scenario could happen. The idea behind an OSRP is to be prepared for the worst possible incident because, no matter how low the probability of occurrence, it is still within the realm of possibilities.

6.5 Air Monitoring

6.5.1 Reasons for Air Monitoring

There is one reason for air monitoring – that is to prevent workers and neighboring communities from being exposed to chemical toxins that are in oil, ingredients that make up dispersants and other chemicals that may be applied during countermeasures, and byproducts of incomplete combustion should in-situ oil slick burning be applied as a countermeasure to the spill.

For workers, OSHA protocol calls for area sweeps prior to conducting spill response activities. By performing such sweeps, the Incident Commander or HSO can determine the most appropriate level of respiratory protection. It is critical that baseline measurements be taken and or an assessment be made of the air quality prior to commencing clean-up activities. As an example, during the Exxon Valdez oil spill workers in shoreline cleaning operations cleaned rocks and boulders using high-pressure water jets. Many workers reportedly did not wear respirators. The activity generated toxic aerosols that workers breathed, making them sick. Many years later, news reports have described long-term chronic ailments. Exxon has reported that there were few, if any, workmen compensation claims filed but also noted that many of the workers employed in those activities were migrant workers, and, hence, any statistics on adverse health impacts from the historic clean-up operation are lost. Nonetheless, the lesson from the Exxon Valdez spill is that, before activities begin, up-most care must be taken to ensure that workers are properly fitted with the appropriate levels of skin and respiratory protection. When in doubt, it is prudent to assume a high-risk situation and to equip workers with the highest levels of protection. This means Level A or B protection, depending on the nature and duration of the response activity.

Additionally, the chemical toxins to be monitored must be determined prior to any response efforts. In the case of crude oil, the chemicals of concern are VOCs (Volatile Organic Chemicals), which include benzene, mixed xylenes, toluene, ethylbenzene, and others. Benzene in particular is of concern because it is a confirmed human carcinogen that the American Petroleum Institute has recommended zero tolerance to since 1948. Should the decision to employ in-situ burning of oil as a countermeasure be made, then it will be important to measure and monitor PAHs (polycyclic aromatic hydrocarbons) and particulate matter.

In the case of particulate matter, particle sizes up to 30 microns are also a concern from the standpoint of human risk exposure. Both PM2.5 and PM10 (particulate matter of size 2.5 and 10 microns, respectively) are within a normal person's respiration range. Therefore, these particles that contain oil, soot from burning, and PAHs will be taken into the lungs and bloodstream.

The EPA[20] notes that there are three primary origins of PM2.5: primary solid particulate matter that is emitted directly in the solid phase, primary condensable particulate that can be emitted at high temperatures in the gas phase but that condenses into the solid phase upon dilution and cooling in the plume from burning, and secondary particulate that is formed through atmospheric reactions of gaseous SO and NO precursor emissions. The secondary components involve complex chemical and physical interactions among the precursor emissions in the atmosphere. The atmospheric reactions and transport processes that result in secondary particulate formation

[20]Pacific Environmental Services, Inc., Getting Started: Emissions Inventory Methods for PM-2.5, Volume IX: Chapter 1; Prepared for: PM-2.5 Committee, Emission Inventory Improvement Program, U.S. EPA, Washington, D.C. September 1999.

extend over large distances from the sources of the emissions and, hence, may threaten neighboring communities. Both directly emitted and secondary PM2.5 can be transported over large distances.

Larger particulate matter up to about 30 microns in size is also a health risk concern. While these large particles are above the respirable range, they are well within the range where they may be ingested. Particles with diameters up to 30 microns can be deposited in the upper respiratory system and ingested.[21]

Other chemicals of concern include solvents and particulate matter that volatilize or are generated as toxic aerosols from the application of cleaning chemicals and dispersants. For example, the dispersant COREXIT contains 2-butoxy ethanol, which is a suspected carcinogen. Acute (short-term) health effects from overexposure to this chemical include irritation of the skin and eyes with possible eye damage. Inhaling 2-butoxy ethanol can irritate the nose and throat, causing coughing and wheezing. 2-butoxy ethanol can cause nausea, vomiting, diarrhea, and abdominal pain. Exposure can cause headache, dizziness, confusion, lightheadedness, and fainting. Aside from being a suspected carcinogen, 2-butoxy ethanol may damage the developing fetus. There is also limited evidence that 2-butoxy ethanol may damage the male reproductive system (including decreasing the sperm count) in animals and may affect female fertility in animals. Other dispersants and shoreline cleaning chemicals may contain a variety of solvents. The MSDS of products to be employed should be carefully examined for toxic ingredients to determine what chemicals to monitor for.

6.5.2 Direct vs. Indirect Methods

The measurement or monitoring of fugitive emissions generated from clean-up efforts is an absolute necessity in order to protect response and post-response personnel in the field, as well as to ensure that potentially impacted communities are not being exposed to chemical toxins during the emergency clean-up efforts.

---

[21]Cynthia Weidner, John Fitzgerald, and Maureen Vallatini, Real-Time Air Monitoring at Construction and Remediation Sites to Estimate Risks of Contaminated Dust Migration, Massachusetts Department of Environmental Protection, *Presented at the 12th Annual Conference on Contaminated Soils, University of Massachusetts at Amherst, Mass. October 1997.*

The available measurement techniques for assessing fugitive emissions may be characterized as either direct or indirect.

Direct methods involve the physical measurement of flow rates and gas concentrations at the source. Indirect measurement techniques rely on the use of a theoretical or empirical model to back-calculate source strength based on pollutant concentrations at a convenient downwind reference point and on appropriate process-activity and/or meteorological data.

Most instruments that are applicable to spill responses fall into the category of field instrumentation and are the direct type. Typically, the key advantages of direct methods over comparable indirect methods include:

- improved sensitivities or detection limits;
- easier-to-assess uncertainties;
- more accurate and disaggregated results allowing better identification and evaluation of specific emission reduction opportunities;
- fast, low-cost means of determining emissions from a few simple point sources;
- simple, easy-to-operate measurement equipment;
- simple data reduction; and
- the ability to isolate and measure emissions from a target source in the presence of other nearby sources to avoid interference problems.

The key advantages of indirect methods usually include:

- the ability to safely determine emissions from hazardous or difficult-to-access sources;
- measurements of all the emissions so there are no missed sources;
- fast, low-cost means of determining aggregate emissions from a large, complex area or volume sources;
- measurements that can be performed off-site; and
- applicability to a wide variety of sources.

Some of the disadvantages of indirect methods include:

- relying heavily on favorable meteorological conditions and downwind terrain;

- susceptibility to interferences from other nearby sources;
- requires the ability to detect small concentration changes of analytes at very low levels,
- generally requires greater expertise and quality assurance/quality control (QA/QC) measures than direct methods,
- large potential errors depending on circumstances and the technique being applied (e.g., emission estimates for simple point sources may have errors of ± 10 to 25 percent under favorable conditions, while estimates for more complex sources and conditions may easily be in error by several orders of magnitude, and
- the true emission rate is never really known unless appropriate confirmation measurements are made which may be difficult and costly to do on large, complex sources or sites);

6.5.3 Instrumentation and Community Air Monitoring Program

There are a number of instruments and analytical methods to aid in an air monitoring program. One of the most common instruments relied on for an oil spill response is a MultiRAE Plus, which is an instrument that combines a PID (Photoionization Detector) with the standard four gases of a confined space monitor (O_2, LEL, and two toxic gas sensors) in one compact monitor with a sampling pump. (Note that LEL stands for Lower Explosion Limit.)

The MultiRAE Plus is a versatile gas detector. It is most often used by HazMat teams in the United States. The MultiRAE Plus is often converted from a technician instrument in the lab to a personal monitor, a hand-held sniffer, or a continuous operational area monitor.

The MultiRAE Plus detector can be made wireless. This allows real-time monitoring information from the detector to be integrated into an existing AreaRAE system. A wireless, RF (radio frequency) modem allows detectors equipped with Firmware version 1.20 or higher to communicate and transmit readings and other information on a real-time basis with a remotely located AreaRAE base controller up to two miles away.

The instrument is a programmable multi-gas monitor designed to provide continuous exposure monitoring of toxic gases, oxygen, and combustible gases for workers in hazardous environments. It monitors three classes of hazardous gases:

- Organic vapors in the ppm range with a Photo-Ionization Detector (PID) using a 9.8, 10.6 eV, or 11.7 eV gas discharge lamp
- Combustible gases near their lower explosive limits (LEL) with a catalytic bead sensor
- Inorganic compounds, including oxygen deficiency, with the electrochemical sensors

Other features of this instrument include the following:

- Lightweight and compact
- Considered by many as dependable and accurate
- 10 hours monitoring with micro controller
- User friendly, menu-driven, intuitive end-use operation
- Equipped with programmable alarm thresholds such as audio buzzer and flashing display alarms
- A multi-gas monitor, a battery charger, rubber boot, alkaline battery adapter, water and dust filters, and calibration accessories
- Separate alarm limit settings for TWA, STEL, and Low and High alarms
- Capable of large data storage: 20,000 readings (64 hours, 5 channels at 1 minute intervals) in non-volatile memory

Table 6.23 provides typical range, resolution, and response times for the instrument. Table 6.24 reports calibration ranges for factory calibrated units.

The instrument includes an integrated sampling pump, which is a diaphragm pump providing about a flow rate of 300 cc per minute at the high setting with the standard filters in place. A low pump speed of about 200 cc per minute is the factory default setting, which will increase the battery life by about 5% and result in an increased LEL sensor lifetime. The high pump speed setting is required for vapors that are especially reactive or adsorb easily to instrument surfaces. Such vapors include Cl_2, PH_3, NH_3, HCN, and

Table 6.23 Range, resolution, and response times for the MultiRAE.

Gas	Symbol	Range	Resolution	Response Time
Carbon Monoxide	CO	0–500 ppm	1 ppm	20 sec
Oxygen	Oxygen	0–100 ppm	1 ppm	30 sec
Sulfur Dioxide	SO_2	0–20 ppm	0.1 ppm	15 sec
Nitric Oxide	NO	0–250 ppm	1 ppm	20 sec
Nitrogen Dioxide	NO_2	0–20 ppm	0.1 ppm	25 sec
	NO	0–10 ppm	0.1 ppm	60 sec
	NO_2	0–30 %	0.10%	15 sec
Volatile Organic Compound	VOC	0–200 ppm	0.1 ppm	10 sec
	VOC	200–2000 ppm	1 ppm	10 sec
Lower Explosion Limit	LEL	0–100 %	1%	15 sec
Hydrogen Cyanide	HCN	0–100 ppm	1 ppm	60 sec
Ammonia	NH_3	0–50 ppm	1 ppm	150 sec
Phosphine	PH_3	0–5 ppm	0.1 ppm	60 sec

semivolatile organic compounds like diesel fuel and jet fuels. For such compounds it is helpful to remove the water trap filter for normal calibration and operation, which will increase the pump speed to about 350 cc/min.

Worker and community sampling must be an integral component in the OSRP. From a community exposure standpoint, a detailed work plan will be needed to address air monitoring and sampling in communities that could potentially be impacted. The OSRP should identify sensitive communities so that monitoring can be implemented at the time of the incident. When a spill does occur, the plan should first require obtaining data or have at hand normal background air concentrations as a control in order to assess whether elevated levels detected during the spill event are causing a problem. Monitoring in the community prior to oil sheens being spotted close to shorelines is important to establishing background air concentrations. In addition, monitoring will need to be performed for the air surrounding the mitigation activities in

Table 6.24 Factory calibration values for the MultiRAE air monitor.

Gas	Symbol	STEL	TWA	Low	High	Gas Concentration Factory Calibration
Oxygen	O_2	–	–	19.50%	23.50%	20.90%
Combustible Gas	–	–	–	10%	20%	50% LEL
VOCs	–	50	25	50	100	100 ppm isobutylene
Carbon Monoxide	CO	100	35	35	200	50 ppm
Hydrogen Sulfide	H_2S	15	10	10	20	10 ppm
Sulfur Dioxide	SO_2	5	2	2	10	5 ppm
Nitric Oxide	NO	25	25	25	50	25 ppm
Nitrogen Dioxide	NO_2	1	1	1	10	5 ppm
Chlorine	Cl_2	1	0.5	0.5	5	10 ppm
Hydrogen Cyanide	HCN	4.7	4.7	4.7	50	10 ppm
Ammonia	NH_3	35	25	25	50	50 ppm
Phosphine	PH_3	1	0.3	1	2	5 ppm

the community in order to protect the general public. Monitoring specific response activities will be needed in order to ensure that operations are being conducted safely.

Integrated air sampling needs to be conducted in representative locations along an impacted or potentially impacted shoreline using standard protocols for sampling. EPA Method TO-15 for volatile organic compounds (VOCs), total hydrocarbons as gasoline, NIOSH Method 5506 for polycyclic aromatic hydrocarbons (PAHs), and the passive dosimeter method (Radiello) for hydrogen sulfide (H_2S). The sampling plan should identify which EPA accredited laboratory samples are to be sent to for analysis. Monitoring locations will need to be identified in advance and updated on a regular basis based upon site activities, public concern and various other factors; hence the plan should be flexible enough to address these variables.

In addition to fixed site air sampling, real-time air monitoring will be needed. This can be accomplished by employing roaming vehicles throughout a community. Roaming vehicles should be equipped with real-time monitoring equipment for measuring VOCs, benzene, sulfur dioxide, particulate matter, and hydrogen sulfide. Roaming vehicles should be operated in a community 24 hours each day.

Air monitoring can be conducted at selected locations, which will need to be determined at the time of the incident and in relationship to projected impacts. Factors such as changes in wind direction and strength will need to be taken into consideration when conducting such monitoring. Additionally, observations concerning the qualitative presence or absence of odors related to oil should be recorded. Air monitoring should be based on a flexible program that responds to concerns and issues raised by the Unified Command and the community.

Real-time air monitoring especially for VOCs will be critical during oil clean-up activities. The MultiRAE Plus, UltraRAE, and Gastec colorimetric detector tubes are considered the field instruments of choice and should be used in combination. The Multi RAE Plus can be set up to monitor for VOCs using its photoionization detector (PID) and for hydrogen sulfide using its electrochemical sensor. It is critical that real-time monitoring also be conducted for benzene; this can be accomplished using the UltraRAE benzene specific PID or with the Gastec colorimetric tubes. PIDs can also be used to detect and monitor any volatile components that may be emitted from oil sheens. Both detector tubes and the UltraRAE are good instruments for periodic chemical specific analysis of gases in the atmosphere and in the event that elevated levels of VOCs are detected using PIDs.

In addition to chemical vapor monitoring, it will be necessary to conduct real-time particulate matter (PM2.5) monitoring. PM2.5 stands for particulate matter that is 2.5 microns in size and smaller. These particle sizes are in the respirable range and further they are small enough to serve as condensation sites for VOCs. Monitoring for PM2.5 is especially important should in-situ burning be applied as a countermeasure. The instruments used for this purpose are a TSI AM510 or TSI DRX 8534 model.

The following table provides a summary of applicable instruments along with detection limits.

Integrated air sampling can be performed using Minicans™ for EPA method TO-15 analysis in cooperation with an EPA certified

Table 6.25 Real-time monitoring instruments with detection limits.

Instrument	Chemical Toxin	Detection Limits	Detection Limit Units
MultiRAE	VOCs	0.1	ppm
MultiRAE	Hydrogen Sulfide	1	ppm
UltraRAE	Benzene	0.1 but down to 0.05 with correction factor	ppm
Gastec Colorimetric Tubes	Hydrogen Sulfide	0.1	ppm
Gastec Colorimetric Tubes	Sulfur Dioxide	0.01	ppm
Gastec Colorimetric Tubes	Benzene	0.05	ppm
TSIAM510	PM2.5	0.001	mg/m^3
TSI DRX 8534	PM2.5	0.001	mg/m^3

lab. This method is especially useful for quantifying total hydrocarbons as gasoline. The sampling/monitoring program may also be supplemented by short-term grab samples to be analyzed via EPA TO-15 analysis. NIOSH Method 5506 is appropriate for sampling PAHs. Both PAHs and VOCs are chemicals that should be tested for; the former based on the toxicology and chemical composition of crude oil, and the latter most importantly when in-situ burning is being conducted as a countermeasure.

Additionally, hydrogen sulfide exposure can potentially be lethal and hence it is best to use Radiello passive dosimeters and then analyze samples using ATL Application # 59 to supplement real-time monitoring.

In both worker and community air monitoring programs, there must be a rational and well planned quality assurance and control (QA/QC) set of protocols that should be rigorously adhered to throughout the program. These should be established at the onset of the monitoring program for each of the chemicals of concern. The QA/QC protocols should include the use of field blanks and a rational basis for crosschecking instrument calibration and analytical test methods based on a statistically number of blanks. For example, for every 10 samples taken in the field, 1 field blank and calibration check should be conducted. And if there is fewer than the minimum number of samples, then at least one per day should be run. A more aggressive sampling and QA/QC program will call for more blanks.

6.5.4 Odors

One area of concern in air monitoring programs is the lack of attention generally given to the detection of odors. While portable air monitoring instruments help to ensure that individuals are not being exposed to concentrations above legal and recommended safe levels of exposure (i.e., a PEL), the presence of odors below TWAs and PELs can still be harmful to humans. They can cause irritation of the eyes, nasal passages, and throat, dizziness, nausea, loss of appetite, headaches, increased blood pressure, wheezing, and can lead to the development of asthmatic conditions and exacerbation of existing asthma. If we look at the disaster in the Gulf, the air monitoring that has been performed reports almost no exposure to airborne concentrations above PELs for crude oil constituents. However, more than 350 fishermen working in the clean-up of the spill have been sent to hospitals in Louisiana because they complained of various symptoms related to odor exposure. News agencies reported (in an article titled "Gulf oil spill still out to sea, but odor has reached shore"[22]) that odors originating from the oil slick were easily detectable to local residents. In the same spill, the associated odors were strong enough to prompt the EPA to draft an advisory document titled "Odors from the BP Oil Spill," addressing complaints from coastal residents. It must be kept in mind that air monitoring really focuses on determining whether or not there is exposure to chemicals that may have a toxicological effect. In contrast, odors are perceived at concentrations that are, generally, considerably below such levels, and yet they are still harmful. And further, there are no stringent guidelines or practices to control or limit exposures.

From an odor standpoint, crude oil has multiple chemical constituents that have impacts on the sense of smell and our reactions to

[22]Nelson, K., 2010. Gulf oil spill still out to sea, but odor has reached shore. Biloxi Sun Herald, 11, May; http://www.mcclatchydc.com/2010/05/11/93992/gulf-oil-spill-still-be-out-to.html

them. Collectively, the soup of chemicals found in crude oil can produce a rather "pungent" smell, as noted on many MSDSs. Depending on the composition, crude oil scents may range from rotten egg to hydrocarbon and petroleum, all of which are general terms used to describe the characteristic odor of crude oil on MSDSs.

Liu[23], Shusterman[24], Schiffman,[25] and others have reported that odors not only act as indicators of potential risk, but that odors themselves may cause damaging health effects.

Odor, as well as noise, crowding, and heat, are environmental stressors. Odors will trigger negative health effects even in the absence of toxicological irritation (see, for example, the study by Nimmermark[26]). Highly odorous compounds like hydrogen sulfide found in sour crude (which can be detected at concentrations as low as 3 ppb, or 0.004 mg/m^3) have been reported to cause symptoms at levels many orders of magnitude lower than those that are known to cause symptoms by toxicologic and irritative mechanisms.

The Agency for Toxic Substances and Disease Registry (ATSDR) has also noted that odors can cause negative health effects stating, "In public health practice, odors are not only warning signs of potential health effects, but may in themselves be the direct cause of some symptoms in humans" (ATSDR, 2009[27]). ATSDR notes that odor can have negative effects on numerous sites of the body, including mucous membranes, upper and lower respiratory airways, heart and blood vessels, stomach and intestines, the brain, as well as psychological and general well being.

The following table compares the odor threshold to the PEL values of several chemicals found in crude oil. As shown, the so-called

---

[23]Liu, 2007. Liu HL, Lam LT, Wang T, Hou CC, Zeng Q, Feng LH, Chen XM. The health effects of odor emission from domestic renovation on household residents in Tianjin, China. *Journal of Environmental Health* 69(7):46–50.

[24]Shusterman, 2001. Shusterman, Dennis. Odor-associated health complaints: competing explanatory models. *Chem Senses* 26:339–343.

[25]Schiffman, 2005. Schiffman SS, Williams CM. Science of odor as a potential health issue. *J Environ Qual* 34:129–138.

[26]Nimmermark, 2004. Nimmermark, Sven. Odour influence on well-being and health with specific focus on animal production emissions. *Ann Agric Environ Med* 11:163–173.

[27]ATSDR, 2009. "The Health Effects of Environmental Odors." Agency for Toxic Substances and Disease Registry. Available at: http://www.atsdr.cdc.gov/sites/intel/docs/Intel%20Package%20-%20Odors%20Health%20Effects%20Generic%20for%20Communities.pdf

Table 6.26 Comparison between odor thresholds and PELs.

CAS No.	Chemical Name	Odor Threshold, mg/m³	OSHA PEL, mg/m³
7783-06-4	Hydrogen Sulfide	0.004	15
1330-20-7	Xylene, Mixture	0.348	435
91-20-3	Napthalene	0.440	50
71-43-2	Benzene	4.500	3.19
108-88-3	Toluene	8.025	750

safe levels of exposure from a toxicological standpoint are well above odor threshold values except for benzene. But even in the case of benzene, the odor threshold is only slightly above the PEL.

These comparisons show that air monitoring can be misleading and does not necessarily provide sufficient information on exposure. Portable field instruments are most often used to determine action levels required for personal protection based on recommended and/or permissible exposure limits. But these limits do not necessarily safeguard workers or communities from the health risks associated with low concentrations associated with odors.

7

Standard of Care and The BP Oil Spill

7.1 The Impacts

The Macondo well is located in Mississippi Canyon Block 252 of the Gulf of Mexico. BP is the operator and principal developer of the oil field with 65% of interest, while the Anadarko Petroleum Corporation owns 25%, and MOEX Offshore 2007, a unit of Mitsui, owns 10%. The prospect may have held 50 million barrels (2.1×10^9 gallons) of producible reserves of oil.

A regional shallow hazard survey and study was executed within the Macondo area by KC Offshore in 1998. High resolution, 2D seismic data, along with 3D exploration seismic data of the MC 252 was collected by Fugo Geoservices in 2003. BP purchased the mineral rights to drill for oil in the Macondo Prospect at the Minerals Management Service's lease sale in March 2008.

Mapping of the block was carried out by BP America between 2008 and 2009. BP secured approval from Mineral Management Services (MMS) in March 2009 to drill the Macondo Prospect and without MMS requiring use of an acoustic blowout preventer actuation alternative. An exploration well was scheduled to be drilled in 2009.

Figure 7.1 Map showing the location of the well. (map courtesy of Google Maps)

On October 7, 2009 the Transocean Marianas semi-submersible rig commenced drilling, but operations were halted at 4,023 feet below the sea floor on November 29, 2009, when the rig was damaged by Hurricane Ida. The Transocean's Deepwater Horizon rig was leased by BP, and drilling operations resumed in February 2010. Figure 7.1 shows the location of the well.

An explosion on the Deepwater Horizon drilling rig occurred on April 20, 2010. The Deepwater Horizon sank on April 22, 2010 in water approximately 5,000 feet deep and has been located resting on the seafloor approximately 1,300 feet (about a quarter of a mile) northwest of the well. Following the rig explosion and subsea blowout, BP started a relief well using Transocean's Development Driller III on May 2, 2010. The relief well would potentially take up to three months to drill. BP started a second relief well using Transocean's GSF Development Driller II on May 16, 2010. Beginning May 17th, some oil and gas was collected through the riser insertion tube tool inserted to the blowout well, the oil being stored and gas being flared on the board of drillship Discoverer Enterprise. Figure 2 shows the severity of the fire on board the drill rig and the efforts to extinguish the flame.

The British Petroleum Corporation (BP) is the 4th largest oil company in the world. In 2009, sales and other operating revenue

Figure 7.2 As Deepwater Horizon sinks, engulfed with flames fueled by tens of thousands of gallons of burning oil. (Image courtesy of US Coast Guard)

equaled $239 billion. It employs over 80,000 personnel, owns or operates 16 refineries, and is active in 30 countries.

The spill is an unfolding event that, to date, has released between more than 80,000,000 to perhaps as much as 140,000,000 million gallons of crude oil into a region that had some of the more, if not most, diverse wild and marine life in the world. The impact of the spill touches almost every fabric of life. The damage to marine life, wildlife, and coral reefs and the disruption to the ecosystem food chain cannot even begin to be taken into consideration because we simply do not have a sense of how extensive this ongoing event is. Those who have visited the Gulf while this event is unfolding told us that marine life is hovering along shoreline areas in unprecedented populations never seen before awaiting an inevitable, agonizing death as oil slowly makes its way to shoreline. There are various Web sites that are tracking wildlife impact statistics from the spill. See for example a website named Daily Dead Birds (http://dailydeadbirds.com/) which is tracking the number of birds, turtles, and mammals killed by the BP oil spill. The website's data comes from consolidated Unified Area Command fish

Table 7.1 Wildlife impacts after 68 days of spill. Source: Daily Dead Birds (http://dailydeadbirds.com/)

	Dead	Oiled but Alive	Cleaned and Released	Totals
Birds	1,117	791	117	2,025
Sea Turtles	406	88	3	497
Animals	50	1	1	52
Totals	1,573	880	121	2,574

and wildlife collection reports. Table 7.1 gives the statistics after 68 days of the spill.

While the statistics recorded in Table 7.1 are high, some readers may react by saying these are not horrifying. It is not unreasonable to extrapolate this data several orders of magnitude after one considers the extent of the spill.

Literally thousands of miles of shoreline and wetlands will be lost not only for decades, but some sensitive wetland zones will disappear forever. Business and livelihoods for all practical purposes are gone. For families that have made their living in the fishing industry, they will simply have to find other means by which to make livings to support their families. While the younger population may retool to another profession or perhaps wind up working on oil rigs in the region, a question for our government is what do people who are approaching or are in senior citizen age brackets do especially when they do not have sufficient savings to support themselves and are unable to retool to another profession so late in their lives? What additional social-economic burden does this place on our country?

For the tourism and restaurant industries, the outlook is just as dismal. At this point in time, hotel resorts and the restaurants are experiencing at least a 40% depression in their business at a time that is supposed to be their peak business and within the midst of a worldwide recession that has not been seen since the early part of the 20th century. Restaurants are just as devastated. While seafood at this point in time has been given a blessing of being untainted, the stigma alone has depressed the consumer markets where we have seen news reports showing seafood restaurants suffering losses of businesses in excess of 50% because of fear of eating oil along with one's meal.

For people who own property close to shoreline areas that have been impacted and for the millions of citizens that enjoyed access to public beaches, both land values and loss of enjoyment have heavy economic and long lasting socio-economic impacts. Valuation of such losses will not be simple because there will be direct, indirect, and short- and long-range effects over many areas. Some of these include the following:

- Valuing the losses to private property appurtenant to the natural resources
- Valuing the loss of enjoyment by impacted properties
- Valuing the losses to businesses that are dependent on impacted natural resources
- Fiscal losses suffered by taxing authorities

Just to scratch the surface, valuation losses centering on private property will need to consider the following:

- The direct impact of the oil spill, i.e., current losses from an ongoing event as well as post-remediation, which could take many years
- Indirect impacts that cause losses, e.g., impacts to businesses, resorts, sport fishing enterprises, etc.
- Losses that are both short and have long-term effects on valuation
- Losses associated with stigma, which may lead to a loss of market value

These losses will vary with the specific areas impacted. If we take Florida as an example, at present about 569 miles of Florida coast has or is expected to be impacted shortly. That represents 14,400 acres. Waterfront property even in today's depressed market sells for between $2 and 8 million. Assuming an average value of $3 million per acre and perhaps 10% depreciation that is due to the oil spill, then we are talking about $4.3 billion dollars in property damages.

Other areas like Louisiana, Mississippi, and Alabama are likely to face devastating impacts to home values as a result of the spill. "Homes in the immediate path of the oil leak are forecast to decline at least 30% in value."[1]

---

[1]K. Chiu, writing for NuWire Investor, 5.17/10, citing analysis from Hosungpredictor.com

The impacts from the spill to property values will be far reaching. The spill touches virtually every fabric of life. There will be impacts to:

- summer/vacation homes;
- personal property (e.g., pleasure boating);
- residents (loss of ambiance);
- natural resource recreation;

There will also be losses that are personal – they will not be a part of the real estate loss but nonetheless have value and should be compensated for.

We may expect business interruption losses. The following are examples:

- Tourism
- Natural resource-based businesses (fishing, oyster, etc.)
- Loss of revenue/increased costs
- Secondary and tertiary losses that will emerge or be defined
- Short- and long-term effects
- BP franchisers

These will vary depending on geography. Here are just a few of the areas that will be impacted by BP's oil spill:

- In Mississippi, leisure and hospitality account for more than 20% of the employment in the Biloxi-Gulfport area.
- Many Mississippi commercial and sport fisheries have reportedly already been closed for the season and may very well remain so.
- The 2009 Mississippi gaming taxes totaled $171 million, which is 3.7% of the state's tax revenues.
- A University of Central Florida economist, Sean Snaith (Director of UCF's Institute for Economic Competitiveness), provided a forecast on June 8th, in which he stated that the spill could cost 195,000 jobs and nearly $11 billion in revenue to Florida's Gulf Coast counties alone.

- The BBVA Compass Bank of Birmingham has reported that the BP oil spill could immediately drain $191 million (0.11%) from the Alabama economy. It also stated that Baldwin County could suffer out-migration and extended losses of tourism, depending on the severity of the spill and ensuing damage.

The above does not even scratch the surface of the financial losses and communities impacted. At this point in time, we have no sense of public health risks. There is a serious question as to how many of these people will wind up with illnesses from exposures to the toxins in crude oil or will expose family members when they bring their oil soiled clothing and shoes home for laundering. This is certainly the case for the workers that BP has required signed releases that would allow fishing boats to assist in the cleanup while denying crew members respirators, protective clothing, and adequate safety training,

And what about areas of our country outside of the Gulf region? Seafood prices have already increased by 30% in many parts of the country due to the loss of this regional industry. And if anyone believes that oil prices will not increase, then they need to rethink. The industry sector has long had a practice of externalizing its costs to the consumer, and prices will unquestionably increase because of this one incident, if for no other reason than the fact that the moratorium on offshore oil drilling costs about 1 million dollars per day per rig while rigs sit idly off shore. Oil companies will simply recover those losses by raising the prices we pay at the pump.

The impacts of the spill not only affect businesses, property values, wildlife, and marine life, as well as raise a big question on the health and safety of communities. The impacts include denying citizens of state and local services and access to public lands, a large part of which were no less than national treasures. BP's oil has, in essence, annexed thousands of miles of shoreline along the coast of the United States. It is in fact one of the largest acts of trespassing in history.

7.2 The Waxman/Stupak Letter

On June 14, 2010, Senators Henry A. Waxman and Bart Stupak (Chairpersons of the Committee on Energy and Commerce) wrote a letter to the Chief Executive Officer of BP, Mr. Tony Hayward. The following are excerpts from the letter:

- "We are looking forward to your testimony before the Subcommittee on Oversight and Investigations... about the causes of the blowout of the Macondo well and the ongoing oil spill disaster in the Gulf of Mexico. As you prepare for this testimony, we want to share with you some of the results of the Committee's investigation and advise you of issues you should be prepared to address."

- "The Committee's investigation is raising serious questions about the decisions made by BP in the days and hours before the explosion on the Deepwater Horizon."
    - "On April 15, five days before the explosion, BP's drilling engineer called Macondo a "nightmare well." In spite of the well's difficulties, BP appears to have made multiple decisions for economic reasons that increased the danger of a catastrophic well failure."
    - "In several instances, these decisions appear to violate industry guidelines and were made despite warnings from BP's own personnel and its contractors."
    - "... BP repeatedly chose risky procedures in order to reduce costs and save time and made minimal efforts to contain the added risk."

- "At the time of the blowout, the Macondo well was significantly behind schedule. This appears to have created pressure to take shortcuts to speed finishing the well. In particular, the Committee is focusing on five crucial decisions made by BP:
    - (1) the decision to use a well design with few barriers to gas flow;
    - (2) the failure to use a sufficient number of "centralizers" to prevent channeling during the cement process;

- (3) the failure to run a cement bond log to evaluate the effectiveness of the cement job;
- (4) the failure to circulate potentially gas-bearing drilling muds out of the well; and
- (5) the failure to secure the wellhead with a lock-down sleeve before allowing pressure on the seal from below."

- "The common feature of these five decisions is that they posed a trade-off between cost and well safety."

7.2.1 Well Design

Deepwater wells are drilled in sections. The process of erecting a well involves drilling through rock, installing and cementing casing to secure the well bore, and then drilling deeper by repeating the process. On April 9, 2010, BP completed drilling the last section of the well. The final section of the well bore extended to a depth of 18,360 feet below sea level, which was 1,192 feet below the casing that had previously been inserted into the well.[2] According to the Waxman/Stupak letter, BP was faced with an important well design decision, namely how to secure the final 1,192 feet of the well.

- On June 3, Halliburton's Vice President of Cementing, Tommy Roth, briefed Committee staff about the two primary options available to BP. The first option involved hanging a steel tube called a "liner" from a liner hanger on the bottom of the casing already in the well and then inserting another steel liner tube called a "tieback" on top of the liner hanger. The second option involved running a single string of steel casing from the seafloor to the bottom of the well.

---

[2]BP, PowerPoint Presentation, *Washington Briefing, Deepwater Horizon Interim Incident Investigation* at 4 (May 24, 2010).

Halliburton's Mr. Roth informed the Committee that the Liner/Tieback Casing provides a safety advantage over the full string casing because it relies on multiple or redundant barriers to annular flow.[3] The Liner/Tieback method provides four barriers to annular flow[4]:

1. the cement at the bottom of the well,
2. the hanger seal that attaches the liner to the existing casing in the well,
3. the cement that secures the tieback on top of the liner, and
4. the seal at the wellhead.

In the case of a single string of casing, there are just two barriers to the flow of gas up the annular space that surrounds the casing: the cement at the bottom of the well and the seal at the wellhead.

Obviously, the Liner/Tieback method is safer because it has a twofold safety factor over the single string casing method. In short, there is double the protection/safety features that would impede the flow of any hydrocarbons in an uncontrolled manner up the annular space of the well.

A responsible company would have chosen a safer practice, but BP understood that the liner-tieback option also takes more time to install. In fact, it required several additional days to complete. Time is money, and lost time represents lost profits piled onto the losses from an overrun schedule that was under-budgeted.

Internal BP documents identified in congressional hearings show that BP was aware of the risks of the single casing practice. An undated "Forward Plan Review" recommended against the single string of casing because of the risks of a well blowout. According to this document, "Long string of casing... was the primary option,"

---

[3]Briefing by Tommy Roth, Vice President of Cementing, Halliburton, to House Committee on Energy and Commerce Staff (June 3, 2010); Halliburton, Power Point Presentation, *Energy and Commerce Committee Staff Briefing* at 12 (June 3, 2010).

[4]Briefing by Tommy Roth, Vice President of Cementing, Halliburton, to House Committee on Energy and Commerce Staff (June 3, 2010); Halliburton, Power Point Presentation, *Energy and Commerce Committee Staff Briefing* at 12 (June 3, 2010).

but a "liner ... is now the recommended option."[5] BP's management ignored industry safe guidelines, safety recommendations from its consultant, and safety recommendations from its own engineers.

Because the company was behind in schedule and wanted to save money, BP decision makers sidestepped all the recommendations from multiple sources urging them to use a safer technology.

The "Forward Plan Review" document provided four reasons against using a single string of casing:

1. "Cement simulations indicate it is unlikely to be a successful cement job due to formation breakdown."
2. "Unable to fulfill MMS (Mineral Management Services) regulations of 500 feet of cement above top HC (hydrocarbon) zone"
3. "Open annulus to the wellhead, with seal assembly as only barrier"
4. "Potential need to verify with bond log, and perform remedial cement job(s)"

In contrast, according to the document, there were four advantages to the liner option:

1. "Less issue with landing it shallow (we can also ream it down)"
2. "Liner hanger acts as second barrier for HC in annulus."
3. "Primary cement job has slightly higher chance for successful cement lift."
4. "Remedial cement job, if required, easier to justify to be left for later"

The document not only shows that BP ignored the technical recommendations of its own engineers and consultants, but it

---

[5]BP, *MC 252#1 Macondo, TD Forward Plan Review – Production Casing & TA Options*, at 9. (Apr. 2010) (BP-HZN-CEC-221 09). The documents provided to the Committee from BP contain three versions of this document. This one and a second nearly identical version (BP, *MC 252#1 Macondo, TD Forward Plan Review – Production Casing & TA Options* (Apr. 2010)) (BP-HZN-CEC-22025) recommend against a single string casing and in favor of a liner.

willfully violated the law by intentionally not meeting the technical requirements of the Mineral Management Services.

Internal communications between employees of BP confirm that they were evaluating these approaches. On April 14, 2010, Brian Morel, a BP Drilling Engineer, e-mailed a colleague, Richard Miller, about the options. His e-mail notes: "this has been [a] nightmare well which has everyone all over the place."[6]

Despite the risks and warnings, BP chose to install the single string of casing instead of a liner and tieback, and they proceeded to apply for an amended permit on April 15, 2010.[7] BP's application stated that the full casing string would start at 9–7/8 inches diameter at the top of the well and would narrow to 7 inches diameter at the bottom. Astounding to us, this application was literally approved on the same day.[8] We question the competency, the level of due diligence performed, and whether there are conflicts of interest on the part of the personnel of the MMSs responsible for reviewing and approving an operating permit modification within a matter of a few hours.

It is obvious to us that BP decision-makers made the decision to run a single string of casing in order save time and to reduce costs.

On March 25th, Mr. Morel e-mailed Allison Crane, the Materials Management Coordinator for BP's Gulf of Mexico Deepwater Exploration Unit, that the long casing string "saves a lot of time ... at least 3 days."[9] On March 30th, he e-mailed Sarah Dobbs, the BP Completions Engineer, and Mark Hafle, another BP Drilling Engineer, that "[n]ot running the tieback ... saves a good deal of

---

[6]E-mail from Brian Morel, Drilling Engineer, BP, to Richard Miller, BP (Apr. 15, 2010) (BP-HZN-CEC-2 1857).

[7]BP, *Form MMS 123A1123S – Electronic Version, Application for Revised Bypass* (Apr. 15, 2010) (BP-HZN-CECO I8357).

[8]E-mail from Frank Patton, MMS, to Heather Powell, J. C. Connor Consulting ("Modification of Permit to Bypass as Location Surface Lease: 032306 Surface Area: MC Surface Block: 252 Bottom Lease: 032306 Bottom ATea: MC Bottom Block: 252 Well Name: 001 Assigned API Number: 608174116901 has been approved, as of 2010-04-15 14:39:39,0"). (Apr. 14, 2010), Ms, Powell then forwarded the approval to BP, E-mail from Heather Powell, J.C. Connor Consulting, to Mark Hafle, Senior Drilling Engineer, BP (Apr. 15, 2010) (BP-HZN-CEC021033).

[9]E-mail from Brian Morel, Drilling Engineer, BP, to Alli son Crane, Materials Management Coordinator, BP Gulf of Mexico Deepwater Exploration (Mar. 25, 2010). (BPHZN-CEC02 1880).

time/money."[10] On April 15th, BP estimated that using a liner instead of the single string of casing "will add an additional $7–$10 MM to the completion cost."[11] The same document calls the single string of casing the "[b]est economic case and well integrity case for future completion operations."

When BP represented the single string case to the MMS as the best well integrity case, it was not being truthful.

At about the same time period, BP prepared another undated version of its "Forward Plan Review." This alternative version of the document reached a different conclusion than the other version, calling the long string of casing "the primary option" and the liner "the contingency option."[12] Like the other version of the plan review, this version acknowledges the risks of a single string of casing, but it described the option as the "best economic case and well integrity case for future completion operations."

7.2.2 Centralizers

Centralizers are attachments that fit around the casing as it is being lowered into the well to keep the casing in the center of the borehole. If the well is not properly centered prior to the cementing process, there is increased risk that channels will form in the cement that allow gas to flow up the annular space around the casing.

API Recommended Practice 65 states, "If casing is not centralized, it may lay near or against the borehole wall. ... It is difficult, if not impossible, to displace mud effectively from the narrow side of the annulus if casing is poorly centralized. This results in bypassed mud channels and inability to achieve zonal isolation."[13]

On April 15th, BP informed Halliburton's Account Representative, Jesse Gagliano, that BP was planning to use six centralizers on the final casing string at the Macondo well. Mr. Gagliano spent that day performing a computer simulation and analysis of a

---

[10]E-mail from Brian Morel, Drilling Engineer, BP, to Sarah Dobbs, Completions Engineer, BP, and Mark Hafle, Senior Drilling Engineer, BP (Mar. 30,20 I 0) (BP-HPCEC021948).

[11]BP, *Drilling & Completions MOC Iniliale* (Apr. 15,20 I 0) (BP-HZN-CEC02 1656).

[12]BP, *TD Forward Plan Review, Production Casing & TA Options* at 6-7 (undated) (BPHZN-CEC-022 145).

[13]API, Recommended Practice 65-Part 2, *Isolating Potential Flow Zones During Well Construction*, 4.6.5.8. , at 28.

number of cement design scenarios in order to determine how many centralizers would be necessary to prevent channeling from occurring.[14] The results of his analysis showed that with ten centralizers, the modeling resulted in a "moderate" gas flow problem.[15] Mr. Gagliano's analysis showed that it would require 21 centralizers to achieve only a "minor" gas flow problem.[16] Mr. Gagliano informed BP of these results and recommended the use of 21 centralizers.[17]

After running a model with ten centralizers, Mr. Gagliano e-mailed Brian Morel, BP's drilling engineer, and other BP officials, stating that the model "now shows the cement channeling" and that "I'm going to run a few scenarios to see if adding more centralizers will help us or not."[18] Twenty-five minutes later, Mr. Morel e-mailed back stating, "We have 6 centralizers, we can run them in a row, spread out, or any combination of the two. It's a vertical hole, so hopefully the pipe stays centralized due to gravity. As far as changes, it's too late to get any more product on the rig, our only option is to rearrange placement of these centralizers."[19]

On April 16, the issue was elevated by Gregory Walz, BP's Drilling Engineering Team Leader, to John Guide, BP's Well Team Leader. Mr. Walz informed Mr. Guide that "we have located 15 Weatherford centralizers with stop collars ... in Houston and worked things out with the rig to be able to fly them out in the morning." The decision was made because "we need to honor the modeling to be consistent

---

[14]House Committee on Energy and Commerce, Transcribed Interview of Jesse Marc Gagliano, at 26 (June 11 , 20 10).

[15]Halliburton, 9 7/8 " X 7" *Production Casing Design Report* (Apr. 15, 2010) (HAL_DO 1 0592).

[16]Halliburton, 9 7/8 " X 7" *Production Casing Design Report* (Apr. 15, 2010) (HAL_DO 1 0699).

[17]House Committee on Energy and Commerce, Transcribed Interview of Jesse Marc Gagliano, at 8 (June 11 , 2010).

[18]E-mail from Jesse Gagliano, Account Representative, Halliburton, to Mark Hafle, Senior Drilling Engineer, BP, Brian Morel, Drilling Engineer, BP, Brett Cocales, Operations Drilling Engineer, BP, and Gregory Walz, Drilling Team Leader, BP (Apr. 15, 2010) (HAL_DO 1 0650).

[19]E-mail from Brian Morel, Drilling Engineer, BP, to Jesse Gagliano, Account Representative, Hallib1ll10n, Mark Hafle, Senior Drilling Engineer, BP, Brett Cocales, Operations Drilling Engineer, BP, and Gregory Walz, Drilling Team Leader, BP (Apr. 15, 2010) (HAL_DO 1 0648).

with our previous decisions to go with the long string."[20] Mr. Walz wrote, "I wanted to make sure that we did not have a repeat of the last Atlantis job with questionable centralizers going into the hole." In this same email Mr. Walz added, "I do not like or want to disrupt your operations... I know the planning has been lagging behind the operations and I have to turn that around."

In his response, Mr. Guide raised objections to the use of the additional centralizers, writing, "it will take 10 hrs to install them...I do not like this and...I [am] very concerned about using them."[21]

An e-mail from Brett Cocales, BP's Operations Drilling Engineer, shows that Mr. Guide's opinion prevailed. On April 16, he e-mailed Mr. Morel stating, "Even if the hole is perfectly straight, a straight piece of pipe even in tension will not seek the perfect center of the hole unless it has something to centralize it. But, who cares, it's done, end of story, will probably be fine and we'll get a good cement job. I would rather have to squeeze than get stuck...So Guide is right on the risk/reward equation."[22] In short, BP made a calculated risk that the financial rewards were worth the risk of a well blow out.

On April 17th, Mr. Gagliano, the Halliburton account representative, was informed that BP had decided to use only six centralizers.[23] He then proceeded to run a model using seven centralizers and found this would more likely than not produce channeling and a failure of the cement job.[24] His April 18th cementing design report states, "well is considered to have a SEVERE gas flow problem." Mr. Gagliano said that BP was aware of the risks and proceeded with the knowledge that his report indicated the well would have a severe gas flow problem.

---

[20]E-mail from Gregory Walz, Drilling Team Leader, BP, to John Guide, Well Team Leader, BP (Apr. 16, 2010) (BP-HZN-CEC0022433).

[21]E-mail from Gregory Walz, Drilling Team Leader, BP, to John Guide, Well Team Leader, BP (Apr. 16, 2010) (BP-HZN-CEC0022433).

[22]E-mail from Brett Cocales, Operations Drilling Engineer, BP, to Brian Morel, Drilling Engineer, BP (Apr. 16, 2010) (BP-HZN-CEC022670).

[23]House Committee on Energy and Commerce, Transcribed Interview of Jesse Marc Gagliano, at 40–41 (June 11, 2010).

[24]Mr. Gagliano testified before the Congressional Committee under oath that at the time he ran a model with seven centralizers, he knew of BP's decision to use only six. He told the Committee that running a model with seven centralizers demonstrated that the difference between six and seven centralizers would be unlikely to affect the outcome of the modeling.

Mr. Gagliano's findings were no surprise to BP because as noted below, BP's mid-April plan review found that if BP used a single string of casing, as BP had decided to do, "cement simulations indicate it is unlikely to be a successful cement job."[25] Nonetheless, BP ran the last casing with only six centralizers.[26]

7.2.3 Cement Bond Log

A cement bond log is an acoustic test that is conducted by running a tool inside the casing after the cementing process has been completed. The cement bond log determines whether the cement has bonded to the casing and surrounding formations. If a channel that would allow gas flow is discovered, the casing can be perforated, and additional cement must be injected into the annular space to repair the cement job.

Mr. Roth, the Halliburton Vice President of Cementing, informed the Committee staff that BP should have conducted a cement bond log. According to Mr. Roth, "If the cement is to be relied upon as an effective barrier, the well owner must perform a cement evaluation as part of a comprehensive systems integrity test."[27]

Minerals Management Service (MMS) regulations required a cement bond log or equivalent tests at the Macondo well. According to the regulations, if there is an indication of an inadequate cement job, the oil company must:

1. "pressure test the casing shoe;
2. run a temperature survey;
3. run a cement bond log; or
4. use a combination of these techniques."[28]

---

[25]BP, *MC 252#1 Macondo, TD Forward Plan Review – Production Casing & TA Options*, at 9. (Apr. 20 I 0) (BP-HZN-CEC-22I 09).
[26]Briefing by Tommy Roth, Vice President of Cementing, Halliburton, to House Committee on Energy and Commerce Staff (June 3, 2010); Halliburton, PowerPoint Presentation, *Energy and Commerce Committee Staff Briefing* at 12 (June 3, 2010).
[27]Briefing by Tommy Roth, Vice President of Cementing, Halliburton, to House Committee on Energy and Commerce Staff (June 3, 2010); Halliburton, PowerPoint Presentation, *Energy and Commerce Committee Staff Briefing* at 12 (June 3, 2010).
[28]See 30 CFR § 250.428.

Waxman and Stupak note that, in the case of the Macondo well, the Halliburton and internal BP warnings should have served as a warning against a potentially inadequate cement job.

On April 18th, BP flew a crew from Schlumberger to the rig. As described in a Schlumberger timeline, "BP contracted with Schlumberger to be available to perform a cement bond log ... should BP request those services."[29] However, at about 7:00 a.m. on the morning of April 20th, 2010, BP told the Schlumberger crew that their services would not be required for a cement bond log test. As a result, the Schlumberger crew departed the Deepwater Horizon at approximately 11:15 a.m. on a regularly scheduled BP helicopter flight.

The Schlumberger crew was scheduled for departure before pressure testing of the well had been completed, indicating that the results of those tests were not a factor in BP's decision to send the crew away without conducting a cement bond log.

Waxman and Stupak report that BP's decision not to conduct the cement bond log test may have been driven by concerns about expense and time. The cement bond log would have cost the company over $128,000 to complete compared to the cost of canceling the service, which was just $10,000.[30] Moreover, Mr. Roth of Halliburton estimated that conducting the test would have taken an additional 9–12 hours.[31] And, of course, if any defects were identified in the cement casing from the test, additional time would have been required to make repairs to the cement job.

The Committee staff asked an independent engineer with expertise in the analysis of well failure about BP's decision not to conduct a cement bond log. The engineer, Gordon Aaker, Jr., P.E., who is a Failure Analysis Consultant with the firm Engineering Services, LLP, said that it was "unheard of" not to perform a cement bond log on a well using a single casing approach, and he described BP's decision not to conduct a cement bond log as

---

[29]Schlumberger, *Mississippi Canyon Block* 252 *Timeline* (undated) (SLB-EC-000002).

[30]Schlumberger, *Estimated Costs of Equipment/Labor to Perform the Contingent Services Identified by BP and the Actual COS/S Upon Cancellation* (SLB-EC-000909).

[31]Briefing by Tommy Roth, Vice President of Cementing, Halliburton, to House Committee on Energy and Commerce Staff (June 3, 2010).

"horribly negligent."[32] Another independent expert consulted by the Committee, John Martinez, P.E., told the committee that "cement bond or cement evaluation logs should always be used on the production string."[33]

7.2.4 Mud Circulation

The practice of fully circulating drilling mud in the well before cementing is a procedure known as "bottoms up." This practice involves circulating drilling mud from the bottom of the well all the way to the surface.

The "bottoms up" process has several purposes:

1. It allows workers on the rig to test the mud for influxes of gas.
2. It permits a controlled release of gas pockets that may have entered the mud.
3. It ensures the removal of well cuttings and other debris from the bottom of the well, thus preventing contamination of the cement, the consequence of which leads to inadequate curing and failure of the casing.

API 's guidelines recommend a full bottoms up circulation between running the casing and beginning a cementing job. API's practice states that "when the casing is on bottom and before cementing, circulating the drilling fluid will break its gel strength, decrease its viscosity and increase its mobility. The drilling fluid should be conditioned until equilibrium is achieved...At a minimum, the hole should be conditioned for cementing by circulating 1.5 annular volumes or one casing volume, whichever is greater."[34]

BP's April 15th operations plan called for a full bottoms up procedure to "circulate at least one (1) casing and drill pipe capacity, if

---

[32]Briefing by Gordon Aaker, Jr., P.E., Failure Analysis Consultant with Engineering Services, L.P. (Houston), to House Committee on Energy and Committee Staff (June 10, 2010).

[33]E-mail from John Martinez, P.E., an independent production specialist on well bore construction, to House Committee on Energy and Commerce Staff (June 10, 2010).

[34]API, Recommended Practice 65-Part 2, *Isolating Potential Flow Zones During Well Cons/rile/ion*, 4.8.4., at 36–37.

hole conditions allow."[35] Halliburton Account Representative Jesse Gagliano testified it was also "Halliburton's recommendation and best practice to at least circulate one bottoms up on the well before doing a cement job."[36]

According to Mr. Gagliano's testimony, a Halliburton engineer on the rig raised the bottoms up issue with BP. But despite the BP operations plan and the Halliburton recommendation, BP did not fully circulate the mud. To the contrary, BP elected to use a procedure "written on the rig," which Mr. Gagliano "did not get input in." BP's final procedure called for circulating just 261 barrels of mud, which is only a small fraction of the mud in the Macondo well.

Mr. Roth of Halliburton informed the Committee that one reason for the decision not to circulate the mud could have been a desire for speed, as fully circulating the mud could have added as much as 12 hours to the operation.[37] Mr. Gagliano expressed a similar view, saying, "the well probably would not have handled too high of a rate, so it would take a little bit," longer than usual to circulate bottoms up in this case."[38]

7.2.5 Lockdown Sleeve

When a casing is placed in the wellhead and cemented in place, it is held in place by gravity. Under certain pressure conditions, however, the casing can become buoyant, rising up in the wellhead and potentially creating an opportunity for hydrocarbons to break through the wellhead seal and enter the riser to the surface. To prevent this from happening, a casing hanger lockdown sleeve is installed.

On June 8, 2010, Transocean briefed Committee staff on its investigation into the potential causes of the explosion on board the Deepwater Horizon. In the presentation, Transocean listed the lack of a lockdown sleeve as one of its "areas of investigation," Slide

---

[35]BP, *GOM Exploration Wells, Me252 #ISTOOPBPOI – Macondo Prospect 7" X 9 7/8 " Interval, Rev, H.2* at 6 (Apr. 15, 2010) (BP-HZN-CEC-O 1762 1),

[36]House Committee on Energy and Commerce, Transcribed Interview of Jesse Marc Gagliano, at 57 (June 11 , 2010).

[37]Briefing by Tommy Roth, Vice President of Cementing, Hallib1ll10n, to House Committee on Energy and Commerce Staff (June 3, 2010).

[38]House Committee on Energy and Commerce, Transcribed Interview of Jesse Marc Gagliano, at 65–66 (June 11, 2010).

seven of Transocean's presentation[39] asks, "Were Operator procedures appropriate?" A sub point details that the "Operaior did not run lock down sleeve prior to negative test or displacement."[40]

Mr. Roth of Halliburton raised a similar concern in his June 3rd briefing for Committee staff.[41]

In BP's planned procedure for the well, BP describes two options involving the lockdown sleeve. BP was seeking permission from the MMS to install the final cement plug on the well at a lower depth than previously approved. If permission was granted, BP's plan was to displace the drilling mud in the riser with seawater and install the cement plug prior to installation of the casing hanger lockdown sleeve. BP's alternative plan, if MMS did not approve the proposed depth of the final cement plug, was to run the lockdown sleeve first, before installing the cement plug at a shallower depth.[42]

On April 16th, Brian Morel, BP's drilling engineer, emailed BP staff saying, "We are still waiting for approval of the departure to set our surface plug... If we do not get this approved, the displacement/plug will be completed shallower after running the LDS."[43] (LDS stands for the lockdown sleeve.)

7.3 Standard of Care

The term "standard of care" refers to the degree of prudence and caution required of an individual who is under a duty of care. It concerns the degree of caution that a reasonable person should exercise in a given situation so as to avoid causing injury. It is the watchfulness, attention, caution, and prudence that a reasonable person under the same or similar circumstances would exercise.

---

[39]Transocean, PowerPoint Presentation, *Deepwater Horizon Incident – Internal Investigation: Investigation Update – Interim Report* at 7 (June 8, 2010).

[40]Transocean, PowerPoint Presentation, *Deepwater Horizon Incident – Internal Investigation: Investigation Update – Interim Report* at 7 (June 8, 2010).

[41]Briefing by Tommy Roth, Vice President of Cementing, Halliburton, to House Committee on Energy and Commerce Staff (June 3, 2010).

[42]BP, *GOM Exploration Wells Mc* 252 *#ISTOOBP01 – Macondo Prospect 7" x 9- 7/8" Interval* at 8 (Apr. 15, 2010) (BP-HZN-CECOI7621).

[43]E-mail from Brian Morel, Drilling Engineer, BP, to Ronald Sepulvado et al. (Apr. 16, 2010) (BP-HZN-CEC022821).

If a person's actions do not meet a standard of care, then the acts are considered negligent. Consider a hypothetical reasonable person or entity that provides an objective by which the conduct of others is judged. A reasonable person is not an average person or a typical person but a composite of the community's judgment as to how the typical community member should behave in situations that might pose a threat of harm to the rest of the community, workers, or the environment. Even though the majority of people or companies in a community may behave in a certain way, that alone does not establish the standard of conduct of the reasonable person or entity. For example, a majority of people in a community may jaywalk, but jaywalking may still fall below the community's standards of safe and responsible conduct.

The operation of any offshore oil rig requires the operator to follow industry-accepted best practices as well as meet statutory requirements for safety inspections, reporting, and recordkeeping. But beyond the statutory requirements, the very nature of deep offshore well drilling is dangerous. There are far too many historical incidents in which there have been spills, fires, explosions, and fatalities. For any company that engages in this field of activity, it is imperative that the best available technologies be employed and not simply the best economically available technologies. There is absolutely no excuse for relying on equipment, technologies, practices, and procedures that in any way trade off the costs for protection of workers, the public, and the environment.

The American Petroleum Institute (API) has spent decades developing best management practices in order to protect the lives of people that work on rigs and in refineries, to insulate and protect the public, and presumably to protect the environment. The recent U.S. congressional investigations show that BP ignored these practices and simply substituted corporate greed for public and worker safety. Given the above considerations and assuming the congressional fact-findings to be accurate, BP may face criminal charges for negligence at some point in time. It must also be kept in mind that, in addition to the environmental and property damages, eleven personnel were killed in the oil rig explosion. Their deaths might possibly be linked to those actions that contributed to the explosion and crude oil release.

Drilling with the Deepwater Horizon started on February 6, 2010. The Deepwater Horizon rig was expensive. Transocean charged BP approximately $500,000 per day to lease the rig. In addition, there were contractors' fees that made this an even more expensive operation. BP was operating under a $96 million budget based on a 51 day schedule to make the well operational.[44,45] The Deepwater Horizon was supposed to be drilling at a new location as early as March 8, 2010, but the Macondo well took considerably longer to complete than BP engineers had scheduled and budgeted for. By April 20, 2010, the day of the blowout, the rig was already 43 days overdue for its next drilling location. This delay cost BP as much as $21 million in leasing fees alone, not to mention contractor and other charges. These factors placed BP's organization under pressure to accelerate efforts to bring the Macondo well on stream.

Congressional hearings have acknowledged that the budget and cost pressures set the context for the series of decisions that BP made in the days and even hours before the blowout. BP's actions show that it cut corners, sidestepped safety procedures and cement integrity testing, ignored risk-based simulations that predicted the bypassing of hydrocarbons in the well due to an inadequate number of centralizers, ignored warnings from the contractors that advised that the design changes being made and installation guidelines being bypassed were unsafe and outside of industry-accepted standards, ignored API standards for mud circulation, and lacked preventive maintenance on the last line of defense – the blowout preventer.

But BP's negligence seems to go even further. BP downplayed the risks of a blowout and in doing so may have fraudulently obtained its license and permit to operate the Macondo well. It never planned for a worst-case scenario of a blowout, nor was it prepared with proper equipment, trained personnel, and other resources to respond to an emergency of the nature that has now unfolded. It also relied on chemical dispersants, which it has neither qualified as being best available to combat the spill. Nor at any point did it bother to consider the negative impacts from employing these chemicals. Its actions have further likely placed an unskilled and untrained workforce into the field to assist in the clean-up without providing adequate training or protective equipment.

---

[44]According to the terms of the contract, the daily rate would range from $458,000 in March 2008 to $5 17,000 in September 2010. *See* Transocean, *Transocean Fleet Update*, fn. **II** (Apr. 13, 2010) (online at http://www.deepwater.com/fw/mainiFleet-Update-Report-58.html).

[45]BP, *GOM Exploration Wells Mc 252 #i – Macondo Prospect Well information* (Sept. 2009) (BP-HZN-CEC0087 14).

7.3.1 Well Design

The Waxman/Stupak letter documents the following:

> "One day before the blowout, BP installed the final section of steel tubing in the well. BP had a choice of two primary options: it could lower a full string of 'casing' from the top of the wellhead to the bottom of the well, or it could hang a 'liner' from the lower end of the casing already in the well and install a 'tieback' on top of the liner. The liner-tieback option would have taken extra time and was more expensive, but it would have been safer because it provided more barriers to the flow of gas up the annular space surrounding these steel tubes. A BP plan review prepared in mid-April recommended against the full string of casing because it would create 'an open annulus to the wellhead' and make the seal assembly at the wellhead the 'only barrier' to gas flow if the cement job failed. Despite this and other warnings, BP chose the more risky casing option, apparently because the liner option would have cost $7 to $10 million more and taken longer."

7.3.2 Mud Circulation

The Waxman/Stupak letter documents the following:

> "... wells are generally filled with weighted mud during the drilling process. The American Petroleum Institute (API) recommends that oil companies fully circulate the drilling mud in the well from the bottom to the top before commencing the cementing process. Circulating the mud in the Macondo well could have taken as long as 12 hours, but it would have allowed workers on the rig to test the mud for gas influxes, to safely remove any pockets of gas, and to eliminate debris and condition the mud so as to prevent contamination of the cement... BP decided to forego this safety step and conduct only a partial circulation of the drilling mud before the cement job."

7.3.3 Centralizers

The Waxman/Stupak letter documents:

- "When the final string of casing was installed, one key challenge was making sure the casing ran down the center of the well bore.... the American Petroleum Institute's (API's) recommended practices explain, if the casing is not centered, "it is difficult, if not impossible, to displace mud effectively from the narrow side of the annulus," resulting in a failed cement job. BP ignored this industry standard.

- Halliburton, the contractor hired by BP to cement the well, warned BP that the well could have a "severe gas flow problem" if BP lowered the final string of casing with only six centralizers instead of the 21 recommended by Halliburton. BP ignored the risk assessment made by its consultant.
- BP rejected Halliburton's advice to use additional centralizers. In an e-mail on April 16, a BP official involved in the decision explained: "it will take 10 hours to install them... I do not like this." Later that day, another official recognized the risks of proceeding with insufficient centralizers but commented: "who cares, it's done, end of story, will probably be fine." BP valued profits and reduced costs more than safety.

7.3.4 Cement Bond Log

The Waxman/Stupak letter documents:

- "BP's mid-April plan review predicted cement failure, stating "Cement simulations indicate it is unlikely to be a successful cement job due to formation breakdown."
    - Despite this warning and Halliburton's prediction of severe gas flow problems, BP did not run a 9- to 12-hour procedure called a cement bond log to assess the integrity of the cement seal.
    - BP had a crew from Schlumberger on the rig on the morning of April 20 for the purpose of running a cement bond log, but they departed after BP told them their services were not needed.

An independent expert consulted by the Committee called this decision "horribly negligent.""

7.3.5 Lockdown Sleeve

The Waxman/Stupak letter documents the following:

"Because BP elected to use just a single string of casing, the Macondo well had just two barriers to gas flow up the annular space around the final string of casing: the cement at the bottom of the well and the seal at the wellhead on the sea floor. The decision to use insufficient centralizers created a significant risk that the cement job would channel and fail, while the decision not to run a cement bond log denied BP the opportunity to assess the status of the cement job...These decisions would appear to make it crucial to ensure the integrity of the seal assembly that was the remaining barrier against an influx of hydrocarbons...BP did not deploy the casing hanger lockdown sleeve that would have prevented the seal from being blown out from below."

7.3.6 Blowout Preventer

Figure 7.3 is an illustration of the Blowout Preventer, which is a critical, if not the most important, safety device used in drilling operations. Its purpose is to avert disasters. It is described by the industry as a failsafe system but also the last line of defense in averting a full-blown well failure. A blowout preventer (BOP) is equipped with giant shears, which are designed to cut and seal off the well's main pipe. The control pods are effectively the brains of the blowout preventer and contain both electronics and hydraulics.

The main components illustrated in Figure 7.3 each serve an important function:

- Annular Preventers are intended to create a seal around the drill pipe or seal off an open wellbore when there is no pipe.
- Control Pods are the brains. They receive electrical signals from the rig and direct movement of hydraulic fluid. The upper portion has electrical components and the lower portion is equipped with hydraulic valves. Only one pod is activated at a time.
- The Blind Shear Ram cuts the drill pipe and is intended to completely seal the well in the event of a blowout.
- The Casing Shear Ram is designed to cut the drill pipe or casing in an emergency when the rig needs to disconnect from the well quickly.
- Accumulators store fluid sent from the rig. During an emergency, pressurized fluid from canisters is intended to provide force to power the blind shear arm.

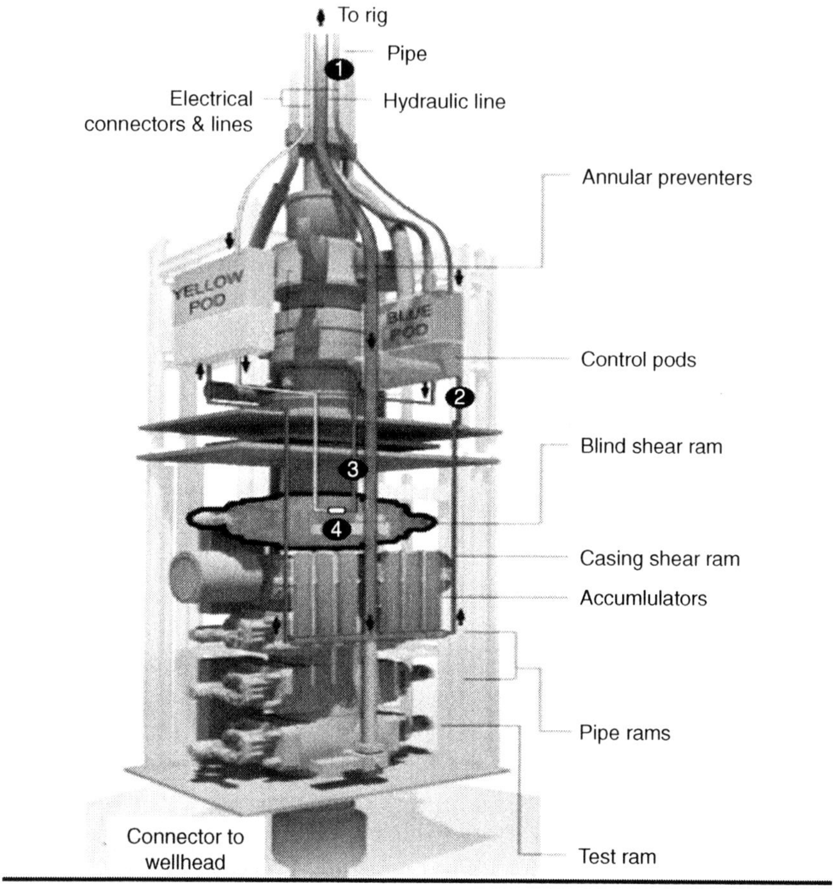

Figure 7.3 Major components of the blowout preventer.

- Pipe rams are supposed to seal off the space between the outside of the drill pipe and the well bore and keep the pipe centered.
- The Test Ram is used to test the rams above it.

The press has reported the following[46]:

- The blowout preventer had a significant leak in a key hydraulic system. This leak was found in the

[46]Stupak Stunner: Oil Well's Blowout Preventer Had Leaks, Dead Battery, Design Flaws, by Joe Romm Published May 13 2010 by Climate Progress, Archived May 13, 2010.

hydraulic system that provides emergency power to the shear rams (the devices that are supposed to cut the drill pipe and seal the well). The blowout preventer was manufactured by Cameron. A senior official at Cameron was interviewed by the Press, and it was disclosed that there was a large leak coming from a loose fitting and likely not caused by the explosion.

- The blowout preventer had been modified. It has an underwater control panel. BP spent a day trying to use this control panel to activate a variable bore ram on the blowout preventer that is designed to seal tight around any pipe in the well. When they investigated why their attempts failed to activate the bore ram, they learned that the device had been modified. A useless test ram, not the variable bore ram, had been connected to the socket that was supposed to activate the variable bore ram. An entire day's worth of precious time had been spent engaging rams that closed the wrong way. BP reported to the Press that the modifications on the BOP were extensive. After the accident, they asked Transocean for drawings of the blowout preventer. Because of the modifications, the drawings they received didn't match the structure on the ocean floor. BP said they wasted many hours figuring this out.
- The Press has reported that its sources have stated that the BOP is not powerful enough to cut through joints in the drill pipe. A Transocean document provided for the Congressional Committee showed that most blind shear rams are designed to shear effectively only on the body of the drill pipe and not threaded parts. Threaded joints between the sections of drill pipe make up about 10% of the length of the pipe. Since the shear rams cannot cut through the joints, then the device would succeed in cutting the drill pipe only 90% of the time. The manufacturer Cameron has confirmed for the Press that it is not powerful enough to cut through the joints in the drill pipe.
- The emergency controls on the blowout preventer may have failed. The blowout preventer has two

emergency controls. One is called the emergency disconnect system or EDS. BP officials have admitted that the EDS was activated on the drill rig before the rig was evacuated. But the Cameron official has reported to the Press that they doubted the signals ever reached the blowout preventer on the seabed. Cameron officials believed the explosion on the rig destroyed the communications link to the blowout preventer before the emergency sequence could be completed. In short, the emergency controls may have failed because the explosion that caused the emergency also disabled communications to the blowout preventer.
- The BOP also has a "deadman switch" which is supposed to activate the blowout preventer when all else fails. But according to Cameron, there were multiple scenarios that could have caused the deadman switch not to activate. One is human oversight – the deadman switch may not have been enabled on the control panel prior to the BOP being installed on the ocean floor. Another is lack of maintenance – the deadman switch won't work if the batteries are dead. The deadman switch is connected to two separate control pods on the blowout preventer. Both rely on battery power to operate. When one of the control pods was removed and inspected after the spill began, the battery was found to be dead. Another reason may be a design flaw. The deadman switch activates only when three separate lines that connect the rig to the blowout preventer are all severed: the communication, power, and hydraulic lines. Cameron believes the power and communication lines were severed in the explosion, but it is possible that the hydraulic lines remained intact, which would have stopped the deadman switch from activating.

The last scenario has raised a controversial issue centering on the acoustic shut-off switch BP failed to install in order to save $500,000. This is a remote device that can trigger the hydraulic shears on the BOP. Some countries require this as an additional safety precaution for BOPs. Because MMS does not require this in their regulations, it is not mandatory in the United States.

Stupak's comments in June hearings went on to say that "...the world is wondering what went wrong to allow explosive gas to shoot out of the drill pipe on the Deepwater Horizon causing the explosion. We heard Chairman Waxman discuss theories of what may have gone wrong in the well (down hole as they call it) and what went wrong on the rig. I would like to take a few minutes to discuss issues related to the blowout preventer (BOP) which was the "fail safe system" to cut off the flow of oil and gas to the rig."

In his testimony, Lamar McKay, the President of BP America, stated that blowout preventers are "intended to ... be fail-safe." But that didn't happen. The blowout preventer used by the Deepwater Horizon rig failed to stop the flow of gas and oil, the rig exploded, people were killed, and an enormous oil spill is now threatening the Gulf Coast.

7.3.7 Emeregncy Response Preparedness

Emergency preparedness plans are intended to direct personnel to the proper procedures for responding to incidents. BP did not have an adequate emergency response plan for its Deepwater Horizon drill rig.

In testimony[47] before the House Transportation and Infrastructure Committee, the former Director of the Minerals Management Service (MMS) said the reason that BP did not complete an Oil Spill Response Plan (OSRP) was that BP reported a decrease in the rig's "worst-case scenario discharge" from 250,000 barrels a day to 160,000.

BP calculated the volume itself, in which the lower figure exempted the company from having to do a site-specific response plan. Instead, the company prepared a plan for the Gulf of Mexico region, where the company has many wells. In short, BP prepared a cut and paste OSRP that was never intended to address a significant spill event from its Deepwater Horizon rig. BP operates about 70 wells in the Gulf. BP also has more than 600 leases in the Gulf.

Leasing and drilling operations on the Federal offshore are governed by laws and regulations that are intended to ensure safe operations and preservation of the environment. The MMS has the

---

[47]Statement of S. Elizabeth Birnbaum, Director Minerals Management Service Department of the Interior Before the House Committee on Transportation and Infrastructure U.S. House of Representatives, May 19, 2010.

authority to enforce compliance with these regulations and periodically updates rules to reflect advances in technology and new information – or at least that is what their Web site states. Changes in MMS regulations may result from outside recommendations or from MMS's ongoing review of technology and investigation of incidents in offshore operations.

The authority for MMS to regulate oil spill planning for affected facilities is derived from the Oil Pollution Act of 1990 (OPA-90) and Executive Order 12777. Direction to lessees regarding federal oil spill planning, preparedness, and response requirements is found in 30 CFR Part 254 (Oil Spill Response Requirements for Facilities Located Seaward of the Coastline). These regulations became effective on June 23, 1997 and require that all designated operators of oil handling, storage, or transportation facilities located seaward of the coastline submit an Oil Spill Response Plan (OSRP) to MMS for approval.

Each OSRP must be consistent with the National Oil and Hazardous Substances Pollution Contingency Plan (the NCP), which falls under the jurisdiction of the National Response Team, which is the interagency body given oversight responsibility under the Oil Pollution Act and with the appropriate Regional Contingency Plan. The NCP provides the organizational structure and procedures for preparing for and responding to discharges of oil and releases of hazardous substances, pollutants, and contaminants (see Chapter 6).

Regional Contingency plans fall under the National Response Team through the appropriate Regional Response Team. The regulation does not relieve the operator of the responsibility to take all actions necessary to immediately abate the source of a spill and remove any spills of oil. Pursuant to 30 CFR 254, a lessee is required to submit an OSRP to the MMS for approval before or concurrent with submitting an exploration plan or Development Operations Coordination Document (DOCD).

The OSRP is supposed to define the availability of spill containment, clean-up equipment, and trained personnel in the event of an emergency.

- The OSRP must ensure that full response capability can be deployed during an oil spill emergency.
- The OSRP must include specifications for appropriate equipment and materials, their availability, and the time needed for deployment.

- The OSRP must also include provisions for varying degrees of response effort, depending on the severity of a spill.
- The Oil Pollution Act requires that the OSRP identify and ensure the availability of private personnel and equipment necessary to respond to a worst-case discharge.

A Regional OSRP is designed to cover multiple facilities or leases of a lessee or operator that are located in close enough proximity to be served by the same response equipment and personnel, have similar modeled spill trajectories and worst-case discharge scenarios, or have the potential to affect the same ecological or socioeconomic resources.

The MMS has the responsibility to review and approve OSRPs every two years unless there is a significant change that requires that the plan be revised immediately. Types of changes that would trigger a review include:

- a change to the plan that significantly reduces the ability to respond to a spill,
- a change in the worst-case discharge scenario, or
- a change in oil spill removal organizations cited in the plan.

According to the MMS, an OSRP must demonstrate that an operator can respond quickly and effectively whenever oil is discharged from its facility. The operator must immediately carry out the provisions of the plan whenever there is a release of oil from a facility. The owner/operator must also carry out the training, equipment testing, and periodic drills described in the plan, and these measures must be sufficient to ensure the safety of the facility and to mitigate or prevent a discharge or a substantial threat of a discharge.

BP's Regional OSRP that covered the Deepwater Horizon was first issued on December 1, 2000 and was last revised on June 30, 2009. BP's Regional ORSP anticipated a worst-case discharge scenario of 250,000 barrels per day. BP's estimate for a worst-case discharge in their exploration plan for the well being drilled by the Deepwater Horizon rig was 160,000 to 162,000 barrels per day. Because the worst-case discharge estimate for the Deepwater Horizon fell below the levels indicated in BP's Regional OSRP, BP was not required to submit a site-specific OSRP.

The worst-case discharge scenario section of the OSRP provides calculations that report how much oil can be spilled from the facility and how the owner/operator would respond to such a spill. In the case of a well blowout, the owner or operator must describe how they would respond to the spill for thirty days. Regional OSRPs list the highest worst-case discharge for all facilities listed in the plan. If a facility covered by a regional OSRP has a larger worst-case discharge than that currently listed in the OSRP, then the plan must be revised with this new worst-case discharge scenario, or a new site-specific plan must be prepared. BP saved itself a lot of expense, time, and effort when it prepared its calculations and misrepresented to the MMS that its worst-case scenario was below the threshold required for a site specific OSRP. We think these actions could very well be argued as being fraudulent, especially since to operate a facility the owner/operator must certify in writing to the MMS Regional Supervisor that it has the capability to respond, to the maximum extent practicable, to a worst-case discharge or a substantial threat of such a discharge. It is also supposed to verify that it has a contract in place for the necessary private personnel and equipment to respond to the discharge.

A briefing[48] paper that the House Transportation and Infrastructure staffers prepared highlighted inadequacies at MMS that may have contributed to lax safety enforcement. The paper also stated that BP was permitted to self-certify itself as safe. The Committee criticized the MMS for being "too cozy" with the industry. In addition, the briefing paper disclosed that evidence is emerging that indicates MMS was aware of the potential failures of blowout preventers and stated that accident reports from MMS "show that the devices have failed or otherwise played a role in at least 14 accidents, mostly since 2005." Moreover, according to an Associated Press investigation, since 2005, the MMS had conducted at least 16 fewer inspections aboard the Deepwater Horizon than it should have under its own inspection policy, which requires monthly inspections. It is reasonable to ask why the MMS skipped at least 25 percent of the monthly inspections they were supposed

---

[48]U.S. House of Representatives Committee on Transportation and Infrastructure, Briefing Paper title "Hearing on Deepwater Horizon Oil Spill Prevention and Response Measures and Natural Resource Impacts," Washington, D.C., May 17, 2010.

to perform on Deepwater Horizon, which this same agency called an "industry model for safety" in 2009.

BP's OSRP did not plan for equipment to kill the Macondo well blow out. As part of its emergency response (not its OSRP), BP implemented what has been referred to as the "Top Kill" approach. Equipment had to be fabricated weeks after the incident and moved to a location near the blowout preventer in order to work on killing the well from the top. Manifold and bypass lines were in place to provide access to valves on the blowout preventer. A so-called "junk shot" of shredded fibrous material was injected into the blowout preventer through these lines. The objective was for the material to travel up the blowout preventer and clog the flow of the well at the pinch point. Once the pressure is controlled, heavy fluids and cement were to be pumped down the well to kill it. The containment dome BP named the "top hat" is apparently the only lesson that BP learned from the fatal Ixtoc I Oil Spill in 1979 in which a rig exploded, caught fire, sank, killed workers, and released millions of gallons of oil into the Gulf of Mexico. In the Ixtoc incident, the containment dome was named nicknamed the Sombrero. After 290 days, a relief well plugged the well with cement and the spill finally came to an end. We wonder why BP expected the Top Hat to work any better than the Sombrero? We also wonder why the Top Hat had to be fabricated after the Macondo well blew out? Again, according to MMS requirements, the OSRP must ensure that full response capability can be deployed during an oil-spill emergency, and the OSRP must include specifications for appropriate equipment and materials, their availability, and the time needed for deployment.

The Congressional briefing document we cited states, "It is important to note that this technology has never been done at this water depth. Significant technical and operational challenges must be overcome for it to be successful." To old engineers like the lead author, this sounds like an experiment. An OSRP is an emergency preparedness plan. It is a document that defines the technologies, procedures, equipment, and all the other resources that are known to be effective in responding to a spill. The British Petroleum Corporation should have developed and tested the Top Hat beforehand. These are clearly not the actions of a responsible corporation in our opinion.

Oil Spill Response Plans (OSRP) are intended to evaluate and determine ways to quickly mitigate disasters caused by oil spills and explosions on drill rigs. Inspections are supposed to ensure that conditions and practices are sufficient and adequate to limit the risk of such disasters from occurring. The OSRP is supposed to make provisions for specialized training of emergency and post-emergency response personnel. The information that has surfaced thus far in Congressional hearings, in the press, and in BP's own certified documents provided to the MMS shows that none of these were performed. The testimony by MMS officials shows that the agency itself was remiss and did not perform inspections, did not verify the OSRP, and simply allowed BP to certify its own plan, which is tantamount to allowing the company to write itself a blank check at the expense of taxpayers and the safety of the public and with a disregard for the environment.

7.3.8 Contractor Training and Worker Protection

News media has reported the following:

- "...the first statewide health survey in Louisiana tracking the health impact of oil spill shows more than 75 workers have been treated for symptoms." (CBS[49])
- "Seven weeks into the oil spill crisis, there is still confusion and uncertainty about how to protect the people cleaning up the mess," reports CBS News medical correspondent Jennifer Ashton. Commercial fisherman George Jackson was overcome by fumes while inspecting boom. BP gave him no protective gear. "They didn't supply us with nothing," says Jackson. "Now they're starting to disperse gloves and hazmat suits."

Chemicals and vapors, both from oil and chemical dispersants pose serious health risks. They can inflame the nose, throat, and lungs and can aggravate asthma. Some products contain ingredients that may be carcinogens. These chemicals and vapors can cause liver and kidney damage and can have long-term effects on the central nervous system. Crude oil itself contains benzene, which is a confirmed human carcinogen. Since 1948, the American Petroleum Institute[50] has

---

[49]http://www.cbsnews.com/stories/2010/06/07/eveningnews/main6558403.shtml

[50]API Toxicological Review for Benzene, American Petroleum Institute, Dept. of Safety, New York, NY, 1948.

reported in a toxicological study that there should be zero tolerance for worker exposure to benzene. When absorbed in the bloodstream, oil compounds like benzene, toluene, xylene, and ethylbenzene can cause headaches and dizziness and could have long-term toxic effects on the brain, liver, and kidneys after prolonged, direct contact.

The Occupational Safety and Health Administration demanded that BP not overwork crews. "We are very concerned about hours worked," (OSHA Assistant Secretary of Labor David Michaels)... "Seeing ads on the internet for 14 hours, seven days a week. We raised this with the Coast Guard. We need to address the number of hours working."

BP's oil spill has shut down fishing operations. The only means by which fisherman can bring income into their homes is by working for BP in the oil spill cleanup. But, in order to do so, we have learned that BP requires workers to sign releases absolving the individuals of any harm caused during the course of the clean-up. The company then sends a flotilla of ships of untrained workers without respirators, chemical protective clothing, and training to skim oil. This defeats the entire purpose and indeed makes a mockery of the process of preparing an OSRP. The OSRP must ensure that full response capability can be deployed during an oil-spill emergency, and the OSRP must include specifications for appropriate equipment and materials, their availability, and the time needed for deployment. BP should have had adequate numbers of emergency and post-emergency cleanup personnel trained and prepared for a worst-case scenario. It clearly has not done so. Its actions not only place the people it has contracted to assist in the clean-up effort in harm's way but the families of these men as well. Without proper segregation of work clothing and OSHA procedures for decontamination, men soiled with oil and chemical dispersants bring these same toxins into their homes and expose family members. We believe BP's actions are irresponsible.

7.3.9 Use of Dispersants

The congressional briefing document states, "In response to the Deepwater Horizon oil spill, BP has employed the use of historic amounts of chemical dispersants, which have been applied to oil on the water's surface and delivered at the point of the oil's release from the well head, approximately 5,000 feet below the water's surface...Recent scientific reports have suggested that the increased use of dispersants may pose a greater long-term threat to the natural resources of the Gulf of Mexico region than other control measures because (1) dispersant transfers the oil (and associated toxic chemicals) from the surface of the water to the water column, (2) the use of dispersant can fundamentally change the chemical makeup of the dispersed oil and its potential impact on the marine environment, and (3) the potential impact of the chemicals contained within the dispersant itself may not be fully understood."

Discussions in Chapters 4 and 6 examine the dispersant Corexit that BP has been applying. Comparisons to other products listed in the NCP approved product list show that this dispersant is less effective as a dispersant and more toxic to marine life than other commercially available products. Additionally, the discussions we have provided in Chapter 6 show that Corexit contains carcinogenic ingredients, which is contrary to the non-carcinogenicity warnings on the product's MSDS.

Our research has not revealed any testing that BP has conducted on its own prior to or during the Gulf disaster that screens for the most effective and least toxic dispersant. As far as we can tell, it simply pulled a product off the shelf and has decided to apply it without performing any risk assessment. Again, the OSRP must ensure that full response capability can be deployed during an oil-spill emergency, and the OSRP must include specifications for appropriate equipment and materials, their availability, and the time needed for deployment.

In short, the OSRP is supposed to be prepared before a disaster and the materials, including chemical dispersants and their negative impacts from application, should be known and defined as a part of the plan. As far as we can tell, BP has done nothing to provide safeguards against, let alone define, the long-term impacts that Corexit may have on the Gulf's marine environment.

7.3.10 BP's Corporate Culture and Day of Reckoning

In conducting the June Congressional hearings, Rep. Stupak made the following statement:

> "Three years ago – almost to the day – this subcommittee held a hearing into British Petroleum's disasters at Texas City and on the North Slope of Alaska. The 2005 Texas City

refinery explosion resulted in the deaths of 15 workers and injured more than 170 people. As a result of that accident and BP's failure to correct potential hazards faced by employees at Texas City, OSHA has twice slapped BP with record setting fines totaling more than $100 million. Several reports criticized management at the Texas City facility including BP's own 2007 *Report of the Management Accountability Project*, which stated "a culture that evolved over the years seemed to ignore risk, tolerated non-compliance, and accepted incompetence.

"In March of 2006, BP discovered their pipeline on Alaska's North Slope had spilled more than 200,000 gallons of oil on the tundra, making it the largest spill in North Slope history. Our hearings discovered that significant cost cutting measures resulted in decreased maintenance, and inspections of the pipeline and BP's management culture deterred individuals from raising safety concerns.

"Since our last hearing, BP has experienced continued problems on the North Slope. September 29, 2008 an 8-inch high-pressure gas line at the Y-Pad location "separated" sending 3 pieces of pipe to the tundra. One segment of the pipe landed 900 feet from the pipeline. Roughly 30 minutes later a second and unrelated incident occurred on the S-Pad where there was a gas release.

January 15, 2009, an ISC cleaning pig became lodged and lost in a 34-inch Oil Transit Line during de-oiling, allowing gas to pass around the pig and travel through Skid to Pump Station, causing a significant venting of gas to the atmosphere and the complete shutdown of the Trans Alaska Pipeline for a period of time.

October 10, 2009 at the Central Compressor Plant, low-pressure flare staging valves were stuck closed, causing gas to travel to the backup low-pressure flare valves, which activated causing the gas to vent to the atmosphere, which could have caused an explosion.

November 29, 2009, an 18-inch three-phase common line near the Lisburne Production Center carrying a mixture of crude oil produced water, and natural gas ruptured spraying its contents over an estimated 8,400 square feet area."

After the Alaska pipeline and Texas refinery disasters, BP promised to make safety its number one priority. The company's safety and environmental track records speak for themselves. This is a company that already has several criminal records of decision against it.

While a $100 million dollar fine sounds like an enormous penalty, the company's annual gross revenues in 2009 were $239 billion – that's $60 billion dollars per quarter. A $100 million dollar fine for criminal violations is a joke. Such a fine is not a deterrent that might be expected to change reckless and unsafe behavior, especially when it appears to be imbedded within the very fabric of how the corporation conducts its business. Such a fine is not much different than an average citizen receiving a $200–300 moving violation – the ticket may sting, but a reckless driver will continue to drive recklessly until his license is taken away or the person does some jail time.

The civil penalty alone for the oil release into the Gulf could amount to $289 million per day. Various experts have reported that the cost of the expected myriad criminal suits could bring BP's total bill to $62.9 billion.[51] As many readers are aware, a $20 billion escrow account was established in which the U.S. Government had BP set aside to respond to victim's claims. This figure is, of course, pale in comparison to what the final ticket may cost for the fiasco, but fortunately it is an evergreen fund. The terms agreed between the U.S. government and BP to set up the escrow fund do not preclude any criminal law suits. Also, the agreement to create the fund does not pre-empt people from using the courts to resolve disputes with BP over the spill, which will likely be the only reasonable approach to bringing some level of closure to the thousands of victims.

Misdemeanor convictions under environmental laws could produce large fines under general federal criminal statutes. The Alternative Fines Act allows the federal government to request twice the gain or loss associated with an offense if the Justice Department shows that a crime was committed. But in recent years court decisions have been readily overturned in favor of industry. And proving a criminal case beyond misdemeanor crimes under federal environmental laws is difficult. While the standard for proving environmental misdemeanors can be relatively low, acts of negligence alone simply lead to misdemeanor penalties under the Clean Water Act. While federal and state prosecutors would probably prefer, given the severity of the ecological crisis caused by the spill, to seek tougher penalty charges, they carry a higher standard of proof. The government would have to prove malicious intent and that BP knew its actions would lead to the mega release of oil on the ocean floor.

---

[51]http://www.cbsnews.com/stories/2010/06/17/business/main6590719.shtml

Certainly BP's prior criminal plea agreements (e.g., its guilty plea in the 2005 refinery explosion that killed 15 people in Texas City, Texas) may be used to support arguments that the Deepwater Horizon disaster is not a unique occurrence but rather the result of a corporate culture that allows schedule and budget pressures to dominate in its risk/reward model for conducting business. But given the corporate veil of protection and the widespread investor holdings in the company, we are skeptical that criminal charges would ever reach the executive chambers of the corporation. Rather, criminal charges would apply to the company as an entity, and based on its history, it is not at all clear that BP may really care.

Some legal experts do argue that criminal indictments could have substantial ripple effects on the company's fortunes. Any company that is indicted risks being blacklisted from future sales contracts with the government under procedures officially known as suspension and debarment. However, the point of debarment under the law is not punishment. Rather, it is aimed to protect the government from poor suppliers. Therefore, at best this would restrict BP from conducting some level of business with the federal government as a contractor for a period of time.

While the public, and presumably the lawmakers, in this country will demand much more information and disclosure of not only BP's decisions and actions taken but those of other contractors and suppliers that may have contributed to the Deepwater Horizon explosion and subsequent spill, it seems rather clear from the limited information released to the public to date that BP made multiple decisions that collectively increased the risk of a well blowout, subsequent explosion, and a release of toxic crude oil of unprecedented amounts into the Gulf. The company's past actions and the activities leading to the Deepwater Horizon mega spill appear to be a reflection of a company that places cost savings and profits above all other considerations. Noted earlier, an email from one of BP's personnel references a risk/benefits formula applied by the company in which it discounted the potential for cement casing failure. It ignored industry best practices, and it sidestepped its legal requirements for performing a cement log bonding test. The record to date shows that, in order to save money, BP cut corners and minimized and even discounted the risks of a catastrophic well blow out. The events unfolding are not an Act of God or nature, and the system failure mechanisms that led to the failure of the BOP and hydrocarbon bypassing of the well casing were not unique and unforeseeable.

There were plenty of warnings, past incidents within the industry, and opportunities to avert the catastrophic release of oil.

At this point in time, the public has not been given full disclosure of all the contractors and partners associated with the Deepwater Horizon rig and the roles they may have played that possibly contributed to the events beginning on April 20th. We do know that BP was in the driver's seat, and they were ultimately in charge of daily operations and in making operational decisions and design alterations. We also know that there were failures on the part of the MMS to provide adequate standards and enforcement, and seemingly there were conflicts of interest through ongoing support of oil industry activities.

In the months and years ahead, the impacts on communities, wildlife, marine life, businesses, and properties will be studied, debated, and litigated over. For some and possibly many areas, what once was, will be no more. We firmly believe that the growing anger and even contempt on the part of American citizens towards British Petroleum are well justified. But we also recognize that some temperance will be necessary. Let us not lose sight of the facts that oil is still lifeblood to our way of life, and that the oil industry has played a dominant role in the economies of the region. The future of this entire industry sector for the region may very well hang in the balance because of the actions of one company.

7.3.11 Mineral Management Services and the Role of Industry

What we find to be remarkable is that the Minerals Management Services (MMS) knew and understood that BOPs have design flaws and that they are not fail safe systems at all.

- Between 1992 and 1998 the MMS knew that there were 319 failures of blowout preventers found in US offshore drilling, which translates into an average of 45 failures per year.[52]

[52]Holand, Per, Reliability of Subsea BOP Systems for Deepwater Application, Phase II DW, Report Prepared for the Minerals Management Service, Project No. 384240, ISBN 82-14-01661-4, July 11, 2007.

- The MMS also knew that between 1992 and 2006 there were at least 39 blowouts off the US coastline, 38 of them in the Gulf of Mexico because it wrote about these failures.[53]
- Between 2007 and 2009, there were 19 blowouts, all of which occurred in the Gulf of Mexico and all of which the MMS documented.[54]
- The largest accidental oil spill in history was a Gulf of Mexico exploratory rig blowout. On June 3, 1979, the exploratory well IXTOC I blew out and ignited, burning down the platform. Divers later activated the blowout preventer to no avail. The well continued to spill oil at a rate of 10,000 to 30,000 barrels per day until it was finally capped on March 23, 1980. It's somewhat surprising that BP and the rest of the oil industry did not learn some lessons from this event.
- A major deepwater blowout followed by a two-month spill occurred in 2009. In "one of Australia's worst oil disasters," a PTTEP oil rig blew out in the Montara oil field on August 21, 2009. Efforts to control the leaking rig set it on fire on November 1st, two days before the leak was finally plugged. Official estimates of the leak rate were five times higher than those of the oil company.[55]

Apparently, the MMS knows quite a lot about the failures of BOPs because it used an awful lot of taxpayer dollars to conduct studies over the past 10 years identifying major design flaws. MMS knows that blowout preventers have "critical" design and operational flaws. There are at least three major studies we found that were commissioned by the MMS to study the causes of BOP failures. Despite identifying the design flaws, this agency took virtually no actions in terms of writing new guidelines and standards for their

[53]David Izon, E.P. Danenberger, Melinda Mayes, Absence of Fatalities in Blowouts Encouraging in MMS Study of OCS Incidents 1992–2006 , Minerals Management Service, Published in D R I L L I N G C O N T R A C T O R Journal, July/August 2007.

[54]"Loss of Well Control – Statistics and Summaries 2006–2010," Mineral Management Services web site: http://www.mms.gov/incidents/blowouts.htm

[55]See full story on Wikipedia: http://en.wikipedia.org/wiki/Montara_oil_spill

design and operation. The following are some of the design flaws that the MMS has known about.

Deep water pressure flaw – In the fall of 2006, West Engineering Services of Brookshire, Texas, turned over to MMS officials a study on the effects of pressure on BOPs. Among its key findings was the fact that high deep-water pressure could severely damage the critical gaskets and seals on BOPs' hydraulic ram valves, causing them to leak and fail in an emergency. One type of hydraulic ram valve (shear ram) is designed to prevent a situation like the one in the Gulf. In the event of a catastrophic failure, the shear rams are supposed to stop the flow of oil by cutting and crumpling the pipe between them. The Deepwater Horizon's shear rams failed, though it's not yet clear why. Many BOP gaskets (the Deepwater Horizon's included) are designed to handle up to 15,000 pounds per square inch (psi) of internal pressure, the pressure of the oil and natural gas pressing outward on the BOP. But they are not mandated to handle external water pressure, which can equal more than 2,000 psi in deep water. Moreover, stress on seals and gaskets could be exacerbated if the pressure inside the well drops dramatically, meaning pressure is higher outside than inside. The MMS sponsored study noted that neither the MMS nor the American Petroleum Institute (API) had any specific standards dictating how much external pressure the seals and gaskets must be able to withstand. The maximum allowable external pressure is never published and indeed may not even be known by the manufacturer, the report states. If differential pressure is applied to a component not designed to withstand it, there could be serious consequences for well control; the deeper the water the greater the risk. Some manufacturers have upgraded seals and gaskets voluntarily. But the lack of specific federal standards resulted in a lack of uniformity in seal quality and their maintenance. Minerals Management Service officials have reported in news media that the agency's safety research division did act on the 2006 report by sharing it with headquarters staff charged with writing regulations, regional field staff who enforce the regulations, and industry organizations. Technical papers were presented at conferences with the findings. But almost four years after the study findings, there are no federal or industry external water pressure standards for BOP closure device seals and gaskets.

Test ram vulnerability – Test rams are a relatively new feature in the offshore oil and gas industry and are currently on only a few deepwater drill rigs' BOPs. They are designed to streamline costly and time-consuming hydraulic test procedures required under federal regulations. One would think that the industry would seek to develop more reliable and safe systems, but in keeping with BP's corporate culture, apparently cost-driven incentives take precedence. The Deepwater Horizon was one of the few rigs whose BOP had a test ram. However, these devices don't help in an emergency, and they may obscure emerging dangers. That is what happened in one of the "safety critical" failures recorded in a 2009 report. One of the BOPs studied experienced a critical failure when a leak developed at the wellhead connector, compromising its ability to maintain the correct pressure. On rigs with test rams, the leak, regardless of size, would not have been identified, the report noted. The report strongly criticized test rams because they obscured leaks and took up space on a BOP that otherwise could have been used for an actual ram. Critics in Congress said the test ram retrofitted onto the Deepwater Horizon impaired the BOP's redundancy and, therefore, the reliability of the BOP.

Lack of safety alerts – A 2009 study recommended a "safety alert" concerning the threat of failure of a BOP valve (the annular). Annular valves are like a large doughnut made of rubber and steel that can be mashed into the pipe to seal a well. The study recommended an alert that would derate, or lower, the estimate of how much pressure the annulars could handle when special large-bore drill pipe was being used. The report notes that failures have occurred while using this drill pipe size with standard [annular] elements. Despite this knowledge, the MMS not once issued any industry alert. The 2009 study is actually a program that was based on a partnership between MMS and 17 oil companies and other industry groups. The study and other documents related to it also reference four senior MMS officials who participated in the report.

Failures to cut pipe – A 2004 study commissioned by the MMS suggested that changes in industry practices made shear rams increasingly prone to failure in deep water, a finding first reported by The Wall Street Journal. Among the problematic changes are higher well-bore pressures and greater drill pipe thickness. Only 3 of 14 newer deepwater drilling rigs were found able to shear pipe at their maximum rated water depths. Additionally, only half of those rigs' operators required a shear-ram test during commissioning or acceptance. This shows a lack of preparedness in the industry to shear and seal a well with the last line of defense against a blowout. The industry has been increasingly compounding the problem by using thicker, harder-to-cut pipes in deeper water. The shear rams could also fail if a thicker "tool joint" in the drilling pipe were between the rams when a disaster stuck. Once again, the MMS did not issue safety alerts or prepare any new standards or requirements for shear ram design to ensure they were powerful enough to cut the thickest pipe.

BP along with the rest of the industry interacts closely with the MMS. The industry itself has conducted cooperative studies with the MMS. Hence, we doubt MMS kept its reports' and findings' best kept secrets. It has, in fact, defended itself by responding to news media that it presented technical papers and attended industry conferences publishing the findings of studies it commissioned. Furthermore, its own web site reports statistics and reasons for past blowouts. BP and the rest of the industry were certainly aware of these design flaws and the probability of BOP failures. With such statistically significant failure rates of 45 per year and well-documented system flaws, we wonder how BP could publish the following statements:

- "The sort of occurrence that we've seen on the Deepwater Horizon is clearly unprecedented." – BP spokesman David Nicholas [AP 4/30/10][56]
- "BP did not build containment devices before disaster because it "seemed inconceivable" the blowout preventer would fail." – BP spokesman Steve Rinehart [AP 5/2/10]

During the past decade, MMS's approach has been to set a performance goal but not to dictate any specific requirements or regulatory standards. The President's 30-day safety review of offshore drilling offers a laundry list of measures that will likely lead to tougher requirements. We wish we could say better late than never, but the extent of the mega spill in the Gulf is potentially so devastating that we find possible legislative toughening too little too late.

These findings show that our own government has failed to act in a responsible manner. The very agency that is supposed to protect the public and the environment failed to provide oversight and write standards and did not take any enforcement action. Its actions are even

---

[56]http://www.cbsnews.com/stories/2010/04/30/national/main6449241.shtml

more irresponsible when one considers that it shared a close working relationship with the very industry it is supposed to regulate.

But MMS's irresponsible behavior was not alone. In ongoing investigations being conducted by the House Energy and Commerce Committee, Rep. Bart Stupak reported to the public and news media that BP's blowout preventer that failed to stop an oil leak in the Gulf of Mexico had a dead battery in its control pod, leaks in its hydraulic system, a "useless" test version of one of the devices that was supposed to close the flow of oil, and a cutting tool that wasn't strong enough to shear through joints that made up 10 percent of the drill pipe.

Summarizing testimony by Mike Saucier, a regional supervisor for field operations for the federal Minerals Management Service, which oversees offshore drilling, Capt. Hung Nguyen of the U.S. Coast Guard has said, "So, MMS approves the design of the well, but they don't check what type of pipe is used. And we have a study some time ago about whether a shear ram would cut a certain pipe (to shut off a well in an emergency), but we don't know what was installed here. I don't understand that."

The testimony from Saucier established that the government agency he works for doesn't do any certification of BOPs. Saucer stated in testimony, "We have self-certification of critical equipment and safety notices that are not enforceable..." We don't see any difference between this and issuing oneself a driver's license.

A panel of three Coast Guard officials and three MMS representatives also appeared disappointed to learn from Saucier that tighter rules for monitoring deepwater drilling safety systems were proposed nine years ago but got lost in the shuffle and never were adopted. Saucier testified that new rules were written in 2001 to require secondary control systems for blowout preventers, the stacks of valves and pistons on the seafloor that are supposed to shut a well in an emergency, and that these proposed regulations were never approved. We may never know if the rules had been adopted whether they could have prevented the catastrophe at Deepwater Horizon.

We may also take a poke at the USCG – it turns out the Coast Guard regulations governing drilling on the Outer Continental Shelf have not been updated since 1996.

Based on the information currently available within the public domain, we view the MMS as an agency that is incompetent and corrupt. We can only hope that sweeping overhauls are made to the agency so that it better serves and protects the public in the future.

We now turn an eye towards the industry sector. There is no question that the petroleum industry is in many ways the foundation of modern society. It has revolutionized and benefited almost every conceivable facet of the quality of life not simply with energy but the chemicals that make the products that support life styles on earth. It is also an industry that has developed and applied cutting edge technologies that are unparalleled in exploration. To quote the U.S. Department of Energy[57], "In the past three decades, the petroleum business has transformed itself into a high-technology industry. Dramatic advances in technology for exploration, drilling and completion, production, and site restoration have enabled the industry to keep up with the ever-increasing demand for reliable supplies of oil and natural gas at reasonable prices."

We actually agree with most of this statement except for the part about dramatic advances in site restoration. And frankly, we find the title of the publication offensive in light of the BP spill (title is Environmental Benefits of Advanced Oil and Gas Exploration and Production Technology).

While the industry has made remarkable advances with 21st century technologies for exploration, the technologies, practices, and infrastructure for oil spill responses have remained frozen in time to the days of the Exxon Valdez spill.

The industry has done nothing to advance safety and emergency response. It has placed all of its efforts into investments that enhance volume of production and profitability. As far as making its technologies safer and providing infrastructure and technologies that can cope with disasters, it has placed virtually no effort. It's not hard to understand why – being prepared for an emergency, establishing extensive preventive maintenance programs in place to ensure that BOPs don't fail, and investing in multiple well log casing integrity testing all cost money, and these represent sunk costs. Sunk costs cannot be readily recovered and, in some cases, cannot be recovered at all. When the focus of the industry is on higher and higher profits, the investments needed to systematically reduce the risks of system failures place a burden on the potential financial rewards. We don't

[57]Environmental Benefits of Advanced Oil and Gas Exploration and Production Technology, U.S. Department of Energy, Office of Fossil Energy, October 1999, Washington, DC.

think that BP's risk/reward model is much different than the formula used by the industry in general.

In 1989 when the Valdez incident occurred, the public and lawmakers were outraged. In response to this disaster the Oil Pollution Act was created which in part required the industry to develop emergency response infrastructure to cope with such events. This promoted the creation of the Marine Spill Response Corporation (MSRC), which the oil industry established as a means of limiting its investment into response technologies from the standpoint of each company investing into large resources needed to respond to emergency releases of oil at sea. The MSRC's web site[58] states, "MSRC's mission now includes response to oil spills of any size, shoreline cleanup..."

While the MSRC's mission may be to respond to oil spills of any size, it clearly does not have the capability to cope with the catastrophe in the Gulf. British Petroleum listed this resource as its prime emergency response contractor in its OSRP. MSRC has no infrastructure to cope with a leaking one-mile deep well, let alone respond to the magnitude of the oil released.

The MSRC's Web site also states that it offers response capability intended to help satisfy the "Worst Case Discharge."[59] If that is the case, then why is the United States now accepting assistance from a dozen nations to battle BP's oil spill?

The Marine Spill Response Corporation is located in Herndon, Virginia. It has an operating budget of somewhere around $80 million per year and essentially constitutes the entire industry sector's emergency and post-emergency response infrastructure comprised of a fleet of 15 dedicated Responder Class Oil Spill Response Vessels (OSRVs) designed and built specifically to recover spilled oil. The OSRVs are approximately 210 feet long, have temporary storage for 4,000 barrels of recovered oil, and have the ability to separate oil and water aboard ship using two oil-water separation systems. To enable the OSRV to sustain clean-up operations, recovered oil is transferred into other vessels or barges. In addition to these 15 OSRVs, MSRC also operates two other OSRVs in the San Francisco Bay, CA, four in Southern California, and one in Puget Sound, WA.

---

[58]http://www.msrc.org/
[59]http://www.msrc.org/AboutUs.htm

The MSRC is outfitted with other specialized response vessels and support equipment, including the following:

- 19 oil spill response barges with storage capacities between 12,000 and 68,000 bbls
- 68 shallow water barges
- 600,000 feet of boom
- Over 240 skimming systems
- 6 self-propelled skimming vessels (32 ft. to 58 ft.)
- 7 mobile communications suites comprising telephone and computer connections, and UHF and VHF marine, aviation, and business band radios
- Various small crafts and shallow-water vessels
- One C-130 dispersant aircraft
- One King Air dispersant/spotter aircraft

Yet, this infrastructure has proven to be substantially inadequate to respond to the events in the Gulf.

Both the industry sector and the MMS have understood the shortcomings of BOPs. Our government has allowed lax rules to be in place, which allowed BP to simply self-certify the safety of its own emergency response plan and the design of its well. With more than 7,000 offshore wells in the Gulf region that the MMS has issued licenses to, we question how many of these other facilities have been allowed to prepare inadequate OSRPs that underscore or discount worst-case scenarios.

The industry clearly does not regulate itself. It has not made investments in new technologies to respond to catastrophic oil releases. It has not invested in research and development to make BOPs more reliable. It has not improved on any of the response technologies that were essentially devised makeshift in response to the Valdez incident in 1989. We are still using in-situ burning to respond to oil spills, which is a practice dating back to World War II. While the industry boasts of its use of high-end technologies and bringing environmental benefits along with offshore oil drilling, we suggest that some of the senior executives of major oil companies spend some time walking along the shores of Mississippi, Louisiana, and Florida and speak with some of the people whose lives, properties, and livelihoods have been destroyed by this environmentally friendly technology.

7.3.12 Commentary

At day 93 the mighty roar of the runaway Macondo well was finally silenced with a temporary shut-off valve, and the two relief wells are within feet of the well casing to hopefully seal it off for all times. To date exact figures on how much oil has been released into the Gulf of Mexico is not known. There were certainly opportunities to measure the release's flow rate. But, in fairness, if we worked for BP and were potentially facing fines from the Clean Water Act, which sets a civil penalty of $1,100 for every barrel of oil released, in addition to responsibility for paying for environmental clean-up and economic loss to as much as $4,300 a barrel in fines if the government determines that the spill was the result of gross negligence, we would want to rely on the most conservative estimate possible.

And lets' face it – how do we really know whether or not the oil from the Macondo well has poisoned marine life, which in turn has decimated businesses and caused property damages? After all, since records started being maintained in 1964 there has been a steady stream of oil into the Gulf. In fact, federal records (MMS published statistics 1964–2009) show that oil spills in this region total 517,847 barrels (22,267,421 gallons) with their impacts reaching as far away as the Yucatan Peninsula. By analogy, roughly one barrel fills an average size bathtub – the amount of oil spilled or dumped into the Gulf, excluding BP's oil, could fill an equivalent number of bathtubs and is nearly twice the amount of the Alaskan Exxon Valdez incident of 1989. Fortunately, someone somewhere will fingerprint the oil in the Gulf, and perhaps BP will take responsibility.

We must also take into consideration that the MMS has now been revamped to better serve the public and the environment. They have now been renamed the Bureau of Ocean Energy Management, Regulation, and Enforcement (BOEMRE). Its Web site states, "Our mission is to manage the mineral resources of the Outer Continental Shelf in an environmentally sound and safe manner."[60] Just the name alone gives us confidence that our national treasures and the public are in safe hands. We cannot even imagine any of the following

---

[60]http://www.gomr.mms.gov/

stories,[61] which were released in 2008 about the MMS being even remotely accurate:

- "Government officials handling billions of dollars in oil royalties engaged in illicit sex with employees of energy companies they were dealing with and received numerous gifts from them, federal investigators said..."
- "The allegations of bad behavior involve 13 government employees in Denver and Washington, reports CBS News investigative correspondent Sheryl Attkisson. Those accused are workers who sell U.S. mineral rights to oil companies. Such sales are one of the government's biggest sources of revenue besides taxes."
- "...the Inspector General for the Interior Department says they rigged contracts, and engaged in illegal moonlighting, drugs, sex, and gift-taking from oil company representatives, according to three reports..."
- "The reports revealed startling allegations including that an employee attended a so-called "treasure hunt" in the desert with all expenses paid by an oil producer, and that a former supervisor, who bought cocaine from a colleague then boosted her performance award, had sex with subordinates and steered government contracts to an outside business where he also worked...."
- "The investigations reveal a 'culture of substance abuse and promiscuity' by a small group of individuals 'wholly lacking in acceptance of or adherence to government ethical standards,' wrote Inspector General Earl E. Devaney."
- "Between 2002 and 2006, nearly a third of the 55-person staff in the Denver office received gifts and gratuities from oil and gas companies, the investigators found."

With a seemingly broken government and an industry whose practices are far from being characterized as having a pristine safety

---

[61]http://www.cbsnews.com/stories/2008/09/10/national/main4436263.shtml

record, we are far less optimistic that lessons have been learned and that the events in the Gulf will make offshore drilling any safer than it is today. Should the record stand and BP's actions remain characterized as reckless, irresponsible, and possibly negligent, then there should be more than financial fines and penalties for the damages attributed to the Deepwater Horizon disaster.

Table 7.2 summarizes key decisions and actions taken in the weeks, days, and hours preceding the April 20, 2010 explosion. Collectively, the actions reveal a series of poor decisions, each one contributing to eroding away at the level of safe operation of the well and the backup devices that could have prevented the spill from occurring. Along the way, decisions to take and/or, in some cases, not take actions when faced with alternatives were based upon downplaying the risks or potential negative impacts and electing to take a faster, more expedient, and less costly path to achieving one objective. Each of the poor or questionable decisions incrementally contributed to reducing the overall safe operations of the rig until the point where there was no margin of safety. In almost all of the decisions that lead to contributory causes of system failure, a decision was made, whether conscious or not, based on placing financial rewards at a higher level of priority. We believe this to be reckless and irresponsible. A facility that operates a nuclear power plant or even a refinery for that matter, relies on a rigorous set of protocols and operating procedures that are intended to control a process in a safe manner. When these are sidestepped, the operator runs the risk of losing control of the process and may not only place the assets of the facility in danger of being damaged, but places the safety of employees and the public in jeopardy.

British Petroleum displayed poor judgment not only in the operation of the Deepwater Horizon rig but also in its level of preparedness for a spill. Its Oil Spill Response Plan was grossly inadequate for a release of the magnitude in the Gulf. In short, BP prepared a cut and paste OSRP that was never intended to address a significant spill event from its Deepwater Horizon rig. This saved BP money because, with a cookie cutter OSRP, it did not have to establish contracts and demonstrate broad-based resources to combat a much larger spill. Nor was it required to conduct time-consuming and costly site inspections and training exercises and to have response equipment on standby. It in fact struggled throughout the entire spill incident to try to find ways to kill the well blowout and to manage the mega-spill. A term

Table 7.2 Risk-based decisions made and actions contributing to the Gulf spill.

Component Affected	Options	Decisions Made/Actions Taken	Consequences
Well Design	BP was faced with two options for well designs. The first involved hanging a steel tube called a "liner" from a liner hanger on the bottom of the casing already in the well and then inserting another steel liner tube called a "tieback" on top of the liner hanger. The second option involved running a single string of steel casing from the seafloor to the bottom of the well. The liner/tieback casing relies on multiple or redundant barriers to annular flow. The liner/tieback method provides four barriers to annular flow (the cement at the bottom of the well, the hanger seal that attaches the liner to the existing casing in the well, the cement that secures the tieback on top of the liner, and the seal at the	BP opted for the single string of casing. This decision saved $7-10 million and 3 days in delay of its drilling schedule.	The decision cut the margin of safety by 50%. The liner/tie-back method is safer because there is double the protection/safety features that would impede the flow of hydrocarbons in an uncontrolled manner up the annular space of the well.

	wellhead). In the case of a single string of casing, there are just two barriers to the flow of gas up the annular space that surrounds the casing: the cement at the bottom of the well and the seal at the wellhead.		
Centralizers	Centralizers are attachments that fit around the casing as it being lowered into the well to keep the casing in the center of the bore-hole. If the well is not properly centered prior to the cementing process, there is increased risk that channels will form in the cement that allow gas to flow up the annular space around the cas-ing. API Recommended Practice 65 states, "If casing is not central-ized, it may lay near or against the borehole wall. ... It is difficult, if not impossible, to displace mud effectively from the narrow side of the annulus if casing is poorly centralized. This results in bypassed mud channels and	BP decided to bypass API Recommended Practice 65. It ignored its Halliburton's engineering calculations, which predicted that 21 centralizers were required to minimize gas flow problems. BP's mid-April plan review found that if BP used a single string of cas-ing, as BP had decided to do, "cement simulations indicate it is unlikely to be a successful cement job." BP decided to run the last casing with only six centralizers. BP evacuated the expensive drilling mud from the riser and replaced it with lighter seawater. It saved itself money both on using fewer	This action most probably created severe channeling and gas flow problems in the well. Model simulations ran by Halliburton for 7 centralizers showed that the "well is con-sidered to have a SEVERE gas flow problem". The removal of the mud is likely the single most important factor leading to the explosion. The heavier mud essentially held back any methane gas from escaping from the reservoir. Once the gas was released it simply found a source of ignition, as in the case of the failed Ixtox I well.

Table 7.2 (cont.) Risk-based decisions made and actions contributing to the Gulf spill.

Component Affected	Options	Decisions Made/Actions Taken	Consequences
	inability to achieve zonal isolation..." BP's contractor/consultant ran mathematical model simulations showing that it would require 21 centralizers to achieve only a "minor" gas flow problem.	centralizers and in the recovery of the drilling chemicals (mud). The decisions and actions saved 10 hours on the drilling schedule.	
Well Integrity Tests	If the well casing cement is to be relied upon as an effective barrier to gas flow, a well owner must perform a cement evaluation as part of a comprehensive systems integrity test. A cement bond log is an acoustic test that is conducted by running a tool inside the casing after the cementing process has been completed. The cement bond log determines whether the cement has bonded to the casing and surrounding formations. If a channel that would allow gas flow is discovered, the casing can be perforated and additional cement must be	BP flew a crew from Schlumberger to the Deepwater Horizon rig. BP contracted with Schlumberger to be available to perform a cement bond log. On the morning of April 20, 2010, BP told the Schlumberger crew that their services would not be required for a cement bond log test. As a result, the Schlumberger crew departed the Deepwater Horizon, and no cement bond log test was performed. The cement bond log would have cost BP over $128,000 to complete compared to the cost of canceling the	The consequence of this action was to complete the drilling schedule without having performed a critical integrity test. Its decision to bypass the test permitted the operators to continue operations blindly and without any margin of safety.

	injected into the annular space to repair the cement job. According to MMS (now called the Bureau of Ocean Energy Management, Regulation, and Enforcement (BOEMRE)) regulations, if there is an indication of an inadequate cement job, the oil company must (1) pressure test the casing shoe, (2) run a temperature survey, (3) run a cement bond log, or (4) use a combination of these techniques.	service, which was just $10,000. BP also saved itself a minimum of 9 to 12 hours on its schedule. If any defects were identified in the cement casing from the test, additional time would have been required to make repairs to the cement job.	
Blowout Preventer	The Blowout Preventer (BOP) is a critical, if not the most important, safety device used in drilling operations. Its purpose is to avert disasters. It is the last line of defense in averting a full-blown well failure. A BOP is equipped with giant shears, which are designed to cut and seal off the well's main pipe. The control pods are the brains of the blowout preventer and contain both electronics and hydraulics. The	Transocean (company that owned the Deepwater Horizon rig) failed to repair a leak in the control pod of the blowout preventer. It also permitted a backlog of 390 maintenance jobs as of September 2010. British Petroleum could have insisted that Transocean maintained an aggressive preventive maintenance program. Transocean had an obligation to maintain and service all of the equipment and	This was a contributor to the failure of the BOP, but it is unknown how significant at this time.

Table 7.2 (cont.) Risk-based decisions made and actions contributing to the Gulf spill.

Component Affected	Options	Decisions Made/Actions Taken	Consequences
	operator had the option of choosing a BOP that had additional safety features and to implement procedures to test, service, and maintain the BOP.	instruments and to prioritize the most critical maintenance jobs.	
		The BOP had a significant leak in a key hydraulic system. This leak was found in the hydraulic system that provides emergency power to the shear rams (the devices that are supposed to cut the drill pipe and seal the well). A senior official at Cameron (the supplier) disclosed that there was a large leak coming from a loose fitting and likely not caused by the explosion.	This was a contributor to the failure of the BOP, but it is unknown how significant at this time.
	The BOP had been modified. BP spent a day trying to use an underwater control panel at the time of the spill in order to	A useless test ram, not the variable bore ram, had been connected to the socket that was supposed to activate the	Because of the modifications, the drawings BP received didn't match the structure on the ocean floor. BP said they

	activate a variable bore ram on the blowout preventer that is designed to seal tight around any pipe in the well. When they investigated why their attempts failed to activate the bore ram, they learned that the device had been modified.	variable bore ram. An entire day's worth of precious time had been spent engaging rams that closed the wrong way. BP reported to the Press that the modifications on the BOP were extensive. After the accident, they asked Transocean for drawings of the blowout preventer. But BP had the option of learning all there is about the operational features prior to April 20, 2010. With a piece of equipment that is so critical to safe drilling operations, the operator has an obligation to understand all of the limitations and features of the device before placing it in service.	wasted many hours figuring this out and subsequently lost a day in which more oil spilled into the Gulf.
	The BOP was not powerful enough to cut through joints in the drill pipe. A Transocean document showed that most blind shear rams are designed to shear effectively only on the body of the drill pipe and not threaded parts.	It seems rather incredible that BP did not bother to investigate this operational feature. When major equipment is placed into service at a refinery or chemical plant, it is procured and shaken down for integrity and	This has been flagged as a major possible contributor to the failure of the BOP.

Table 7.2 (cont.) Risk-based decisions made and actions contributing to the Gulf spill.

Component Affected	Options	Decisions Made/Actions Taken	Consequences
	Threaded joints between the sections of drill pipe make up about 10% of the length of the pipe. Since the shear rams cannot cut through the joints, then the device would succeed in cutting the drill pipe only 90% of the time.	reliability. Persons responsible for providing equipment are obligated to explain limitations of any piece of equipment that is being placed into service or relied upon as a critical component in any engineered system. If the devise was known to fail 90% of the time (i.e., unable to cut the drill pipe 90% of the time), then the option is to work with the supplier to make modifications. A simple analogy is no one in their right mind would purchase a car if it was disclosed that the breaks on the vehicle brought one to a full stop only 10% of the time when driving above a certain speed.	

	The emergency controls on the blowout preventer may have failed. The blowout preventer has two emergency controls. One is called the emergency disconnect system or EDS. BP officials have admitted that the EDS was activated on the drill rig before the rig was evacuated. But the Cameron official has reported to the Press that they doubted the signals ever reached the blowout preventer on the seabed. Cameron officials believed the explosion on the rig destroyed the communications link to the blowout preventer before the emergency sequence could be completed. In short, the emergency controls may have failed because the explosion that caused the emergency also disabled communications to the blowout preventer.	BP had the option of installing an acoustic switch. An acoustic switch allows an operator on the rig to activate the BOP rams without the use of the umbilical cord.	While very sophisticated, the system basically performs the safety function of a "deadman" system, where the BOP will slam shut if it loses communication with the rig through the umbilical in the event of a catastrophic event such as the Deepwater Horizon. However, in this particular incident, something most likely jammed the shear rams (see above). So, even if the Deepwater Horizon had had an acoustic system installed, it very likely would not have resulted in a different outcome.

Table 7.2 (cont.) Risk-based decisions made and actions contributing to the Gulf spill.

Component Affected	Options	Decisions Made/Actions Taken	Consequences
	The BOP also has a "deadman switch," which is supposed to activate the blowout preventer when all else fails. But according to Cameron, there were multiple scenarios that could have caused the deadman switch not to activate. One is human oversight – the deadman switch may not have been enabled on the control panel prior to the BOP being installed on the ocean floor. Another is lack of maintenance – the deadman switch won't work if the batteries are dead. The deadman switch is connected to two separate control pods on the blowout preventer. Both rely on battery power to operate. When one of the control pods was removed and inspected after the spill began, the battery was	These are design and operational flaws as well as lack of preventive maintenance. One may argue that BP was not entirely responsible for these factors if they are eventually shown to have contributed to preventing the release of oil. Transocean had the responsibility of maintenance. But as the operator, BP should have insisted on a prioritization of schedules and elevated the importance of inspections on batteries used in the BOP. Should there be design flaws associated with the operation of the deadman switch, then Cameron faces liability and should take every effort to improve its design.	This was a contributor to the failure of the BOP, but it is unknown how significant at this time.

	found to be dead. The battery in the other pod has not been inspected yet. Another reason may be a design flaw. The deadman switch activates only when three separate lines that connect the rig to the blowout preventer are all severed: the communication, power, and hydraulic lines. Cameron believes the power and communication lines were severed in the explosion, but it is possible that the hydraulic lines remained intact, which would have stopped the deadman switch from activating.		

often used synonymously with an Oil Spill Response Plan is an 'Emergency Preparedness Plan'. As the term implies, the plan is supposed to define the exact procedures and equipment to apply to the incident in a formalized plan. An OSRP or Emergency Preparedness Plan (also known as an Emergency Response Plan) is supposed to be created before an incident or release and it must be based on proven responses and technologies so the a proper response can be made. This is NOT a plan that is supposed to evolve or that teams of engineers are chartered with assignments to struggle with approaches to invent new technologies in order to control a catastrophic release. While OSRPs need to be flexible and fluid in an emergency, they are not supposed to be created after the incident has occurred, which is exactly what the British Petroleum Company did in this instance.

We are quite confident that there are optimists out there saying a lot of lessons can be learned from the tragedy in the Gulf, and that in the end, better practices and safeguards will be devised to prevent such an incident from occurring again. There may even be some readers that will argue that BP will become a leaner and more responsible company, having learned major lessons from its follies. We are less optimistic for one fundamental reason; namely, we believe that in the end BP's philosophy of high risk-taking works. The company has a criminal record, and yet it continues to conduct high-risk business including with the U.S. Government. Let's take a moment to look at the company's record on environmental and safety performance:

- Over the past two decades, BP subsidiaries have been convicted no less than three times of environmental crimes in Alaska and Texas, including two felonies. It remains on probation for two of them.
- BP received the largest ever fine for willful work safety violations in U.S. history and is the subject of a wide range of safety investigations. BP's best-known disaster took place in 2005, when an explosion at its refinery in Texas City near Galveston killed 15 workers, injured 180 people, and forced thousands of nearby residents to remain sheltered in their homes. An investigation of the explosion by the U.S. Chemical Safety and Hazard Investigation Board squarely blamed BP for the explosion and offered a scathing assessment of the company.

It reported organizational and safety deficiencies at all levels of the corporation and said management failures could be traced from Texas to London. BP eventually pleaded guilty to a felony violation of the Clean Air Act, was fined $50 million, and sentenced to three years probation.

- The Occupational Health and Safety Administration assessed BP the largest fine in OSHA history, $87 million, after inspectors found 270 safety violations that had been previously cited but not fixed and 439 new violations at its Texas refinery. But BP has great lawyers and has managed to have many of these fines reduced or dismissed, paying a mere $100,000 for safety violations in Texas.
- In Alaska, BP first brought attention to itself more than 20 years ago in the aftermath of the Exxon Valdez oil spill. Exxon was BP's partner in Alaska's Prudhoe Bay oilfield, the nation's largest, and shared in the ownership of the trans-Alaska pipeline system, known as Alyeska and headed then by a BP executive who was on loan to the pipeline company. Alyeska failed to live up to its promises to contain spills.
- BP ran afoul of federal environmental laws in Alaska after it was discovered that from 1993 to 1995 a BP contractor, Doyon Drilling, had illegally dumped hazardous materials down oil well shafts on the North Slope, the giant Alaska oil production area bordered by the Brooks Range mountains to the south and the Arctic Ocean on the north. Doyon pleaded guilty in federal court to a felony violation of the Clean Water Act and was fined $3 million. BP was convicted on February 1, 2000, of failing to report the dumping as soon as it learned about it, which is a felony. BP was fined $500,000, placed on five years probation, and ordered to create a nationwide environmental management program that cost the company about $40 million. But with a company that does more than $280 billion dollars per year in revenues, this is a slap on the wrist and simply a part of its perceived cost of doing business. BP did state in court, "We are committed to ensuring this never happens again."

- But BP was still on probation when new problems surfaced, this time in its North Slope corrosion control program. Despite warnings from a leak detection system, a badly corroded 34-inch-diameter pipeline in Prudhoe Bay lost oil for at least five days before a worker driving down a nearby service road on March 2, 2006 smelled oil and spotted the spill, which covered at least two acres of tundra. At 200,000 gallons, it was the largest spill ever on the North Slope.
- Five months later on August 6, 2006, a second spill of about 1,000 gallons was discovered on another line. Subsequent investigation found the line was riddled with corrosion, with 176 places where more than half the original diameter had been eaten away. Congressional hearings were held to probe the spills focused on claims that BP actively discouraged workers from reporting safety and environmental problems. BP's head of the corrosion unit took the 5th Amendment against self-incrimination during the hearings. In this incident, Rep. Joe Barton suggested BP had decided to "bet the farm" that the pipeline wouldn't fail before Prudhoe Bay would run out of oil, saving it the cost of replacement. He accused the company of fostering a "corporate culture of seeming indifference to safety and environmental issues."
- In 2007, BP pleaded guilty in federal court in Anchorage to another violation of the Clean Water Act for the 2006 spill. This crime was a misdemeanor, but it still cost BP $20 million in fines and restitution and three more years of probation. Prosecutors said the spill occurred because BP was more interested in cutting costs than in maintaining an aging oil field – a theme that seems to resonate in the Gulf of Mexico. But once again, there were assurances because a BP vice president told the judge that the corrosion problems were "out of character" for the company and that BP had learned its lesson.

BP's 2005 Texas City explosion has fostered the harshest accusations against it from the U.S. Chemical Safety and Hazard Investigation Board, which issued a 341-page report in March 2007 two years after the blast, and from a separate commission led by former Secretary of State James Baker III. Both groups faulted BP's management at all levels for overlooking problems. In these reports, the Chemical Safety and Hazard probe concluded "warning signs of a possible disaster were present for several years, but company officials did not intervene effectively to prevent it...Cost-cutting, failure to invest, and production pressures from BP Group executive managers impaired process safety performance at Texas City." As an example, the board cited a blowdown stack where the first explosion occurred when a geyser of flammable liquid erupted from it. The blowdown stack was described by the board as antiquated equipment of unsafe design originally installed in the 1950s.

The Baker panel also concluded that BP safety efforts were impacted by bad management and cost cutting. The panel said that the company had "a false sense of confidence" about safety and didn't always make sure that "adequate resources were effectively allocated" to safety issues.

After the 2005 explosion, BP officials said they created a panel to study safety practices at its site, increased the number of people responsible for safety and environmental issues, and spent more than $1 billion on upgrades and repairs. The company spent millions of dollars on TV ads talking about how the company is a pioneer for efforts to move "beyond petroleum."

The only lesson we see from this history is that crime seems to pay. Companies that have a culture of being lax about safety, placing schedules and cost savings above public safety and preservation of precious natural resources and wildlife, cut corners and allow incompetence and shoddy work to prevail with the knowledge that they risk harm to their neighbors in almost countless ways represent a danger to society. Such companies should really be treated in a manner that is no different than the way society acts towards convicted child molesters and sex offenders. If we can place such individuals on a watch list, warn communities and societies when they move into areas, and closely monitor their behaviors and actions in order to protect our communities, why would we not do the same with companies that act irresponsibly toward our environment, the properties of others, jobs, human life, safety and other businesses, and the wildlife that we as human beings are supposed to be committed to protecting? Industry has an obligation that goes far beyond statutes when it comes to protecting the environment. Indeed, the overriding reason why the U.S. EPA was created some 40 years ago was to protect human life because damaging our environment has a direct impact on human health and sustainability. Companies that act responsibly toward the environment and the safety of their employees do so because they understand that good environmental stewardship is good for business. One should not separate good financial performance from good environmental performance, unless the company simply intends to conduct business in both an illegal and unethical manner. Companies that rely on a risk/reward model that is built on the premise that the higher the risks the greater the rewards should be outlawed or, at the very least, heavily regulated, assuming our policymakers and regulators can come to the realization that that is what they are receiving benefits and salaries for through our taxes and that they should not be receiving industry payoffs and bribes.

If we want to learn any genuine lessons, then a fundamental question for which an answer should be pressed for is how did the former MMS fail to regulate the Deepwater Horizon rig? With the incredibly poor safety track record of British Petroleum, why would the former MMS simply accept BP's words on how safe its operations were, how prepared its OSRP was, and how little the risk was from a blowout and the potentially worst-case scenario? Why would the people who worked, and presumably continue to work, for the former MMS not make regularly scheduled inspections of the rig to ensure that it was being operated in a safe manner?

In the end, there is one more group of contributors to the megaspill in the Gulf – that is we the authors and you the readers. We have listened to endless rhetoric from our policymakers on how dependent our nation, and indeed the world, is on oil. Perhaps it is finally time to change the rhetoric into action by changing our approaches to lifestyles and views towards sustainability.

Index

About the Authors

Nicholas P. Cheremisinoff is a consultant to industry and governments on standard of care as it pertains to environmental management and worker protection. He has more than 35 years of international consulting, business, and applied research & development experience, having worked throughout Eastern Europe, Russia, parts of the Middle East, Latin America, the Far East, Africa, and the United States. He has led and participated in hundreds of pollution prevention and environmental health and safety audits of large industrial complexes ranging from refineries, to coke-chemical plants, pulp and paper mills, steel mills, pharmaceutical and pesticide manufacturing plants, wood treating facilities, and other industrial operations. He has assisted several foreign governments in strengthening environmental laws and trained several thousand regulators and industry personnel on environmental auditing, pollution prevention and safe chemical management through U.S. Agency for International Development, European Union, and World Bank Organization sponsored programs. He is the author, co-author or editor of more than 150 technical books. Dr. Cheremisinoff earned his Ph.D., M.Sc. and B.Sc. in chemical engineering from Clarkson College of Technology.

Anton Davletshin is Senior Project Manager for N&P, Ltd, a firm specializing in standard of care assessments. Anton project manages sampling programs, conducts historical literature reviews and assists in environmental impact assessments that aid N&P staff in forensic reconstruction of legacy pollution problems. His expertise extends to applications of air dispersion simulations using EPA approved AerMod used in the evaluation of community impacts from point and area sources of emissions from within industrial complexes. A graduate of Virginia Polytechnic Institute, he holds a degree in Construction Management.

Also of Interest

Check out these forthcoming and available related titles from Scrivener Publishing

Ethics in Engineering, by James Speight and Russell Foote, March 2011, ISBN 9780470626023. Covers the most thought-provoking ethical questions in engineering.

Concentration Cells: Fabrication Methods and Materials, by Ralph Zito, September 2011, ISBN 9780470639214. The most comprehensive and practical volume on concentration cells available.

The Greening of Petroleum Operations, by Rafi qul Islam, ISBN 9780470625903. A breakthrough treatment of one of the most difficult and sought-after subjects of the modern era: sustainable energy. *NOW AVAILABLE!*

Formulas and Calculations for Drilling Operations, by Robello Samuel, ISBN 9780470625996. The only book every drilling engineer must have, with all of the formulas and calculations that the engineer uses in the field. *NOW AVAILABLE!*

Acid Gas Injection and Carbon Dioxide Sequestration, by John Carroll, ISBN 9780470625934. Provides a complete overview and guide on the very important topics of acid gas injection and CO2 sequestration. *NOW AVAILABLE!*

Advanced Petroleum Reservoir Simulation, by M.R. Islam, S.H. Mousavizadegan, Shabbir Mustafiz, and Jamal H. Abou-Kassem, ISBN 9780470625811. The state of the art in petroleum reservoir simulation. *NOW AVAILABLE!*

Energy Storage: A New Approach, by Ralph Zito, ISBN 9780470625910. Exploring the potential of reversible concentrations cells, the author of this groundbreaking volume reveals new technologies to solve the global crisis of energy storage. *NOW AVAILABLE!*

Zero-Waste Engineering, by Rafiqul Islam, February 2011, ISBN 9780470626047. In this controvercial new volume, the author explores the question of zero-waste engineering and how it can be done, efficiently and profitably.

Fundamentals of LNG Plant Design, by Saeid Mokhatab, David Messersmith, Walter Sonne, and Kamal Shah, August 2011. The only book of its kind, detailing LNG plant design, as the world turns more and more to LNG for its energy needs.

Flow Assurance, by Boyun Guo and Rafiqul Islam, September 2011, ISBN 9780470626085. Comprehensive and state-of-the-art guide to flow assurance in the petroleum industry.